PRAISE FOR THE *MANGA GUIDE* SERIES

"A fun and fairly painless lesson on what many consider to be a less-than-thrilling subject."
—SCHOOL LIBRARY JOURNAL

"This is really what a good math text should be like. . . . It presents statistics as something fun, and something enlightening."
—GOOD MATH, BAD MATH

"The most enjoyable tech book I've ever read."
—LINUX PRO MAGAZINE

"If you want to introduce a subject that kids wouldn't normally be very interested in, give it an amusing storyline and wrap it in cartoons."
—MAKE

"This is a solid book and I wish there were more like it in the IT world."
—SLASHDOT

"Great for anyone wanting an introduction or a refresher on statistics."
—PARENTING SQUAD

"A light, impressively non-oppressive read, especially considering the technical nature of its subject."
—AIN'T IT COOL NEWS

"Makes accessible a very intimidating subject, letting the reader have fun while still delivering the goods."
—GEEKDAD, WIRED.COM

"Definitely way better than trying to comprehend a bland statistics book."
—DR. DOBB'S CODETALK

"I would have killed for these books when studying for my school exams 20 years ago."
—TIM MAUGHAN

"An awfully fun, highly educational read."
—FRAZZLEDDAD

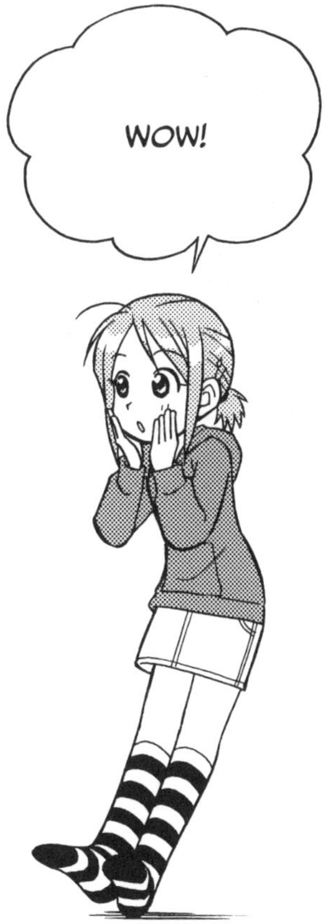
WOW!

THE MANGA GUIDE™ TO MOLECULAR BIOLOGY

THE MANGA GUIDE™ TO
MOLECULAR BIOLOGY

MASAHARU TAKEMURA
SAKURA
BECOM CO., LTD.

Ohmsha

no starch press

This English edition is co-published by No Starch Press and Ohmsha, Ltd.

13 12 11 10 09   1 2 3 4 5 6 7 8 9

ISBN-10: 1-59327-202-2
ISBN-13: 978-1-59327-202-9

SUSTAINABLE FORESTRY INITIATIVE Certified Fiber Sourcing
Label applies to the text stock www.sfiprogram.org

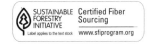

Publisher: William Pollock
Author: Masaharu Takemura
Illustrator: Sakura
Producer: Becom Co., Ltd.
Production Editors: Kathleen Mish and Magnolia Molcan
Developmental Editor: Tyler Ortman
Technical Reviewers: Read Siry, E. Jane Richardson, and Kerri Lendo
Compositor: Riley Hoffman
Proofreader: Cristina Chan
Indexer: Sarah Schott

For information on book distributors or translations, please contact No Starch Press, Inc. directly:
No Starch Press, Inc.
555 De Haro Street, Suite 250, San Francisco, CA 94107
phone: 415.863.9900; fax: 415.863.9950; info@nostarch.com; http://www.nostarch.com/

Library of Congress Cataloging-in-Publication Data

Takemura, Masaharu, 1969-
[Manga de wakaru bunshi seibutsugaku. English]
The manga guide to molecular biology / Masaharu Takemura, Sakura, Becom Co., Ltd.
p. cm.
Includes index.
ISBN-13: 978-1-59327-202-9
ISBN-10: 1-59327-202-2
1. Molecular biology--Comic books, strips, etc. 2. Molecular biology--Popular works. I. Becom Co. II. Title.
QH506.T3513 2009
572.8--dc22

2009025876

CONTENTS

3
DNA REPLICATION AND CELL DIVISION. 91

4
HOW IS A PROTEIN MADE? 131

5
GENETIC TECHNOLOGY AND RESEARCH

PREFACE

Molecular biology is an academic discipline aimed at understanding the behavior of living organisms too small for our eyes to see. Genes play important roles in our world. However, they are not only invisible to our eyes but also difficult to observe even with a microscope.

Researchers of molecular biology such as myself are conducting many experiments every day in laboratories at colleges, research institutes, and corporations. Researchers work to understand the behavior of DNA, proteins, and RNA based on their experiments and to understand this small world using models they create.

Since we cannot see the subjects of our experiments, knowledge in the field of molecular biology tends to be based on experimental data—and there are still many things we do not understand. While this research itself is difficult to pursue, the more difficult task is conveying the world of molecular biology to nonspecialists in an easy-to-understand manner. *The Manga Guide to Molecular Biology* is an attempt to do just that.

The main characters of this book are two college girls, Ami Kasuga and Rin Natsukawa. These two girls are called to a small isolated island owned by Professor Moro for a molecular biology make-up class. Through a virtual reality machine that brings them into the world of molecular biology, they learn a lot, along with help from Marcus, the professor's handsome assistant.

Since the girls aim to grasp the big picture of molecular biology, this book contains many descriptions designed to facilitate readers' understanding of the subject. In other words, the processes of replication of DNA, gene transcription, and protein synthesis are not quite as simple as they seem in this book.

If readers come to feel that the world of molecular biology is more complex and contains many more areas they wish to understand, then more than half of the purpose of publishing this book, I think, will have been achieved.

Having said that, it also needs to be said that molecular biology is a profound academic discipline. It is inevitably linked to other areas of study, including medical science, agriculture, and engineering, as well as basic scientific areas, such as physics, chemistry, and geosciences, not to mention biology. And it is closely related to the daily lives of many people.

Thanks to research results that have increased at an explosive pace from the end of the 20th century and into this century, the field of molecular biology has vastly expanded. It is difficult for a single researcher now to have sufficient knowledge to grasp the entire picture of molecular biology.

This book covers only the basics of molecular biology. If you want to obtain a true picture of molecular biology, I recommend using this book as a beginning and then going on to nurture your interest in the variety of other materials available on this subject.

In conclusion, I would like to take this opportunity to thank all the staff at Ohmsha, Ltd., Mr. Masayoshi Maeda for the wonderful scenarios, Mr. Sakura for creating the amazing representation of the complex world of molecules in cartoon form, and, above all, the readers who have taken up this book.

MASAHARU TAKEMURA
JANUARY 2008

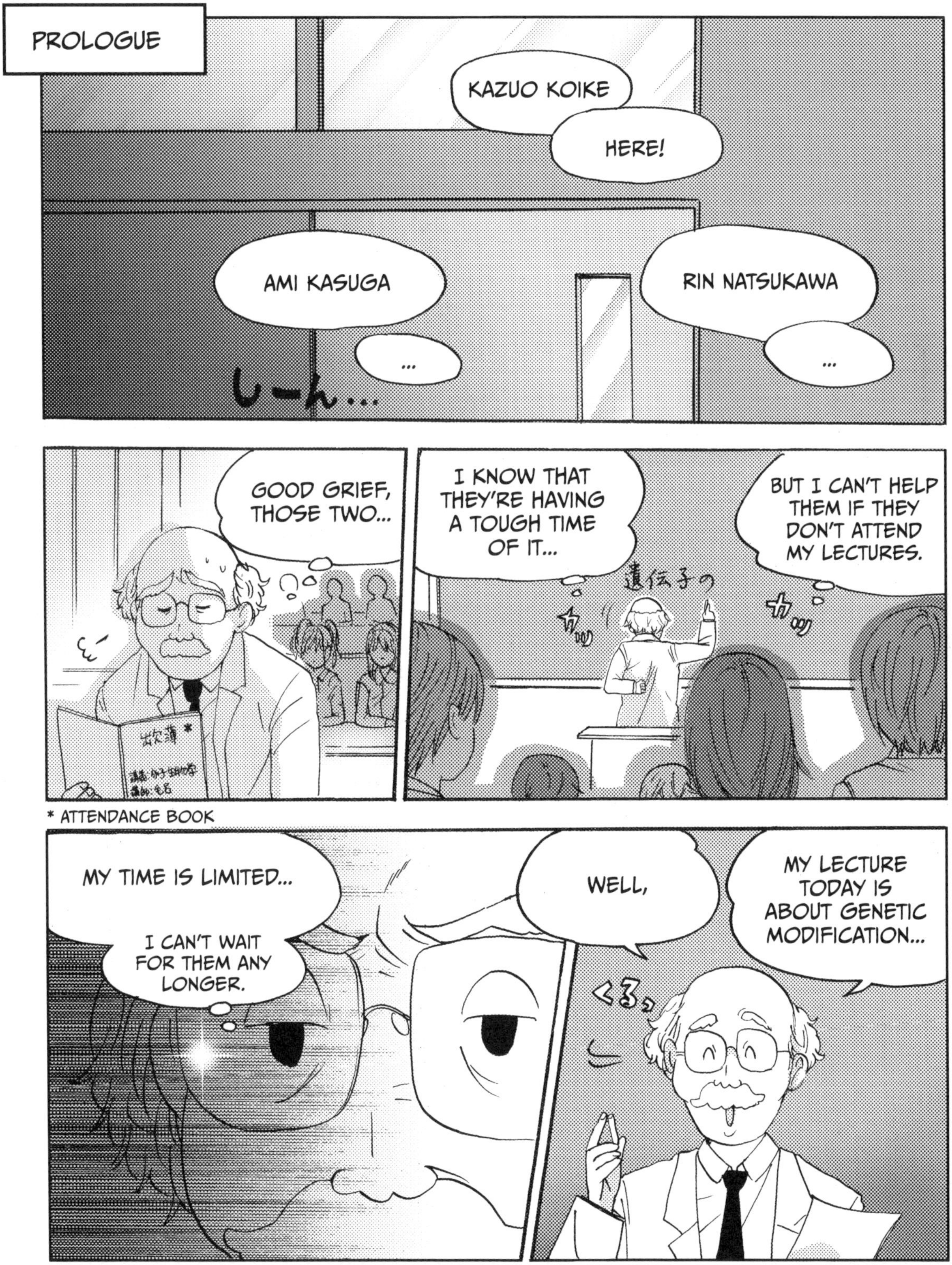
PROLOGUE
KAZUO KOIKE
HERE!
AMI KASUGA
...
RIN NATSUKAWA
...
しーん...
GOOD GRIEF, THOSE TWO...
I KNOW THAT THEY'RE HAVING A TOUGH TIME OF IT...
BUT I CAN'T HELP THEM IF THEY DON'T ATTEND MY LECTURES.
遺伝子の
カッ
カッ
* ATTENDANCE BOOK
MY TIME IS LIMITED...
I CAN'T WAIT FOR THEM ANY LONGER.
WELL,
MY LECTURE TODAY IS ABOUT GENETIC MODIFICATION...
くるっ

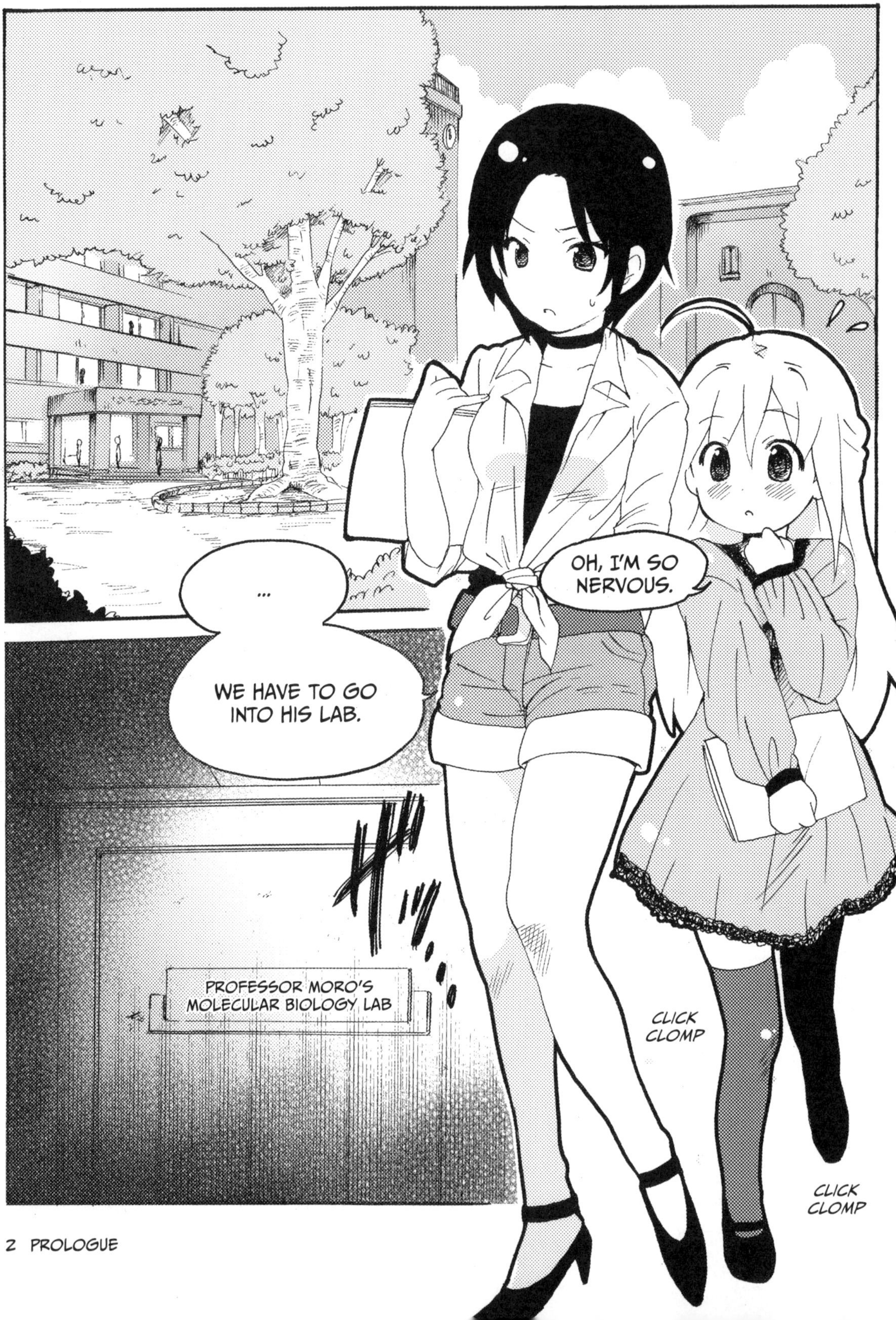
...
OH, I'M SO NERVOUS.
WE HAVE TO GO INTO HIS LAB.
PROFESSOR MORO'S MOLECULAR BIOLOGY LAB
CLICK CLOMP
CLICK CLOMP

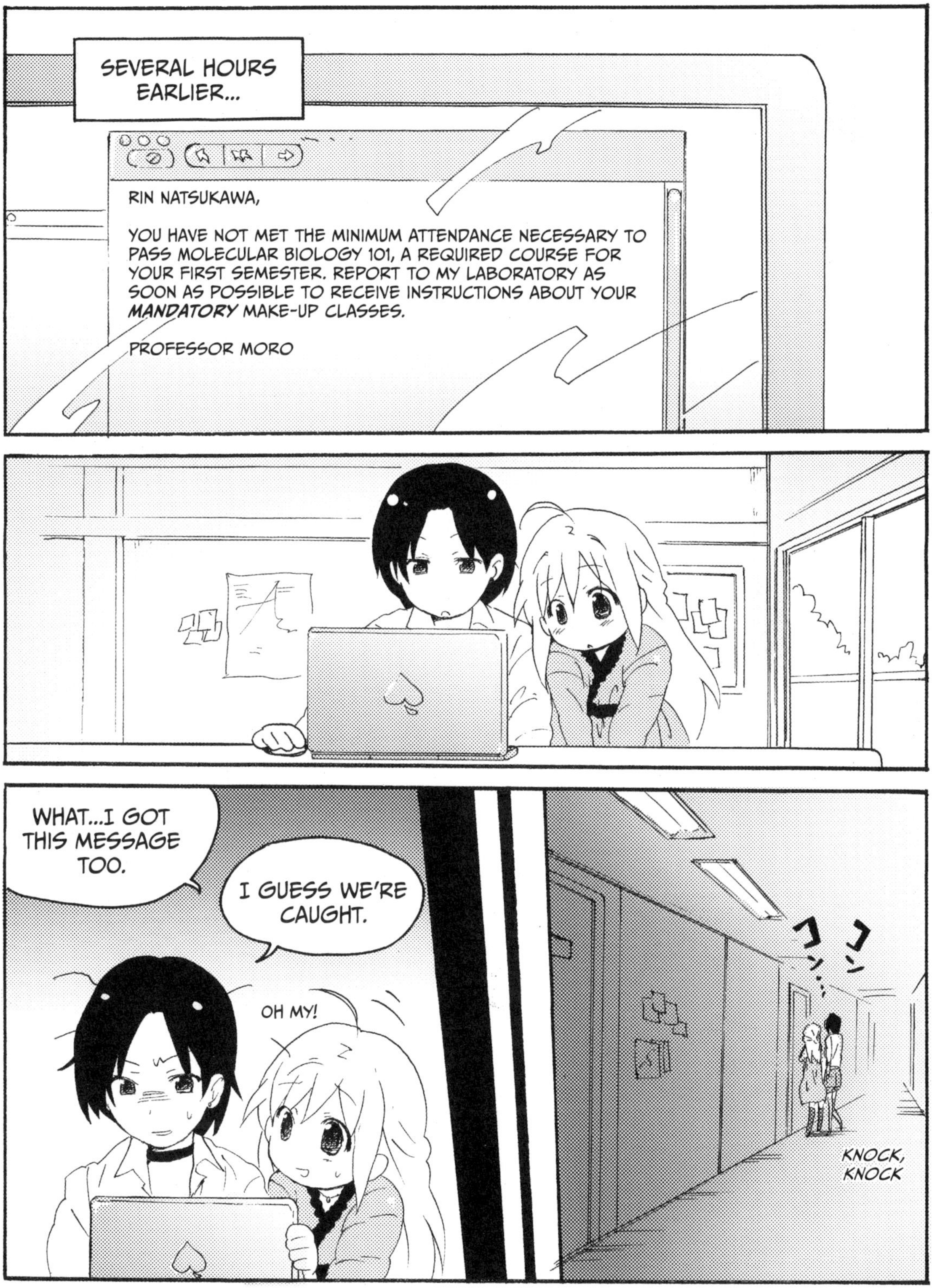
SEVERAL HOURS EARLIER...
RIN NATSUKAWA,
YOU HAVE NOT MET THE MINIMUM ATTENDANCE NECESSARY TO PASS MOLECULAR BIOLOGY 101, A REQUIRED COURSE FOR YOUR FIRST SEMESTER. REPORT TO MY LABORATORY AS SOON AS POSSIBLE TO RECEIVE INSTRUCTIONS ABOUT YOUR MANDATORY MAKE-UP CLASSES.
PROFESSOR MORO
WHAT...I GOT THIS MESSAGE TOO.
I GUESS WE'RE CAUGHT.
OH MY!
KNOCK, KNOCK

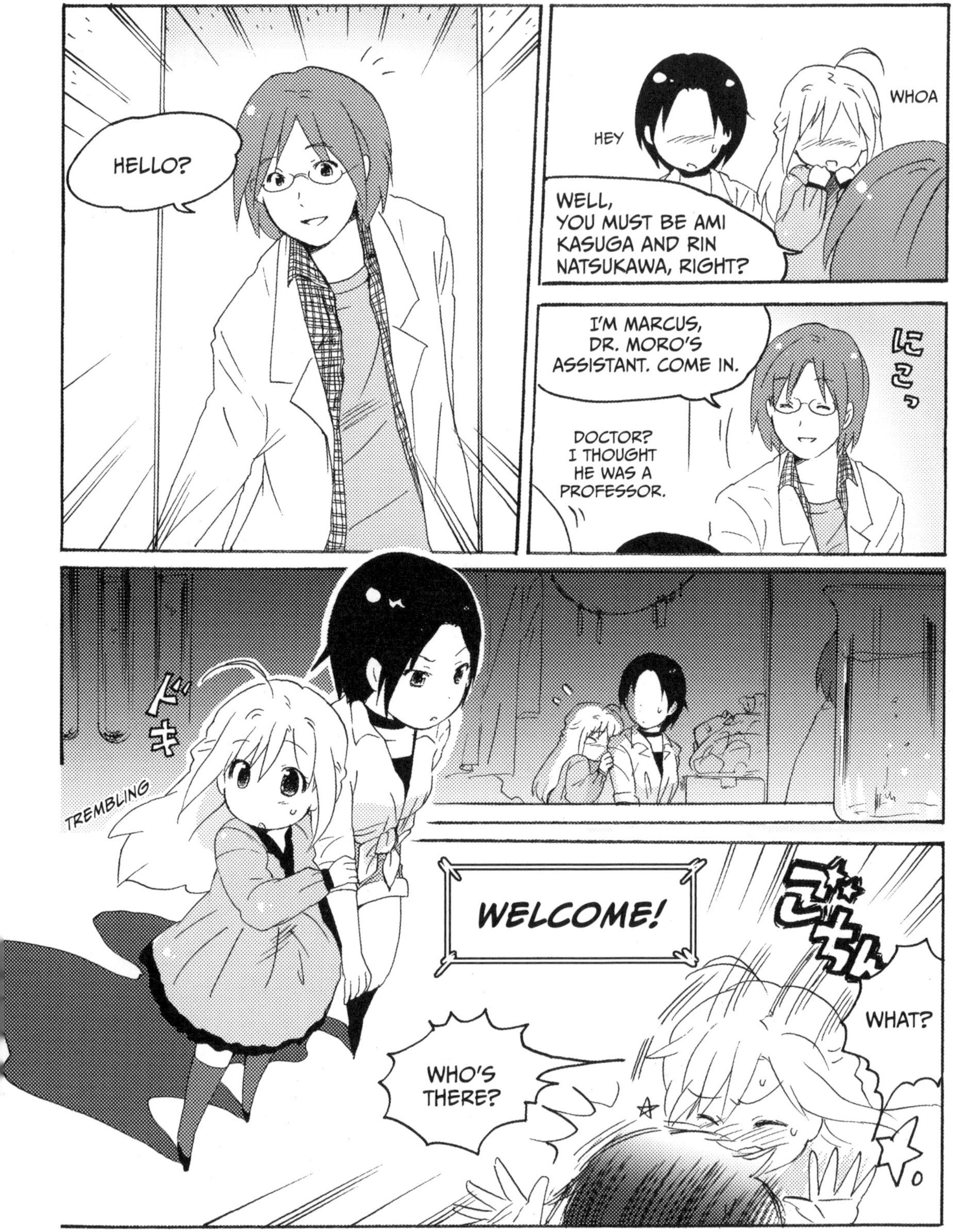
HELLO?
HEY
WHOA
WELL,
YOU MUST BE AMI
KASUGA AND RIN
NATSUKAWA, RIGHT?
I'M MARCUS,
DR. MORO'S
ASSISTANT. COME IN.
にこっ
DOCTOR?
I THOUGHT
HE WAS A
PROFESSOR.
ドキ
TREMBLING
WELCOME!
ごちん
WHO'S
THERE?
WHAT?

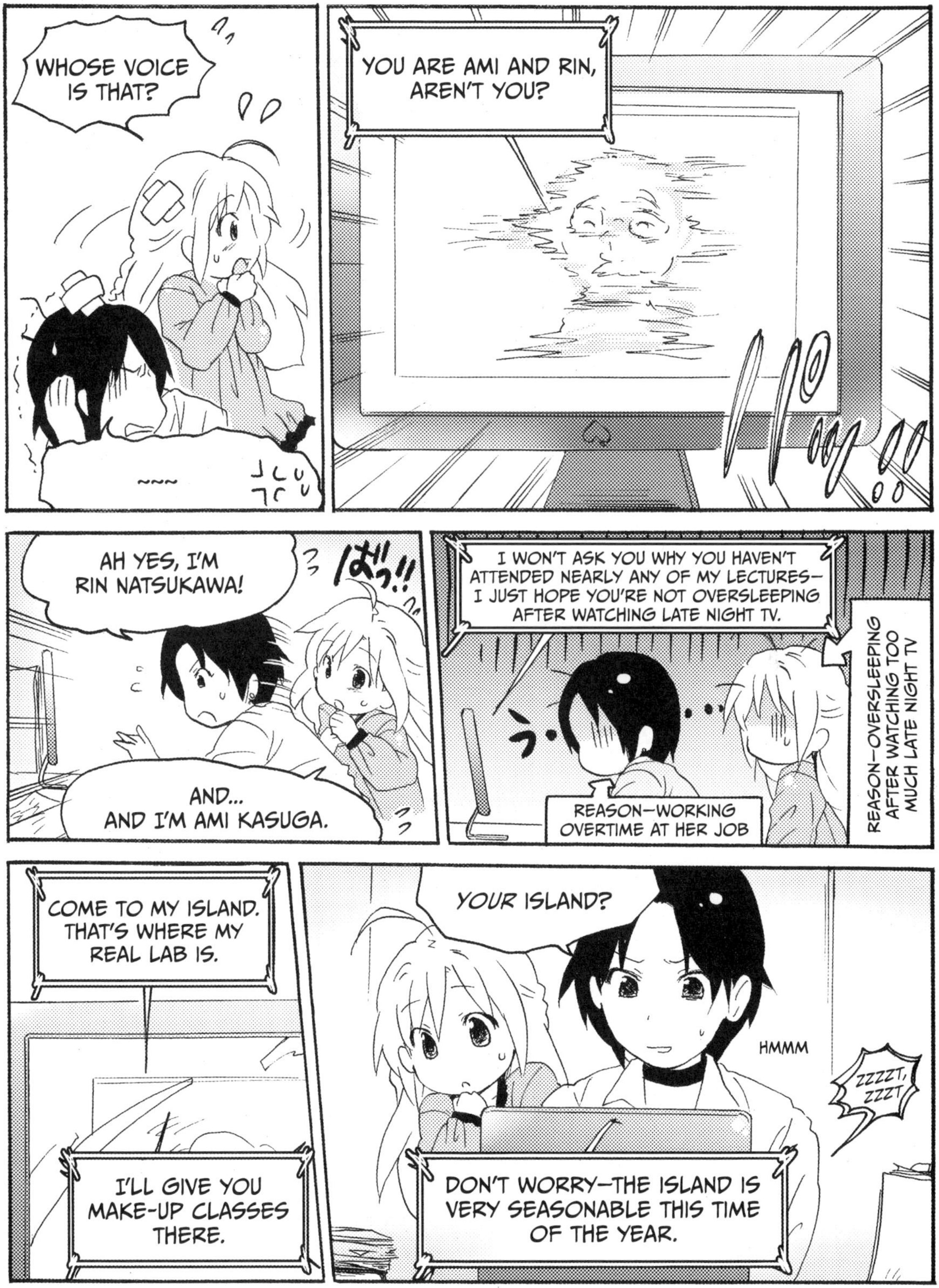
WHOSE VOICE IS THAT?
YOU ARE AMI AND RIN, AREN'T YOU?
AH YES, I'M RIN NATSUKAWA!
ばっ!!
AND... AND I'M AMI KASUGA.
I WON'T ASK YOU WHY YOU HAVEN'T ATTENDED NEARLY ANY OF MY LECTURES— I JUST HOPE YOU'RE NOT OVERSLEEPING AFTER WATCHING LATE NIGHT TV.
う・
REASON—WORKING OVERTIME AT HER JOB
REASON—OVERSLEEPING AFTER WATCHING TOO MUCH LATE NIGHT TV
COME TO MY ISLAND. THAT'S WHERE MY REAL LAB IS.
I'LL GIVE YOU MAKE-UP CLASSES THERE.
YOUR ISLAND?
HMMM
ZZZZT, ZZZT
DON'T WORRY—THE ISLAND IS VERY SEASONABLE THIS TIME OF THE YEAR.

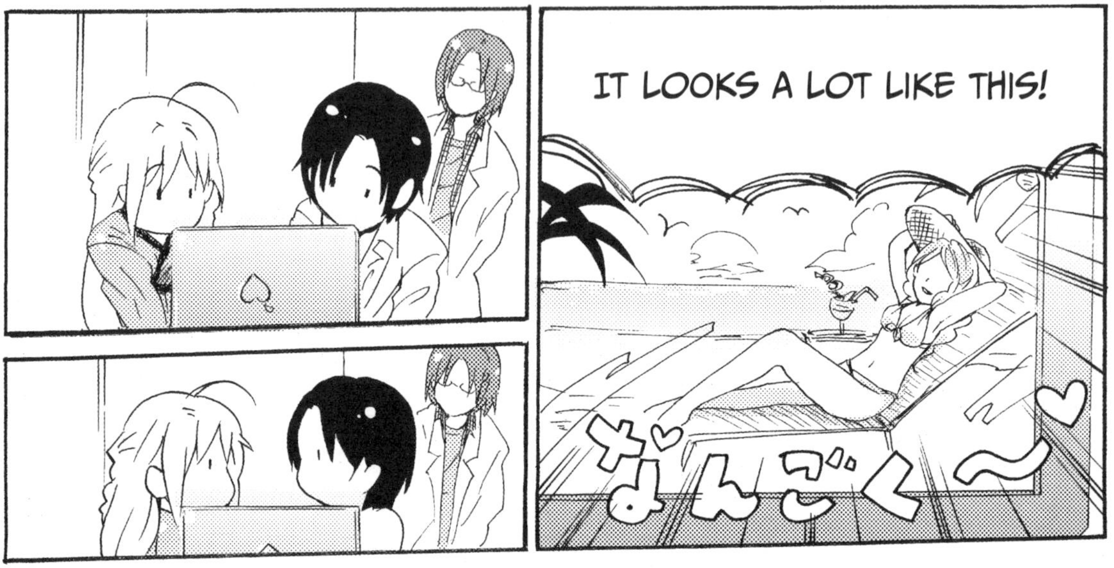
IT LOOKS A LOT LIKE THIS!
なんごく〜

A FEW DAYS LATER...

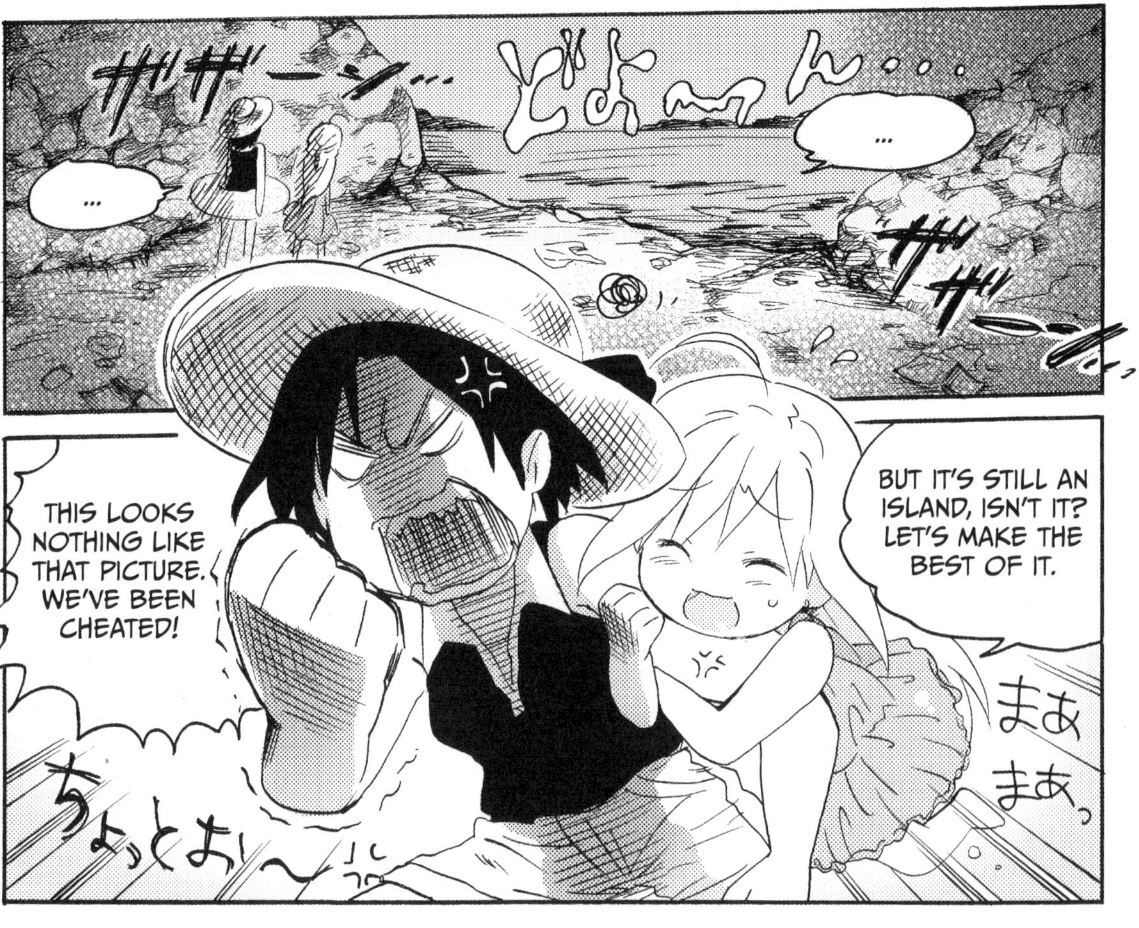
ザザーン…
どよ〜ん…
...
...
ザザーン
THIS LOOKS NOTHING LIKE THAT PICTURE. WE'VE BEEN CHEATED!
BUT IT'S STILL AN ISLAND, ISN'T IT? LET'S MAKE THE BEST OF IT.
まあまあっ
ちょっとお〜

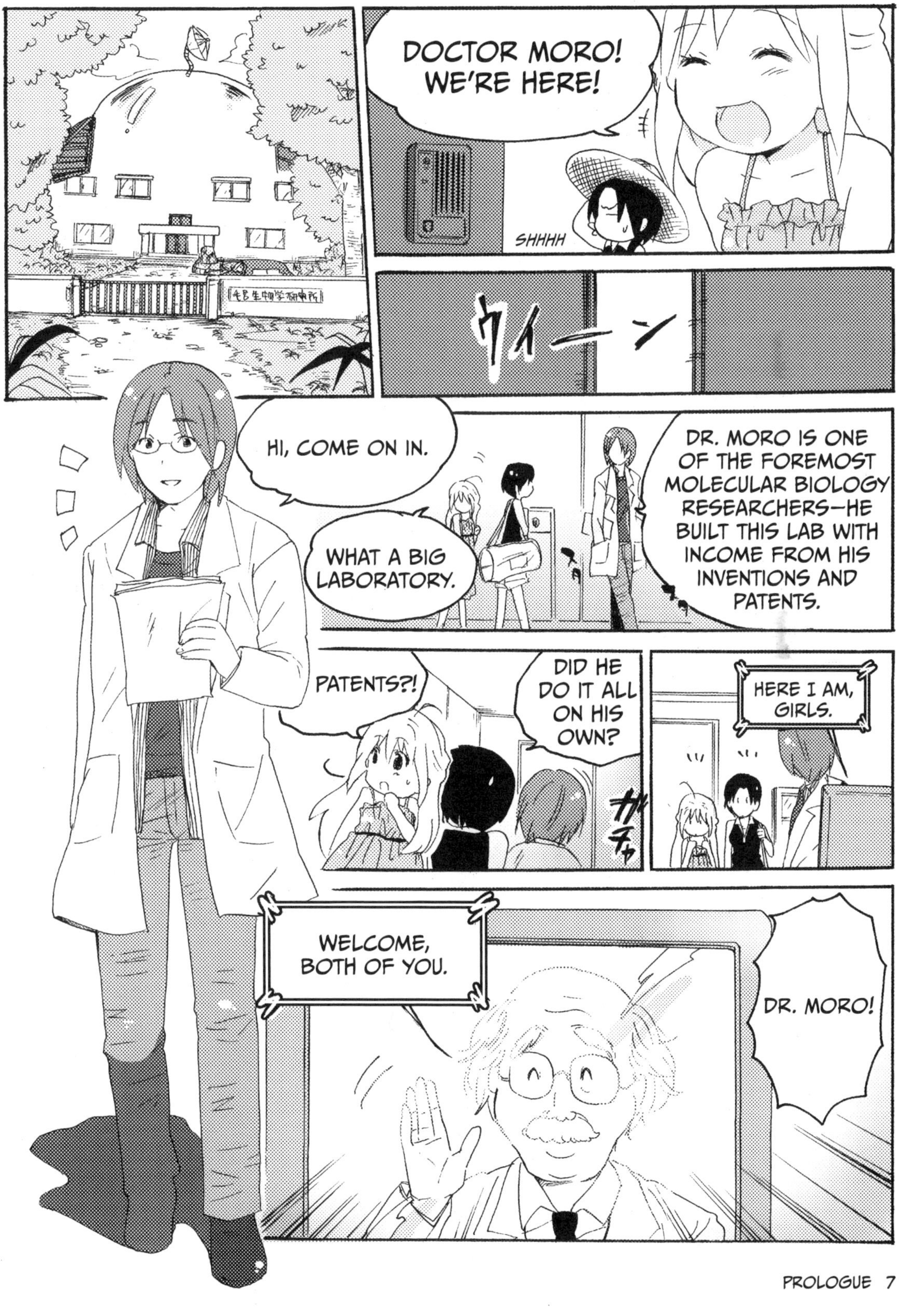
DOCTOR MORO! WE'RE HERE!
SHHHH
ウィーン
HI, COME ON IN.
WHAT A BIG LABORATORY.
DR. MORO IS ONE OF THE FOREMOST MOLECULAR BIOLOGY RESEARCHERS—HE BUILT THIS LAB WITH INCOME FROM HIS INVENTIONS AND PATENTS.
PATENTS?!
DID HE DO IT ALL ON HIS OWN?
ガチャ
HERE I AM, GIRLS.
WELCOME, BOTH OF YOU.
DR. MORO!

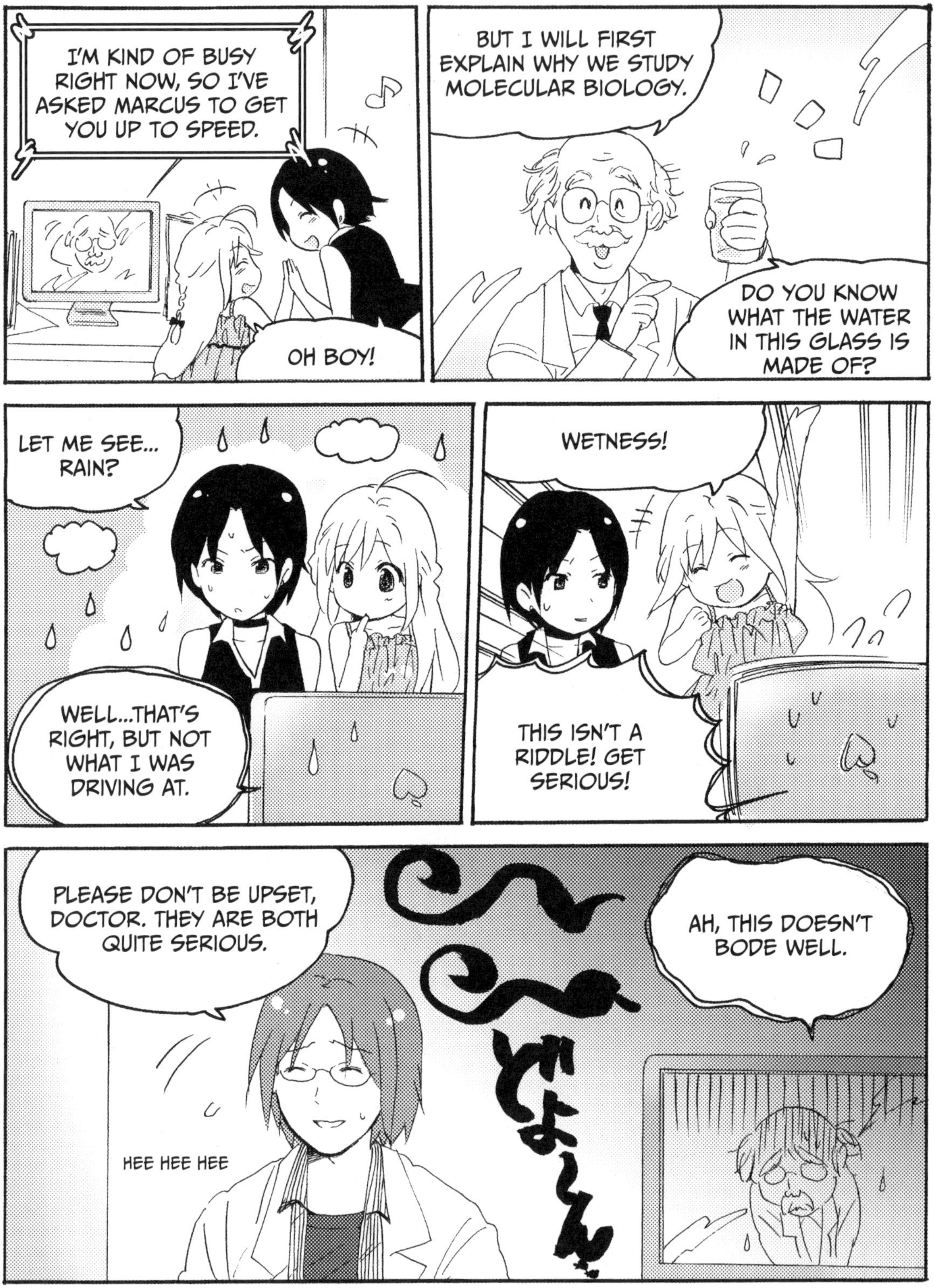
I'M KIND OF BUSY RIGHT NOW, SO I'VE ASKED MARCUS TO GET YOU UP TO SPEED.
OH BOY!
BUT I WILL FIRST EXPLAIN WHY WE STUDY MOLECULAR BIOLOGY.
DO YOU KNOW WHAT THE WATER IN THIS GLASS IS MADE OF?
LET ME SEE... RAIN?
WELL...THAT'S RIGHT, BUT NOT WHAT I WAS DRIVING AT.
WETNESS!
THIS ISN'T A RIDDLE! GET SERIOUS!
PLEASE DON'T BE UPSET, DOCTOR. THEY ARE BOTH QUITE SERIOUS.
HEE HEE HEE
AH, THIS DOESN'T BODE WELL.

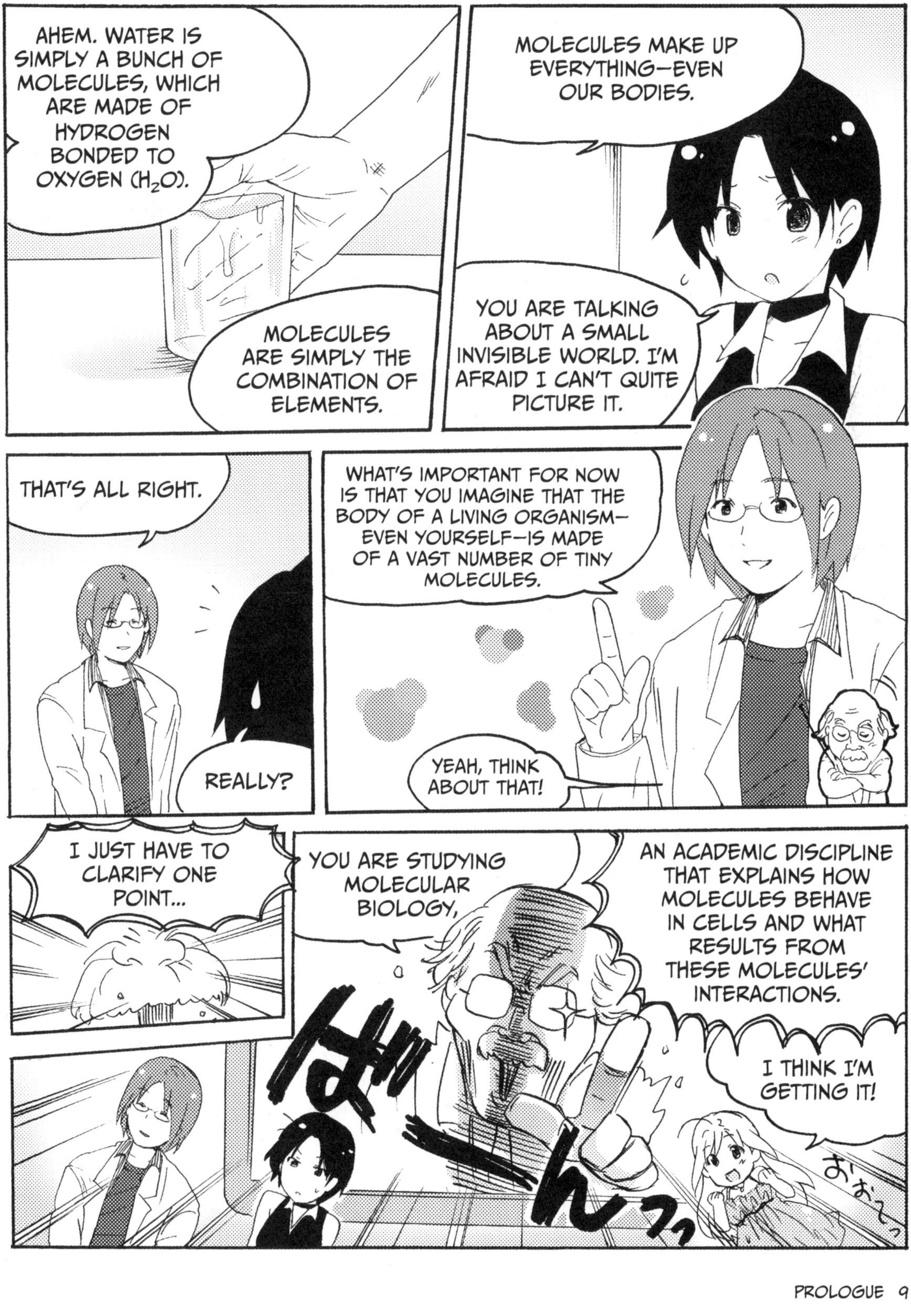
AHEM. WATER IS SIMPLY A BUNCH OF MOLECULES, WHICH ARE MADE OF HYDROGEN BONDED TO OXYGEN (H_2O).
MOLECULES ARE SIMPLY THE COMBINATION OF ELEMENTS.
MOLECULES MAKE UP EVERYTHING—EVEN OUR BODIES.
YOU ARE TALKING ABOUT A SMALL INVISIBLE WORLD. I'M AFRAID I CAN'T QUITE PICTURE IT.
THAT'S ALL RIGHT.
REALLY?
WHAT'S IMPORTANT FOR NOW IS THAT YOU IMAGINE THAT THE BODY OF A LIVING ORGANISM—EVEN YOURSELF—IS MADE OF A VAST NUMBER OF TINY MOLECULES.
YEAH, THINK ABOUT THAT!
I JUST HAVE TO CLARIFY ONE POINT...
YOU ARE STUDYING MOLECULAR BIOLOGY,
AN ACADEMIC DISCIPLINE THAT EXPLAINS HOW MOLECULES BEHAVE IN CELLS AND WHAT RESULTS FROM THESE MOLECULES' INTERACTIONS.
I THINK I'M GETTING IT!

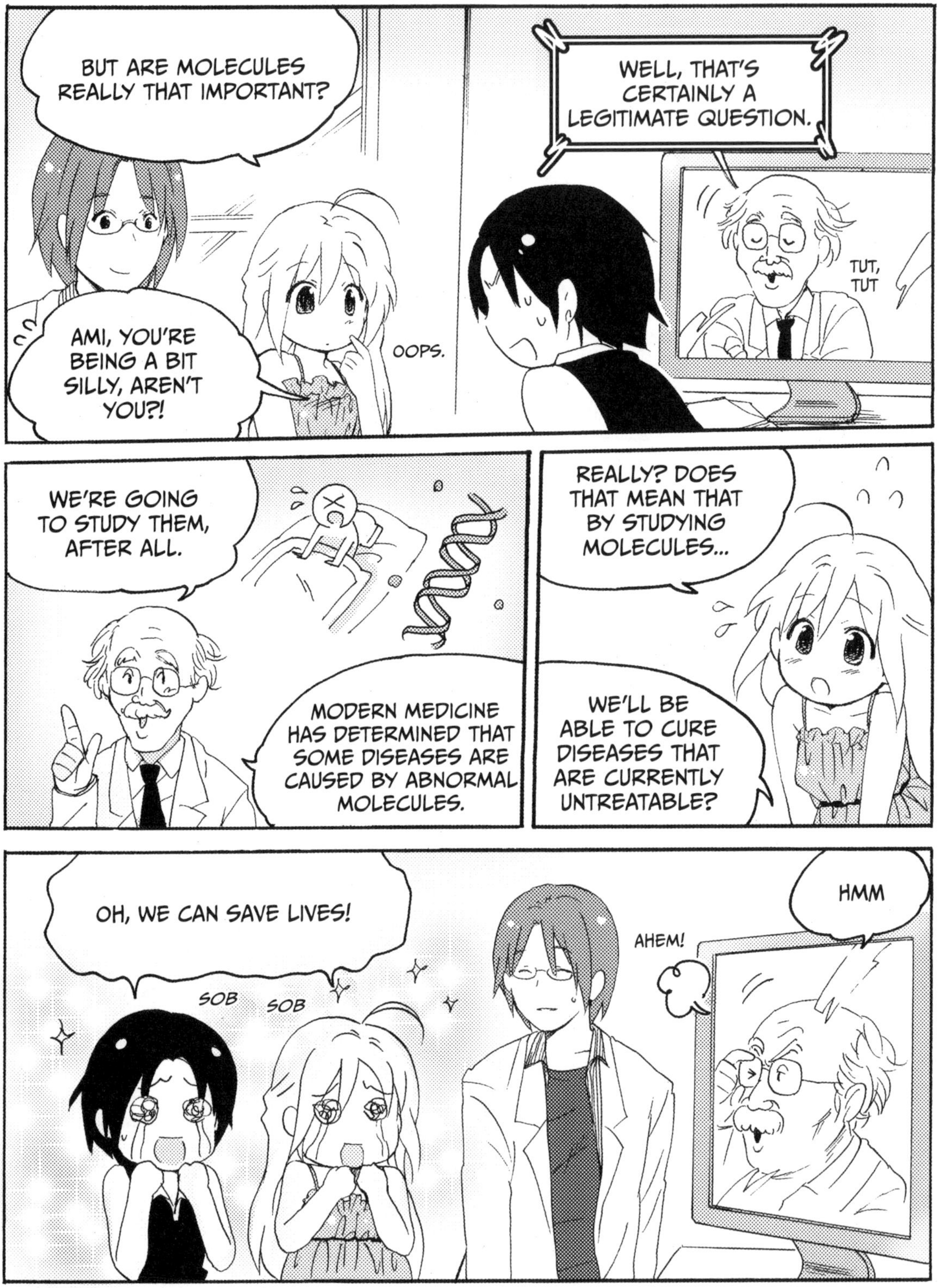
BUT ARE MOLECULES REALLY THAT IMPORTANT?
AMI, YOU'RE BEING A BIT SILLY, AREN'T YOU?!
OOPS.
WELL, THAT'S CERTAINLY A LEGITIMATE QUESTION.
TUT, TUT
WE'RE GOING TO STUDY THEM, AFTER ALL.
MODERN MEDICINE HAS DETERMINED THAT SOME DISEASES ARE CAUSED BY ABNORMAL MOLECULES.
REALLY? DOES THAT MEAN THAT BY STUDYING MOLECULES...
WE'LL BE ABLE TO CURE DISEASES THAT ARE CURRENTLY UNTREATABLE?
OH, WE CAN SAVE LIVES!
SOB
SOB
AHEM!
HMM

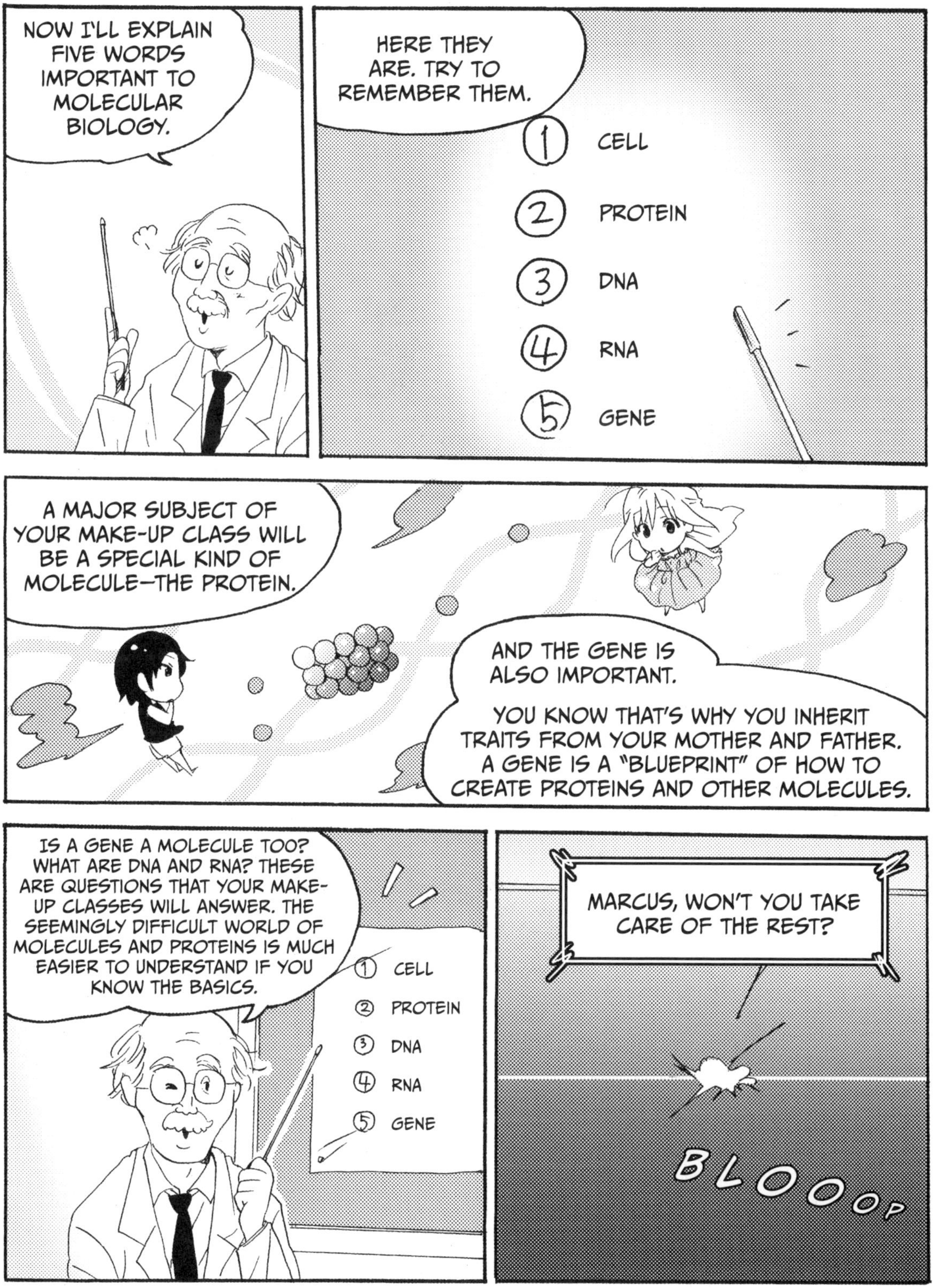
NOW I'LL EXPLAIN FIVE WORDS IMPORTANT TO MOLECULAR BIOLOGY.
HERE THEY ARE. TRY TO REMEMBER THEM.
1 CELL
2 PROTEIN
3 DNA
4 RNA
5 GENE
A MAJOR SUBJECT OF YOUR MAKE-UP CLASS WILL BE A SPECIAL KIND OF MOLECULE—THE PROTEIN.
AND THE GENE IS ALSO IMPORTANT.
YOU KNOW THAT'S WHY YOU INHERIT TRAITS FROM YOUR MOTHER AND FATHER. A GENE IS A "BLUEPRINT" OF HOW TO CREATE PROTEINS AND OTHER MOLECULES.
IS A GENE A MOLECULE TOO? WHAT ARE DNA AND RNA? THESE ARE QUESTIONS THAT YOUR MAKE-UP CLASSES WILL ANSWER. THE SEEMINGLY DIFFICULT WORLD OF MOLECULES AND PROTEINS IS MUCH EASIER TO UNDERSTAND IF YOU KNOW THE BASICS.
1 CELL
2 PROTEIN
3 DNA
4 RNA
5 GENE
MARCUS, WON'T YOU TAKE CARE OF THE REST?
BLOOOOP

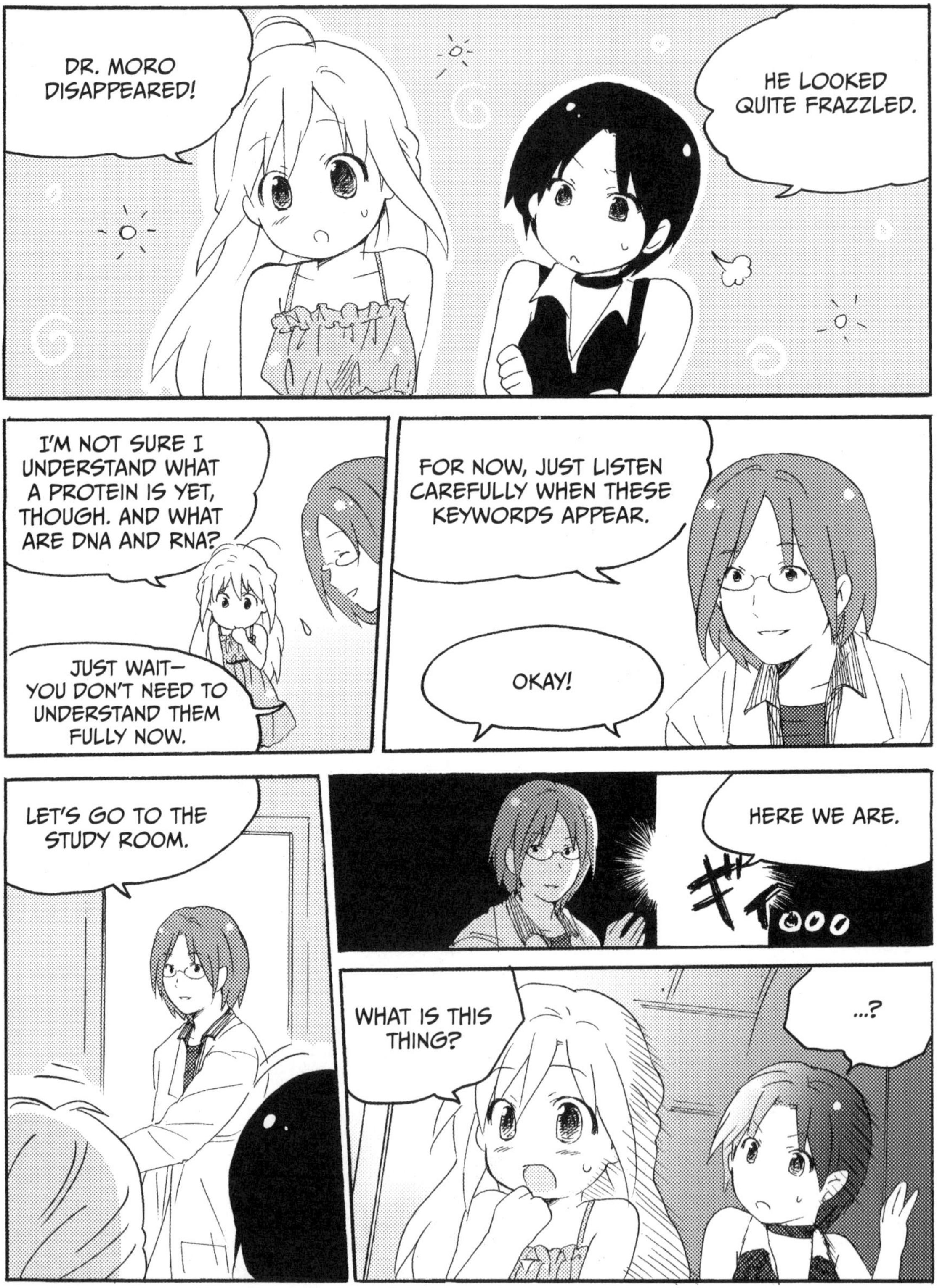

DR. MORO DISAPPEARED!
HE LOOKED QUITE FRAZZLED.
I'M NOT SURE I UNDERSTAND WHAT A PROTEIN IS YET, THOUGH. AND WHAT ARE DNA AND RNA?
JUST WAIT— YOU DON'T NEED TO UNDERSTAND THEM FULLY NOW.
FOR NOW, JUST LISTEN CAREFULLY WHEN THESE KEYWORDS APPEAR.
OKAY!
LET'S GO TO THE STUDY ROOM.
HERE WE ARE.
WHAT IS THIS THING?
...?

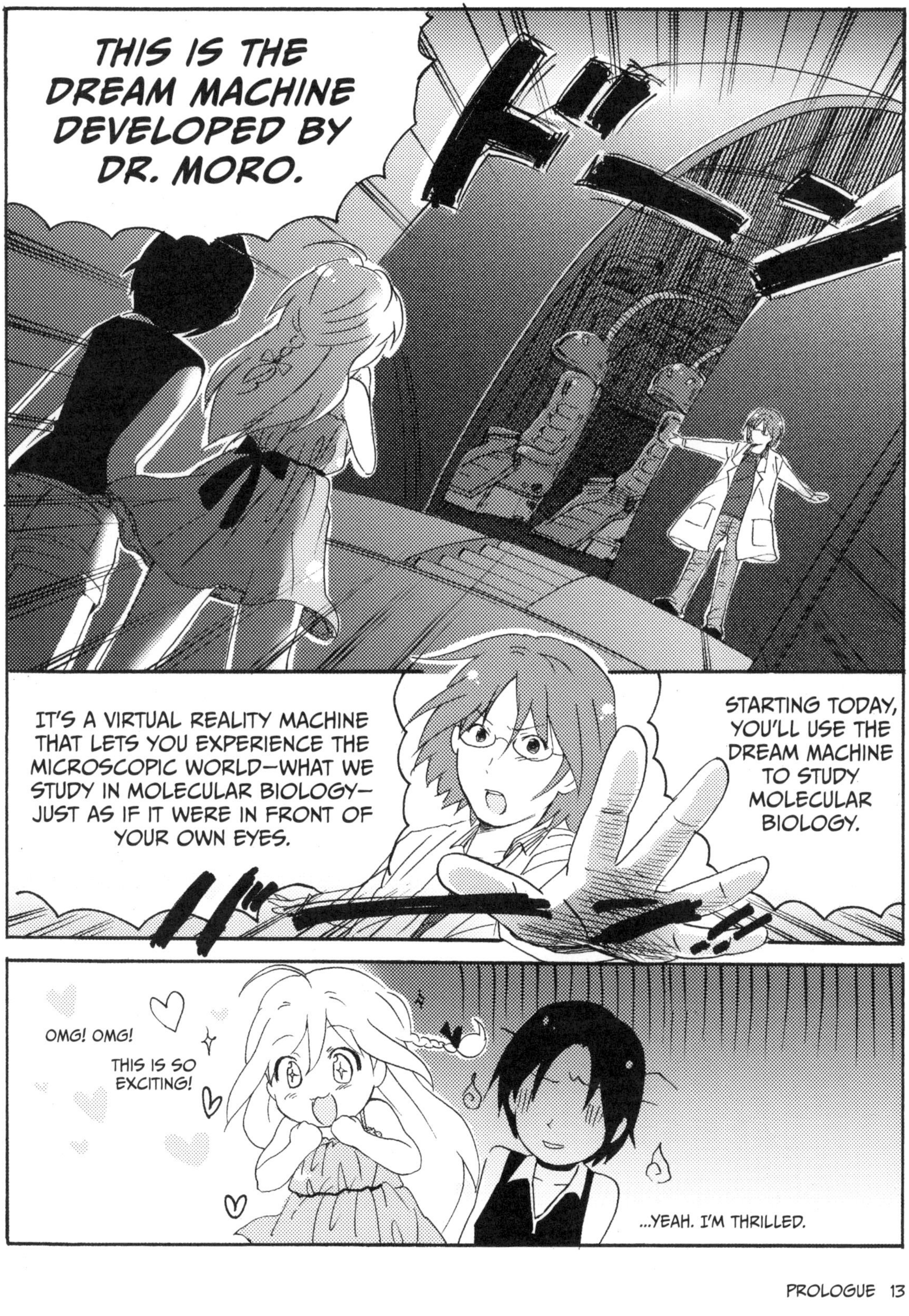
THIS IS THE DREAM MACHINE DEVELOPED BY DR. MORO.
IT'S A VIRTUAL REALITY MACHINE THAT LETS YOU EXPERIENCE THE MICROSCOPIC WORLD—WHAT WE STUDY IN MOLECULAR BIOLOGY—JUST AS IF IT WERE IN FRONT OF YOUR OWN EYES.
STARTING TODAY, YOU'LL USE THE DREAM MACHINE TO STUDY MOLECULAR BIOLOGY.
OMG! OMG!
THIS IS SO EXCITING!
...YEAH. I'M THRILLED.

1
WHAT IS A CELL?

A CELL IS A LITTLE SACK OF LIFE

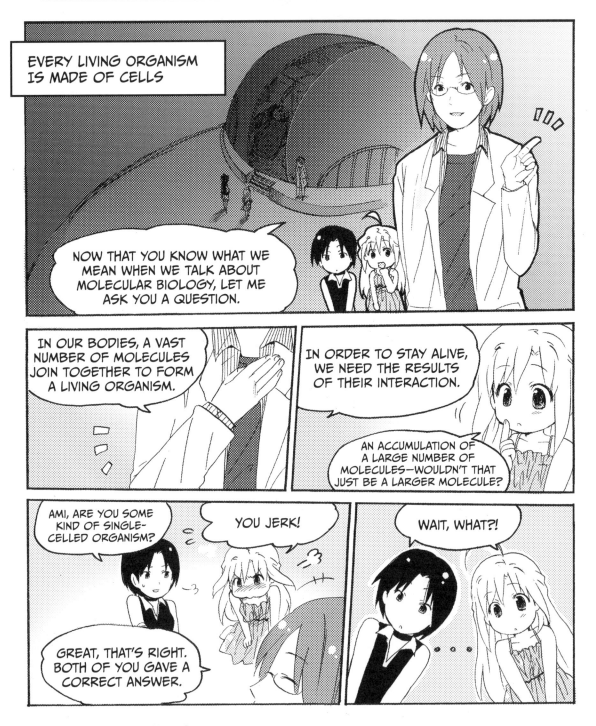

EVERY LIVING ORGANISM IS MADE OF CELLS

NOW THAT YOU KNOW WHAT WE MEAN WHEN WE TALK ABOUT MOLECULAR BIOLOGY, LET ME ASK YOU A QUESTION.

IN OUR BODIES, A VAST NUMBER OF MOLECULES JOIN TOGETHER TO FORM A LIVING ORGANISM.

IN ORDER TO STAY ALIVE, WE NEED THE RESULTS OF THEIR INTERACTION.

AN ACCUMULATION OF A LARGE NUMBER OF MOLECULES—WOULDN'T THAT JUST BE A LARGER MOLECULE?

AMI, ARE YOU SOME KIND OF SINGLE-CELLED ORGANISM?

YOU JERK!

WAIT, WHAT?!

GREAT, THAT'S RIGHT. BOTH OF YOU GAVE A CORRECT ANSWER.

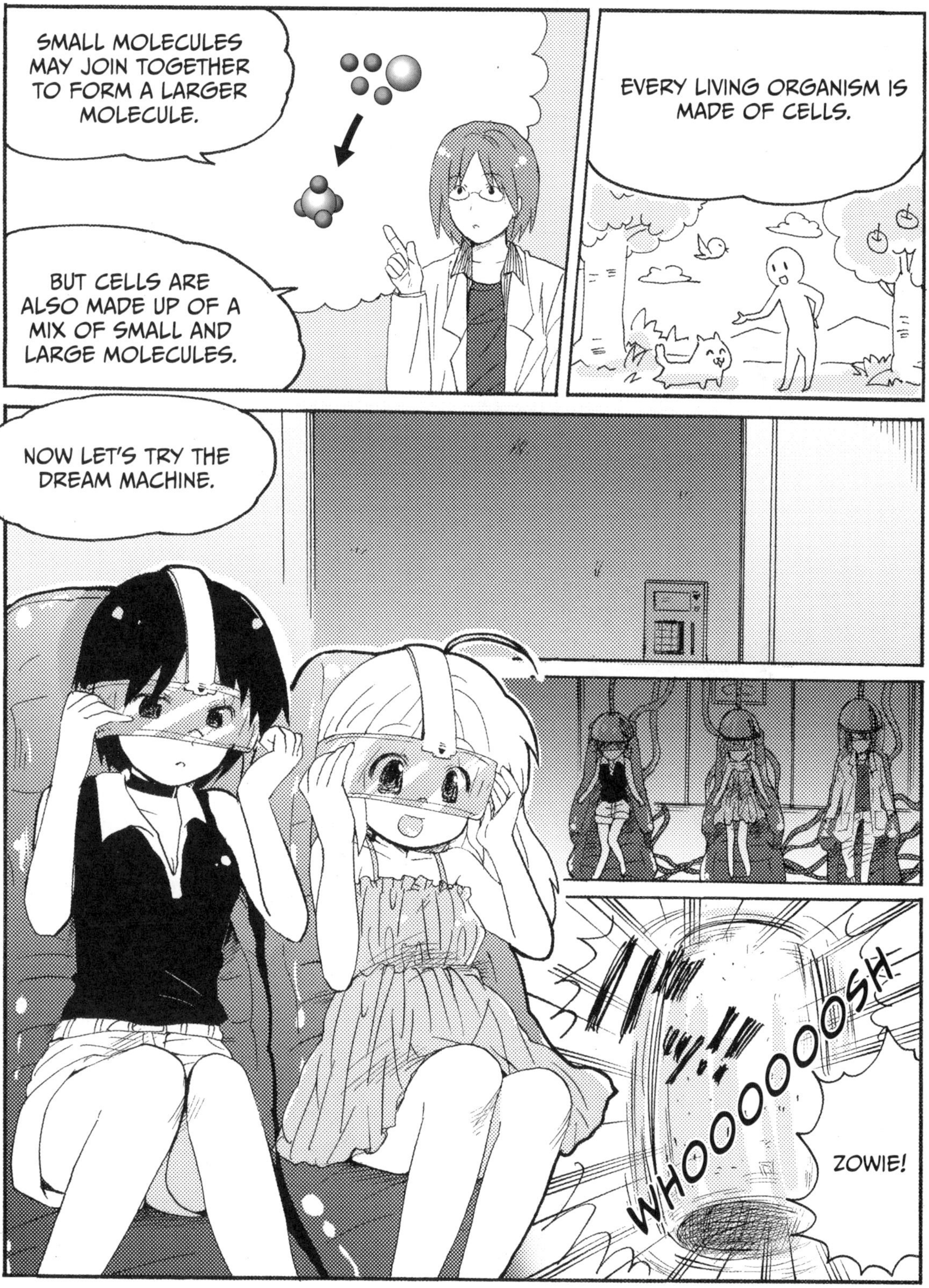
SMALL MOLECULES MAY JOIN TOGETHER TO FORM A LARGER MOLECULE.
BUT CELLS ARE ALSO MADE UP OF A MIX OF SMALL AND LARGE MOLECULES.
EVERY LIVING ORGANISM IS MADE OF CELLS.
NOW LET'S TRY THE DREAM MACHINE.
WHOOOOOOOSH
ZOWIE!

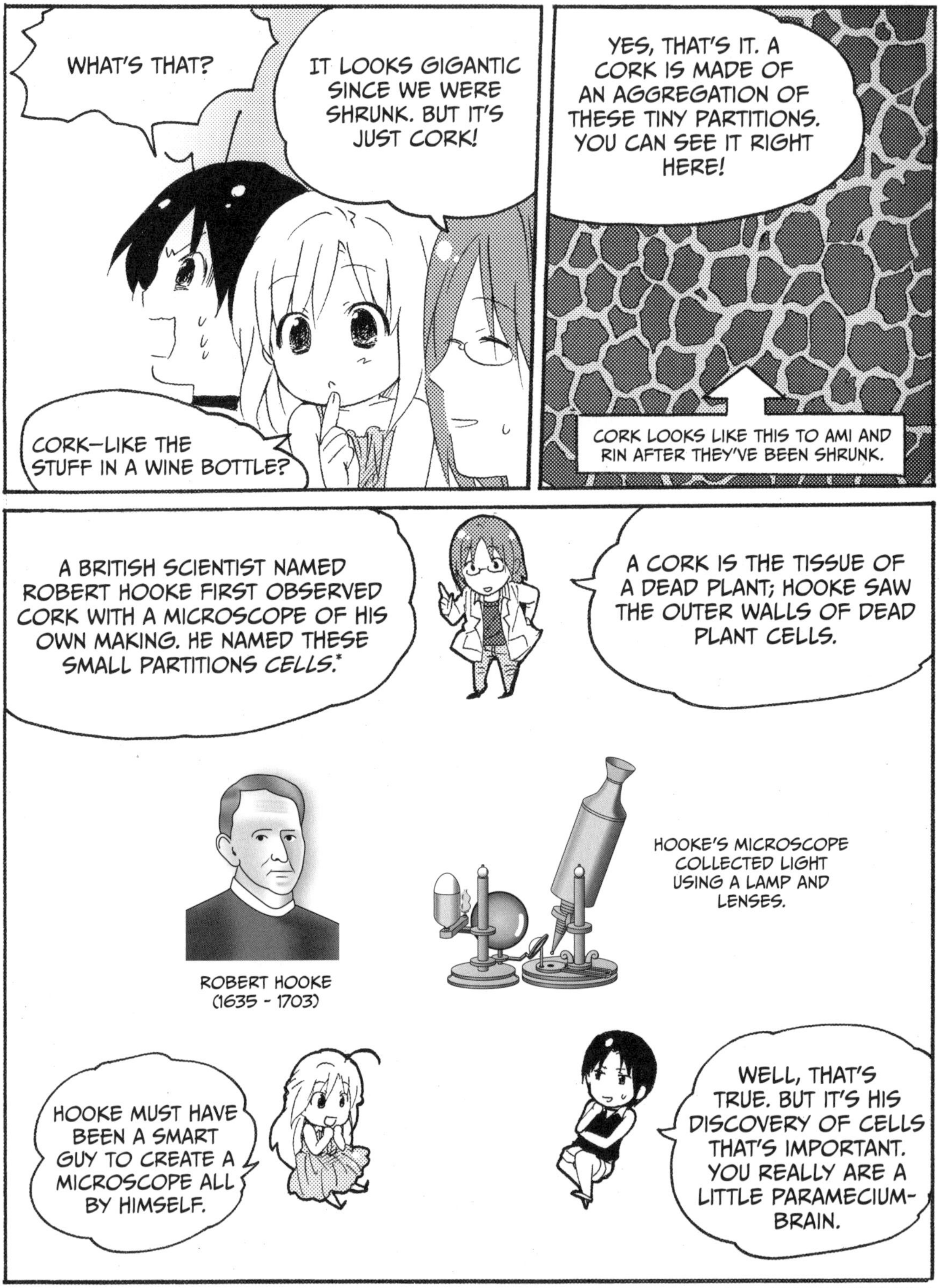
WHAT'S THAT?
IT LOOKS GIGANTIC SINCE WE WERE SHRUNK. BUT IT'S JUST CORK!
CORK—LIKE THE STUFF IN A WINE BOTTLE?
YES, THAT'S IT. A CORK IS MADE OF AN AGGREGATION OF THESE TINY PARTITIONS. YOU CAN SEE IT RIGHT HERE!
CORK LOOKS LIKE THIS TO AMI AND RIN AFTER THEY'VE BEEN SHRUNK.
A BRITISH SCIENTIST NAMED ROBERT HOOKE FIRST OBSERVED CORK WITH A MICROSCOPE OF HIS OWN MAKING. HE NAMED THESE SMALL PARTITIONS *CELLS*.*
A CORK IS THE TISSUE OF A DEAD PLANT; HOOKE SAW THE OUTER WALLS OF DEAD PLANT CELLS.
ROBERT HOOKE (1635 - 1703)
HOOKE'S MICROSCOPE COLLECTED LIGHT USING A LAMP AND LENSES.
HOOKE MUST HAVE BEEN A SMART GUY TO CREATE A MICROSCOPE ALL BY HIMSELF.
WELL, THAT'S TRUE. BUT IT'S HIS DISCOVERY OF CELLS THAT'S IMPORTANT. YOU REALLY ARE A LITTLE PARAMECIUM-BRAIN.

* HE CALLED THESE PARTITIONS CELLS BECAUSE THEY REMINDED HIM OF THE CELLS IN A MONASTERY—THE SMALL ROOMS WHERE MONKS LIVE.

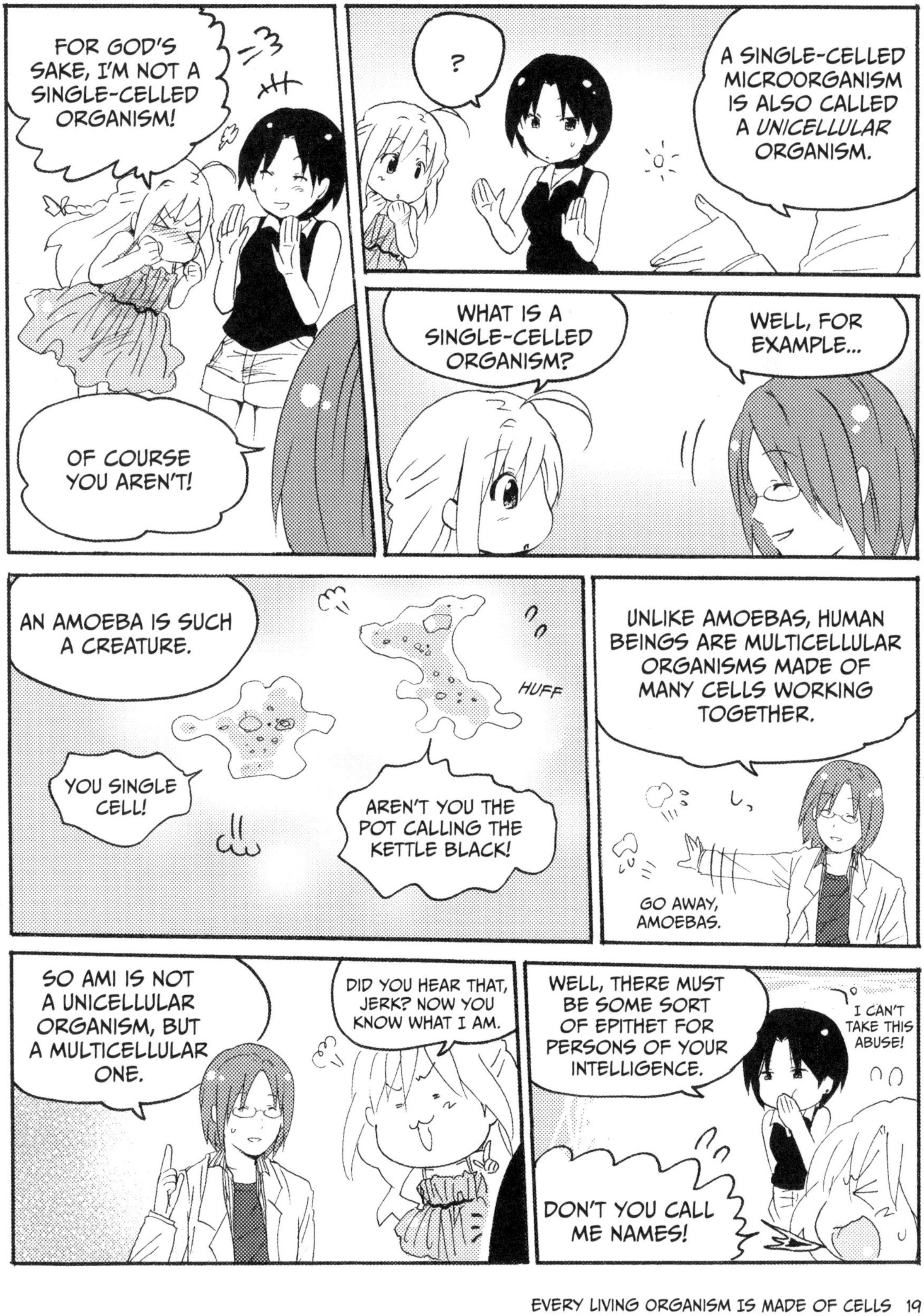
FOR GOD'S SAKE, I'M NOT A SINGLE-CELLED ORGANISM!
OF COURSE YOU AREN'T!
?
A SINGLE-CELLED MICROORGANISM IS ALSO CALLED A *UNICELLULAR* ORGANISM.
WHAT IS A SINGLE-CELLED ORGANISM?
WELL, FOR EXAMPLE...
AN AMOEBA IS SUCH A CREATURE.
HUFF
YOU SINGLE CELL!
AREN'T YOU THE POT CALLING THE KETTLE BLACK!
UNLIKE AMOEBAS, HUMAN BEINGS ARE MULTICELLULAR ORGANISMS MADE OF MANY CELLS WORKING TOGETHER.
GO AWAY, AMOEBAS.
SO AMI IS NOT A UNICELLULAR ORGANISM, BUT A MULTICELLULAR ONE.
DID YOU HEAR THAT, JERK? NOW YOU KNOW WHAT I AM.
WELL, THERE MUST BE SOME SORT OF EPITHET FOR PERSONS OF YOUR INTELLIGENCE.
I CAN'T TAKE THIS ABUSE!
DON'T YOU CALL ME NAMES!

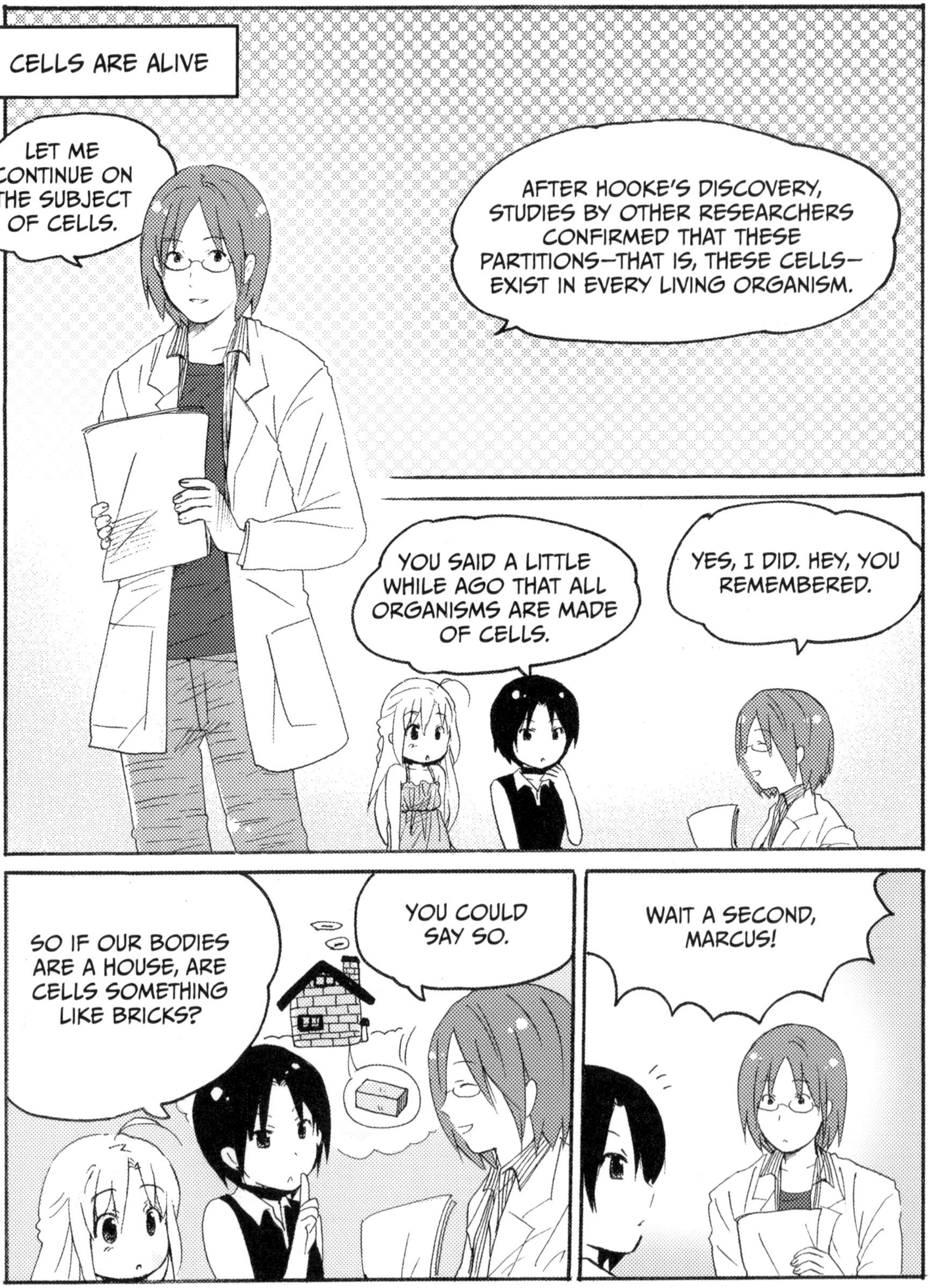
CELLS ARE ALIVE
LET ME CONTINUE ON THE SUBJECT OF CELLS.
AFTER HOOKE'S DISCOVERY, STUDIES BY OTHER RESEARCHERS CONFIRMED THAT THESE PARTITIONS—THAT IS, THESE CELLS—EXIST IN EVERY LIVING ORGANISM.
YOU SAID A LITTLE WHILE AGO THAT ALL ORGANISMS ARE MADE OF CELLS.
YES, I DID. HEY, YOU REMEMBERED.
SO IF OUR BODIES ARE A HOUSE, ARE CELLS SOMETHING LIKE BRICKS?
YOU COULD SAY SO.
WAIT A SECOND, MARCUS!

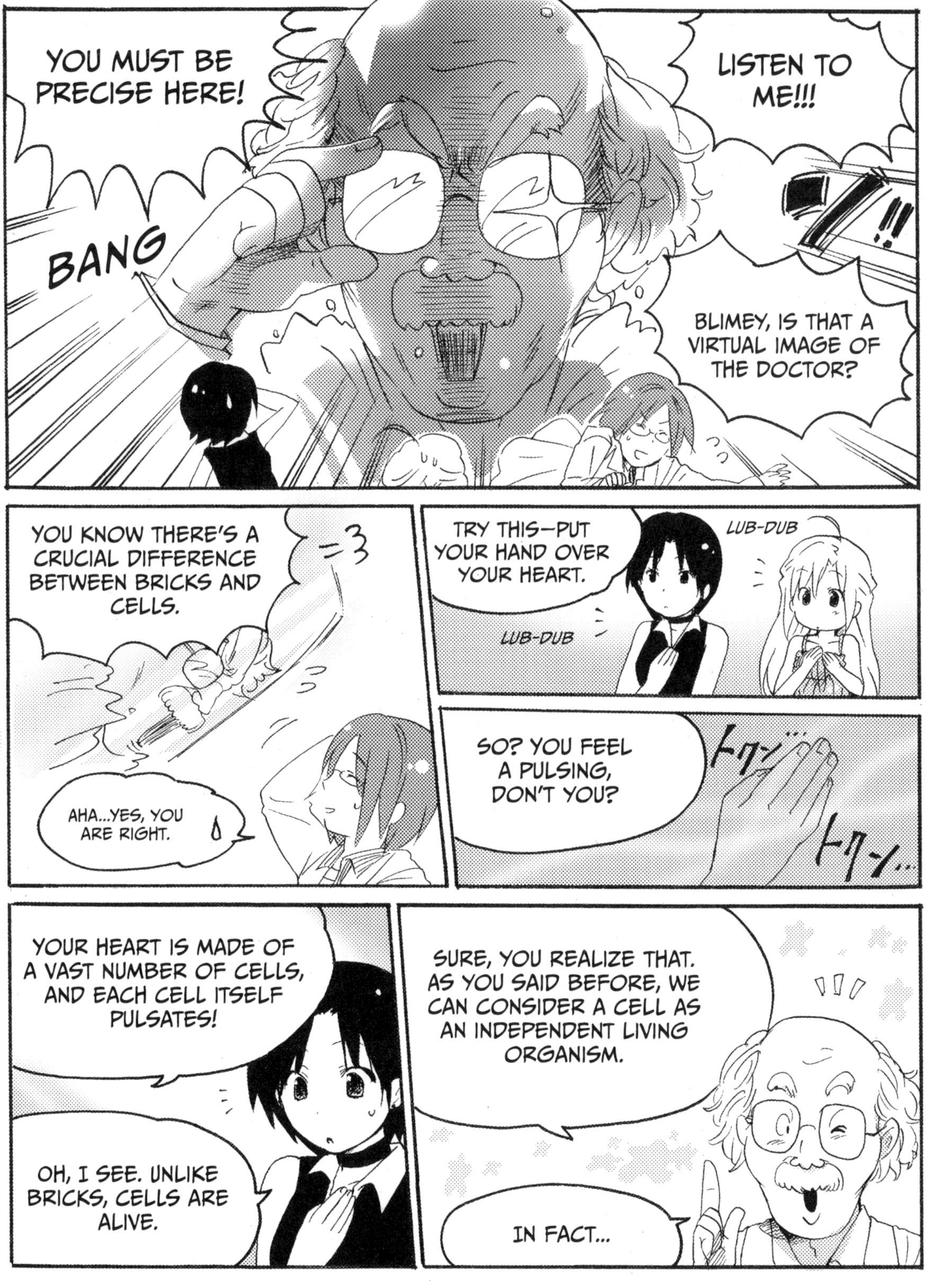
YOU MUST BE PRECISE HERE!
LISTEN TO ME!!!
BANG
BLIMEY, IS THAT A VIRTUAL IMAGE OF THE DOCTOR?
YOU KNOW THERE'S A CRUCIAL DIFFERENCE BETWEEN BRICKS AND CELLS.
AHA...YES, YOU ARE RIGHT.
TRY THIS—PUT YOUR HAND OVER YOUR HEART.
LUB-DUB
LUB-DUB
SO? YOU FEEL A PULSING, DON'T YOU?
トクン…
トクン…
YOUR HEART IS MADE OF A VAST NUMBER OF CELLS, AND EACH CELL ITSELF PULSATES!
OH, I SEE. UNLIKE BRICKS, CELLS ARE ALIVE.
SURE, YOU REALIZE THAT. AS YOU SAID BEFORE, WE CAN CONSIDER A CELL AS AN INDEPENDENT LIVING ORGANISM.
IN FACT...

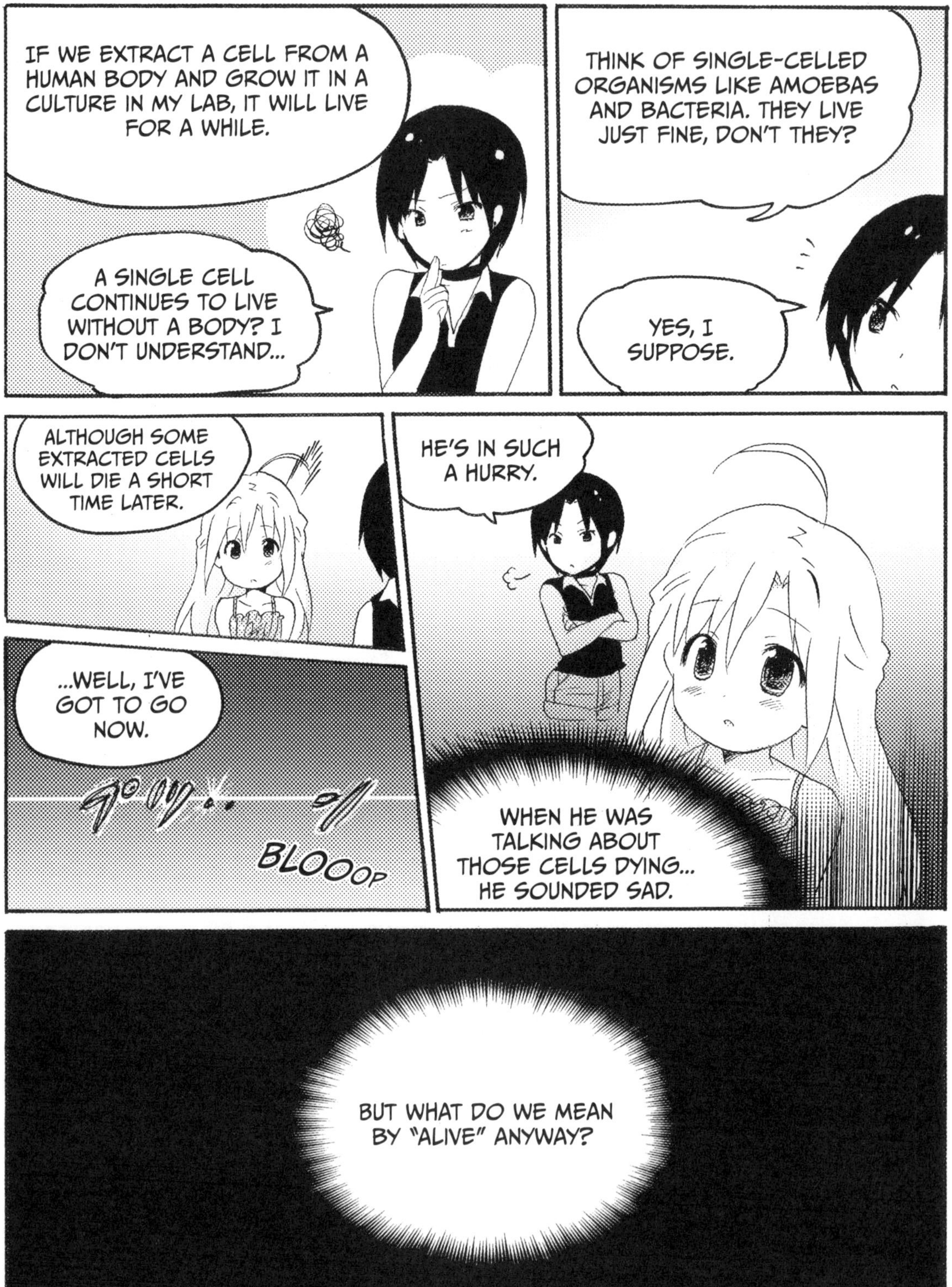
IF WE EXTRACT A CELL FROM A HUMAN BODY AND GROW IT IN A CULTURE IN MY LAB, IT WILL LIVE FOR A WHILE.
A SINGLE CELL CONTINUES TO LIVE WITHOUT A BODY? I DON'T UNDERSTAND...
THINK OF SINGLE-CELLED ORGANISMS LIKE AMOEBAS AND BACTERIA. THEY LIVE JUST FINE, DON'T THEY?
YES, I SUPPOSE.
ALTHOUGH SOME EXTRACTED CELLS WILL DIE A SHORT TIME LATER.
...WELL, I'VE GOT TO GO NOW.
BLOOOP
HE'S IN SUCH A HURRY.
WHEN HE WAS TALKING ABOUT THOSE CELLS DYING... HE SOUNDED SAD.
BUT WHAT DO WE MEAN BY "ALIVE" ANYWAY?

A CELL IS MADE UP OF VARIOUS MOLECULES

A cell is a result of many different molecules acting together. A variety of molecules, large and small, react with each other to form a working "society," which we call a cell.

Large molecules are things like nucleic acids (such as DNA), proteins, lipids (such as fats and cholesterol), and polysaccharides (such as starch). Smaller molecules like water, amino acids, and minerals are also in cells.

Do you remember that Dr. Moro said that proteins play an important role in cellular activity?

A large protein molecule is made up of a number of linked molecules called *amino acids*, which can be subdivided into 20 types. Proteins with various properties are created, depending upon their combination. The structure of a protein determines its function. Each protein carries out its own unique work—and our cells are alive thanks to the work done by these proteins.

Now let's take a closer look at the structure of a cell. The outside of the cell is called the *cell membrane*—it's made of a fatty material called *lipids*.

At its most basic, a cell is simply a cell membrane made of lipids, with various molecules floating inside it.

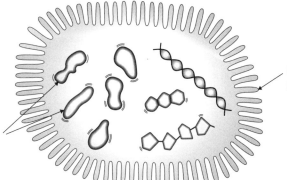

Molecules floating in water

Membrane made of lipids

A number of molecules are floating in a cell.

Glucose is also present in the cell—it is one of the most basic carbohydrates. You must have heard that rice and spaghetti are composed of carbohydrates. Glucose is contained in those foods and functions as an energy source in a cell.

I'VE NEVER SEEN A CELL!

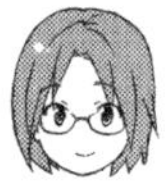

But wait, perhaps you have. So far I've talked about a world that's visible only through a microscope. But you probably don't have a microscope at home. So what can we do?

Just open the door to your refrigerator. You may see a gigantic oblong cell right there—yes, eggs, wonderful eggs! The chicken egg you eat for breakfast is just a single cell!

THE LONGEST CELL IN OUR BODIES

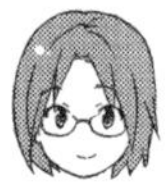

And of course, the human body is made up of cells, just like any other organism. At first glance, you can't find tissue that you can identify as "a single cell." But many different cells exist in our bodies, working together, as organs and other clusters.

Since we can't usually see cells without a microscope, you might think their size might be entirely microscopic. But we have a long, fine cell that's almost equivalent to the length of our bodies! This is the *nerve cell*, which responds to various stimuli, like light, sound, and touch, and is responsible for communicating these messages to the brain. Nerve cells are also known as *neurons*.

A nerve cell consists of a cell body and an axon. An *axon* is like a cable transmitting messages and stimuli through your body—and even to your brain. A number of protrusions called *dendrites* stick out from the cell body and receive messages from other neurons. The nerve cells in our body can have a meter-long axon.

Cell body
Axon
Dendrite

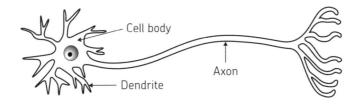

LET'S LOOK INSIDE A CELL

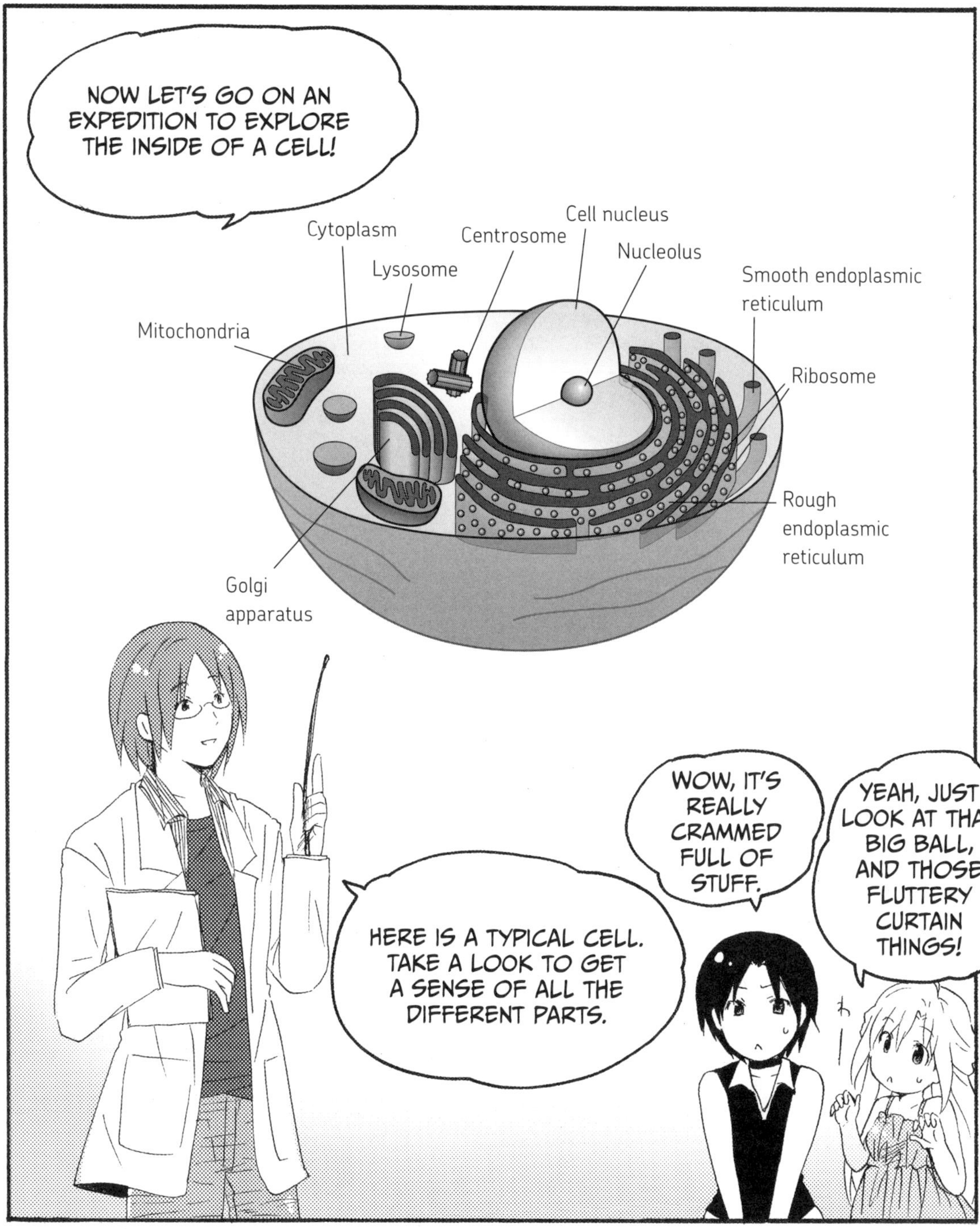
NOW LET'S GO ON AN EXPEDITION TO EXPLORE THE INSIDE OF A CELL!
Cytoplasm
Centrosome
Cell nucleus
Nucleolus
Lysosome
Smooth endoplasmic reticulum
Mitochondria
Ribosome
Rough endoplasmic reticulum
Golgi apparatus
HERE IS A TYPICAL CELL. TAKE A LOOK TO GET A SENSE OF ALL THE DIFFERENT PARTS.
WOW, IT'S REALLY CRAMMED FULL OF STUFF.
YEAH, JUST LOOK AT THAT BIG BALL, AND THOSE FLUTTERY CURTAIN THINGS!

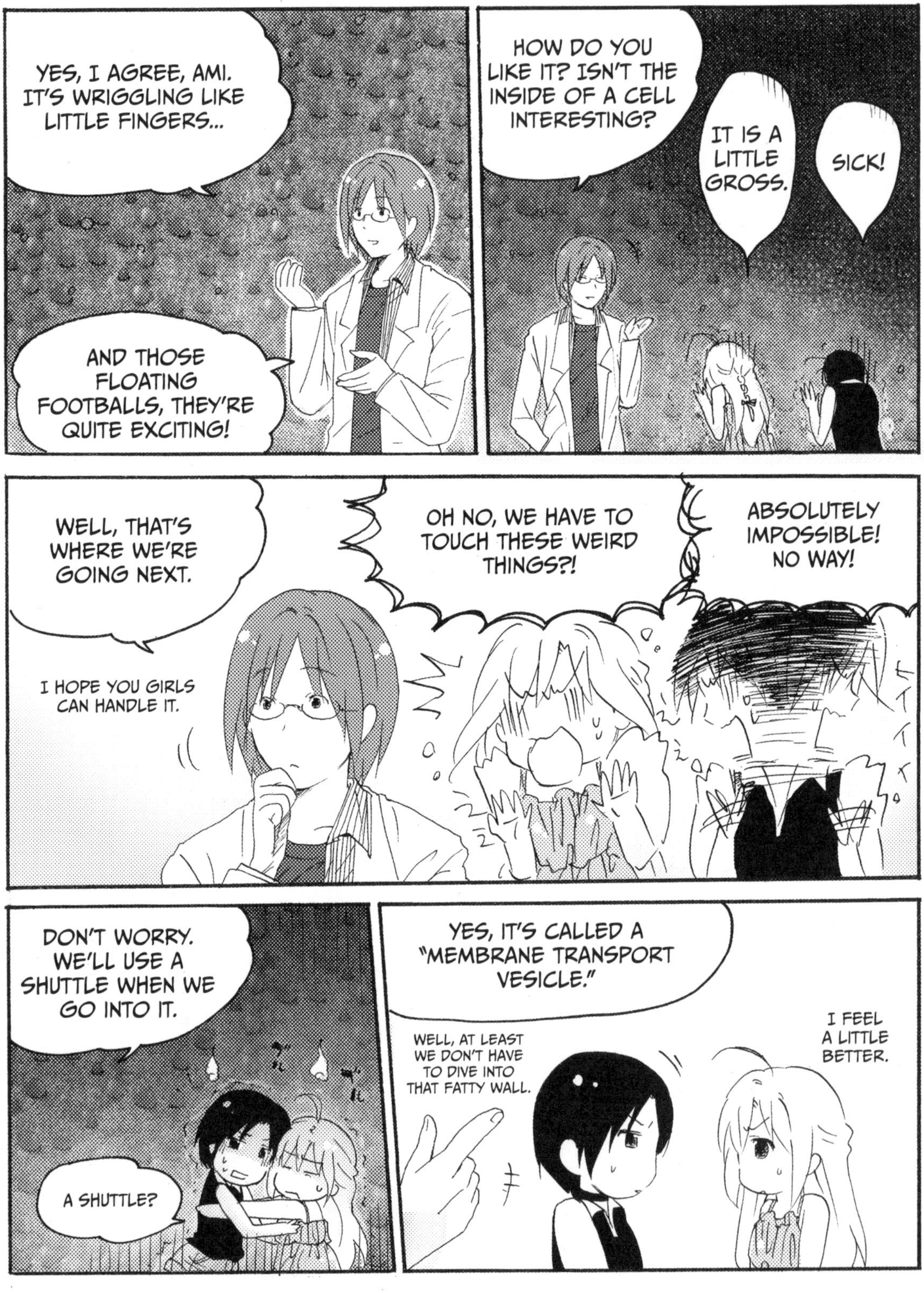
YES, I AGREE, AMI. IT'S WRIGGLING LIKE LITTLE FINGERS...
AND THOSE FLOATING FOOTBALLS, THEY'RE QUITE EXCITING!
HOW DO YOU LIKE IT? ISN'T THE INSIDE OF A CELL INTERESTING?
IT IS A LITTLE GROSS.
SICK!
WELL, THAT'S WHERE WE'RE GOING NEXT.
I HOPE YOU GIRLS CAN HANDLE IT.
OH NO, WE HAVE TO TOUCH THESE WEIRD THINGS?!
ABSOLUTELY IMPOSSIBLE! NO WAY!
DON'T WORRY. WE'LL USE A SHUTTLE WHEN WE GO INTO IT.
A SHUTTLE?
YES, IT'S CALLED A "MEMBRANE TRANSPORT VESICLE."
WELL, AT LEAST WE DON'T HAVE TO DIVE INTO THAT FATTY WALL.
I FEEL A LITTLE BETTER.

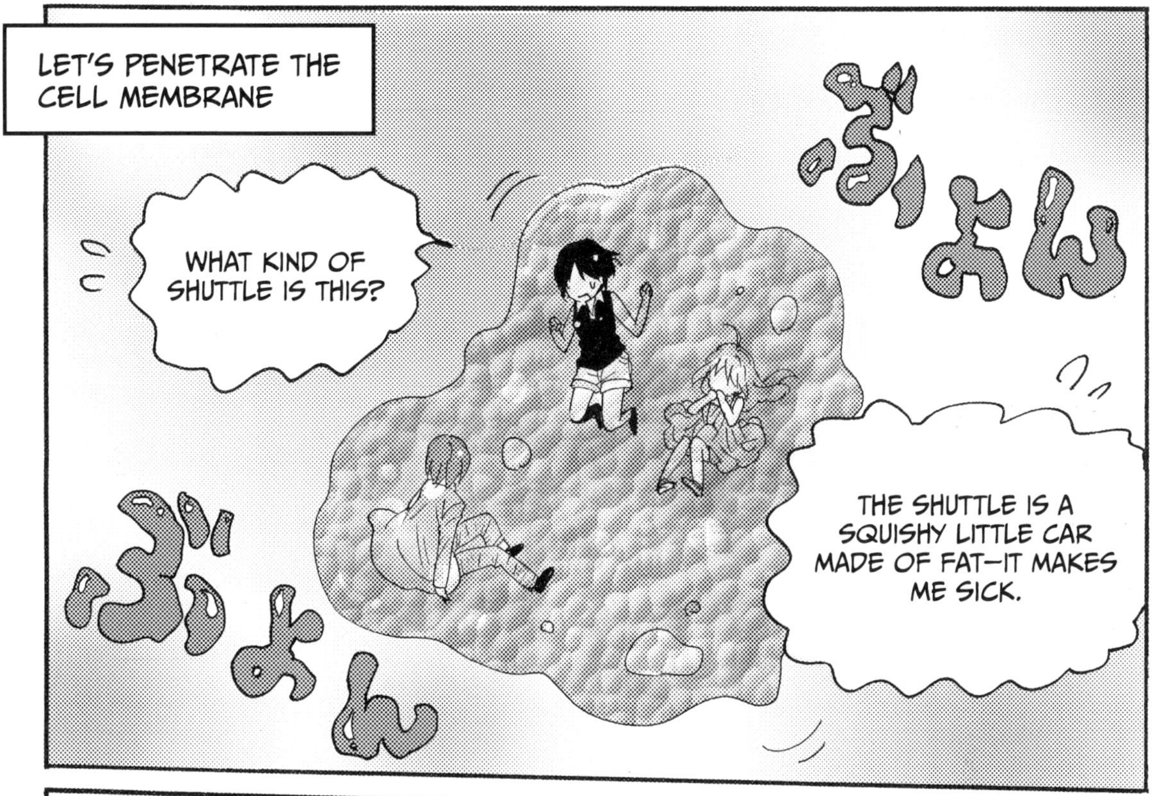
LET'S PENETRATE THE CELL MEMBRANE
ぶよん
WHAT KIND OF SHUTTLE IS THIS?
THE SHUTTLE IS A SQUISHY LITTLE CAR MADE OF FAT—IT MAKES ME SICK.
ぶよん

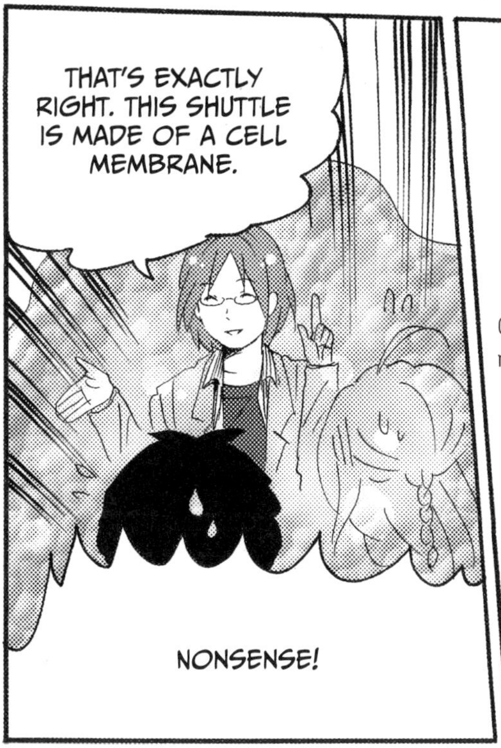
THAT'S EXACTLY RIGHT. THIS SHUTTLE IS MADE OF A CELL MEMBRANE.
NONSENSE!

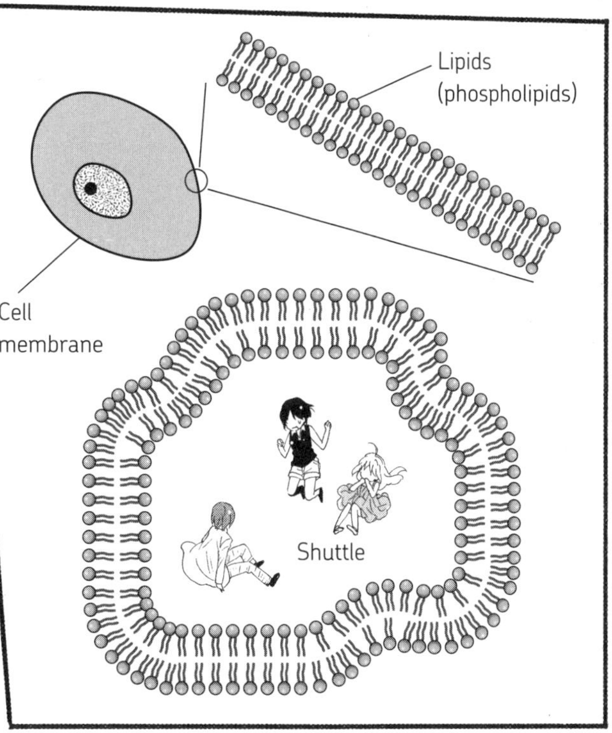
Lipids (phospholipids)
Cell membrane
Shuttle

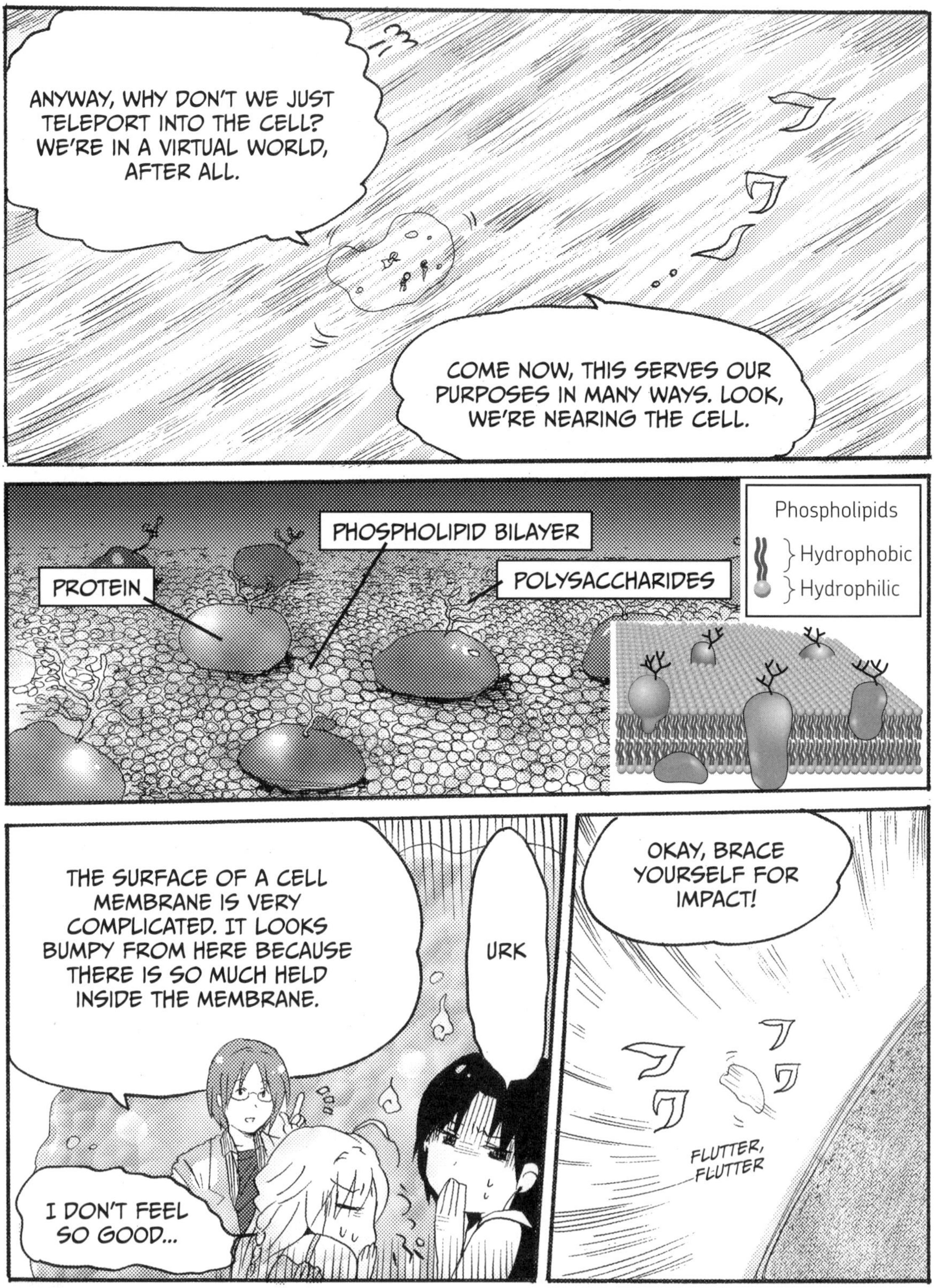
ANYWAY, WHY DON'T WE JUST TELEPORT INTO THE CELL? WE'RE IN A VIRTUAL WORLD, AFTER ALL.
COME NOW, THIS SERVES OUR PURPOSES IN MANY WAYS. LOOK, WE'RE NEARING THE CELL.
PHOSPHOLIPID BILAYER
PROTEIN
POLYSACCHARIDES
Phospholipids
Hydrophobic
Hydrophilic
THE SURFACE OF A CELL MEMBRANE IS VERY COMPLICATED. IT LOOKS BUMPY FROM HERE BECAUSE THERE IS SO MUCH HELD INSIDE THE MEMBRANE.
URK
I DON'T FEEL SO GOOD...
OKAY, BRACE YOURSELF FOR IMPACT!
FLUTTER, FLUTTER

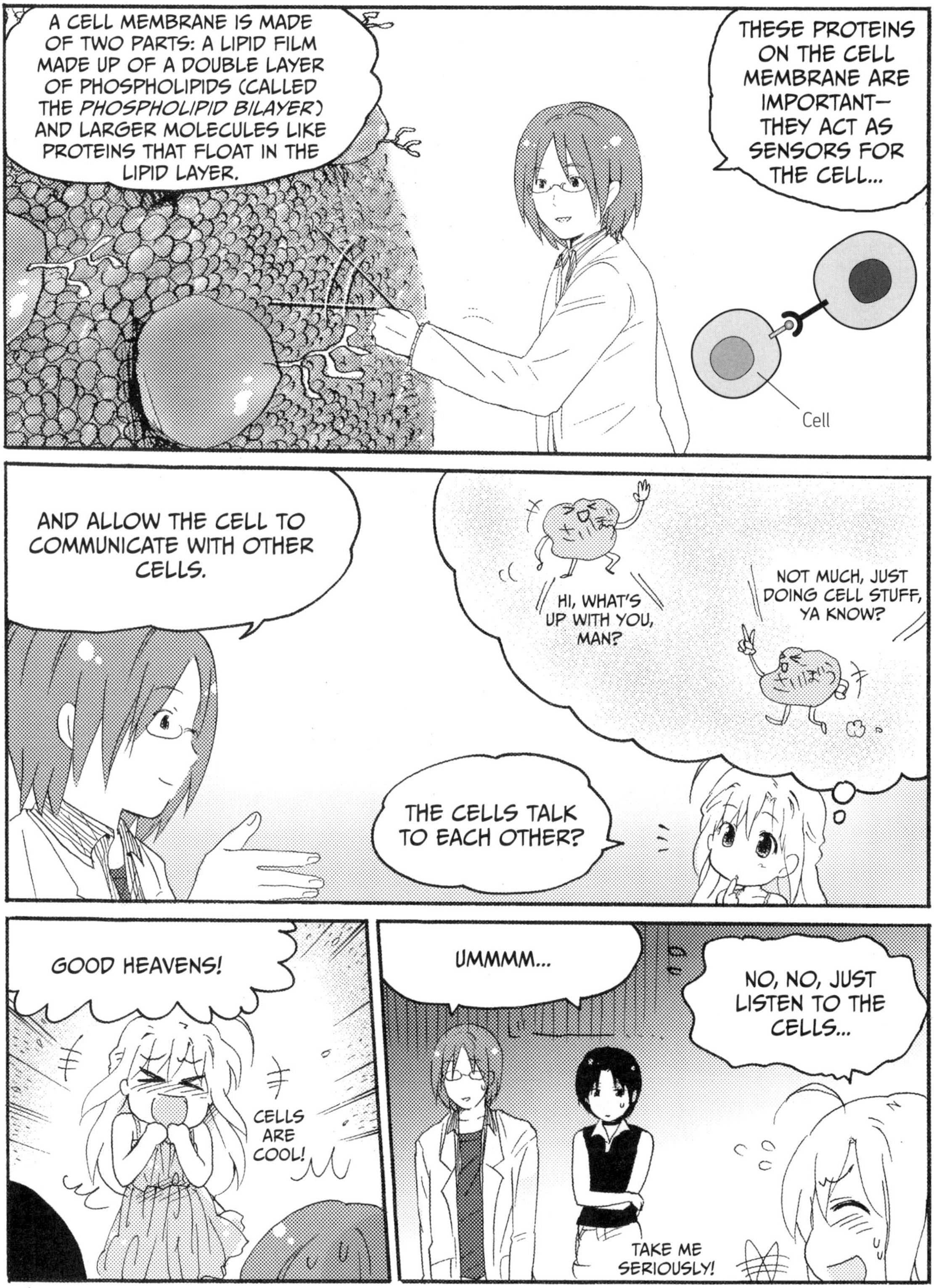

A CELL MEMBRANE IS MADE OF TWO PARTS: A LIPID FILM MADE UP OF A DOUBLE LAYER OF PHOSPHOLIPIDS (CALLED THE *PHOSPHOLIPID BILAYER*) AND LARGER MOLECULES LIKE PROTEINS THAT FLOAT IN THE LIPID LAYER.
THESE PROTEINS ON THE CELL MEMBRANE ARE IMPORTANT—THEY ACT AS SENSORS FOR THE CELL...
Cell
AND ALLOW THE CELL TO COMMUNICATE WITH OTHER CELLS.
HI, WHAT'S UP WITH YOU, MAN?
NOT MUCH, JUST DOING CELL STUFF, YA KNOW?
THE CELLS TALK TO EACH OTHER?
GOOD HEAVENS!
CELLS ARE COOL!
UMMMM...
NO, NO, JUST LISTEN TO THE CELLS...
TAKE ME SERIOUSLY!

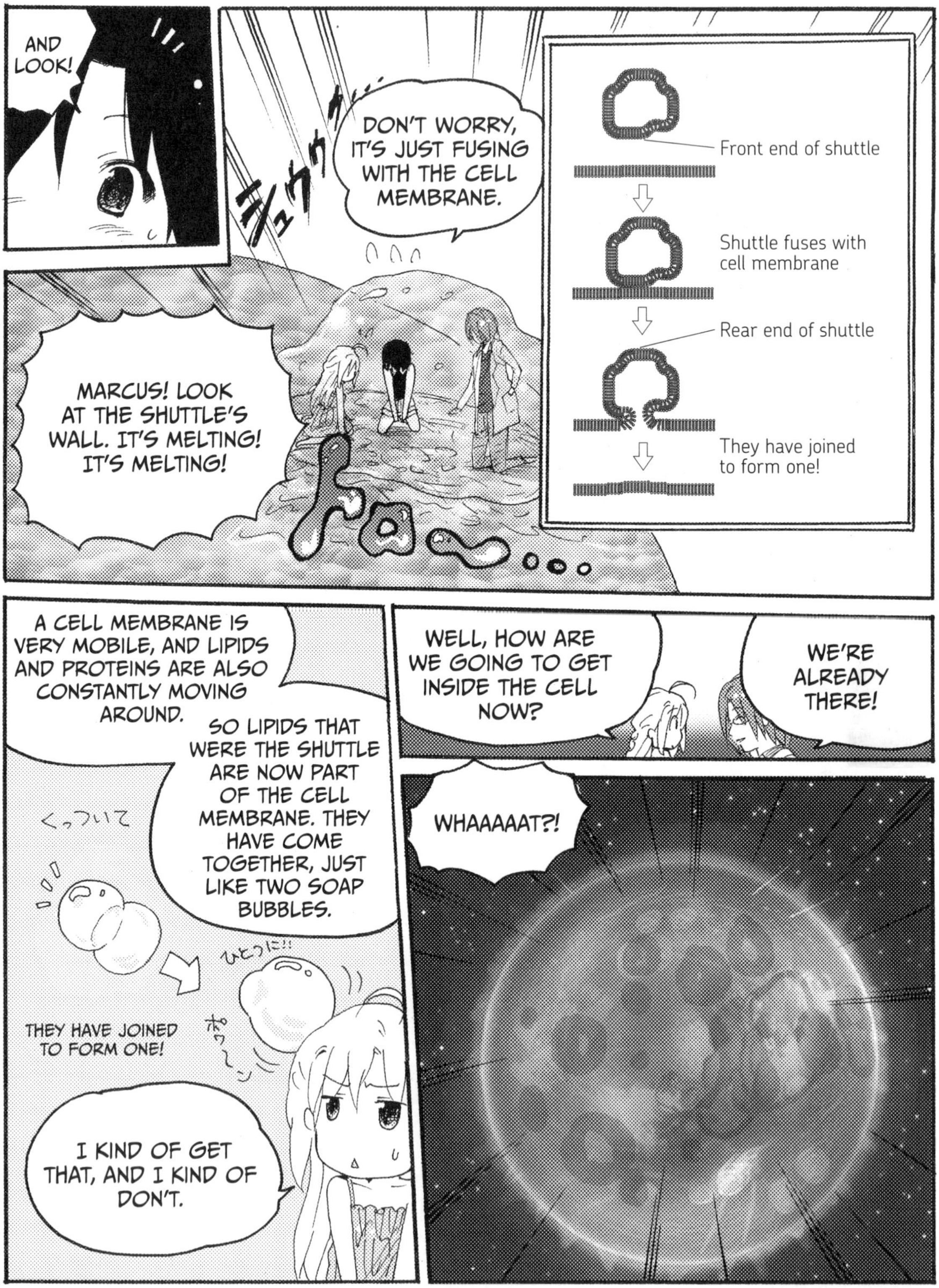
AND LOOK!
シュウゥ...
DON'T WORRY, IT'S JUST FUSING WITH THE CELL MEMBRANE.
Front end of shuttle
Shuttle fuses with cell membrane
Rear end of shuttle
They have joined to form one!
MARCUS! LOOK AT THE SHUTTLE'S WALL. IT'S MELTING! IT'S MELTING!
A CELL MEMBRANE IS VERY MOBILE, AND LIPIDS AND PROTEINS ARE ALSO CONSTANTLY MOVING AROUND.
SO LIPIDS THAT WERE THE SHUTTLE ARE NOW PART OF THE CELL MEMBRANE. THEY HAVE COME TOGETHER, JUST LIKE TWO SOAP BUBBLES.
くっついて
ひとつに!!
THEY HAVE JOINED TO FORM ONE!
I KIND OF GET THAT, AND I KIND OF DON'T.
WELL, HOW ARE WE GOING TO GET INSIDE THE CELL NOW?
WE'RE ALREADY THERE!
WHAAAAAT?!

CELL ORGANELLES

What's this gel that surrounds us?

It's *cytoplasm*, a thick solution of water and dissolved molecules. These molecules are nutrients that the cell needs and leftover waste.

As we float through the cell's cytoplasm, we see a few larger objects floating. The cell is really packed full of these little things. Sometimes an object that resembles a giant blue whale, or a spaceship, or a giant football drifts by us.

As I mentioned, individual cells are living things and must carry out a number of tasks in order to survive. These tasks are performed by these different shapes—which are called *organelles*. Organelle means "little organ." Just like the human body has a heart, brain, and other organs that have specific duties in the body, a cell has "little organs" responsible for different tasks.

That big spaceship was actually a cell organelle.

Hey Marcus, I've been wondering for a while, what are those wall-shaped things overlapping each other in layers?

Let's move in for a closer look. The walls are made of thin films of phospholipid bilayer, just like cell membranes, but they are not actually walls. As we move closer, you can see they have a structure similar to that of a folded ribbon.

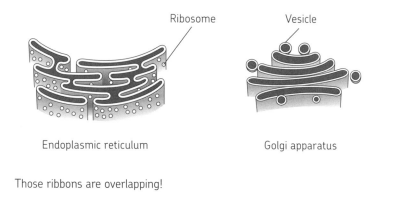

Ribosome   Vesicle

Endoplasmic reticulum   Golgi apparatus

Those ribbons are overlapping!

It's a cell organelle called the *endoplasmic reticulum*. Its surface is covered with many ribosomes, another smaller organelle. These two types of organelles work together to synthesize and process proteins.

Actually, there are many places in the body, like the liver and lymph nodes, where the cells make proteins specifically to be secreted for use by other cells in the body. When these proteins are secreted, a cell organelle called the *Golgi apparatus* functions as a delivery center. The process for packaging and delivering proteins is just like the way we entered the cell—but in reverse! The Golgi apparatus is also known as a Golgi body. Golgi bodies package proteins and other molecules into membrane bags, called *vesicles*. These vesicles deliver molecules made inside the cell to the outside; they can also deliver molecules to other organelles in the cells, like lysosomes, via membrane fusion.

Golgi apparatus

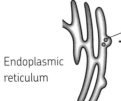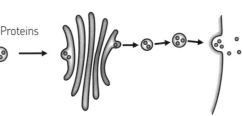

Proteins

Endoplasmic reticulum

Secreted to the outside of the cell

The Golgi apparatus secretes proteins outside the cell.

 Aha! That's why you carried us on such an unusual shuttle.

 That's right—we entered the cell through membrane fusion.

Look, bags of cell membrane are used all over the cell to perform different functions. In a cell organelle called a *lysosome*, large molecules are broken down into smaller sizes. It functions like the digestive system of a cell, by degrading molecules.

There are also bag-like vesicles called *peroxisomes* everywhere inside the cell. They oxygenate harmful substances like bacteria, in order to neutralize them.

Mitochondria are also very important cell organelles. They are a sort of "power plant" for producing the energy necessary for cells to live. Cells need mitochondria to survive!

Cristae
Matrix

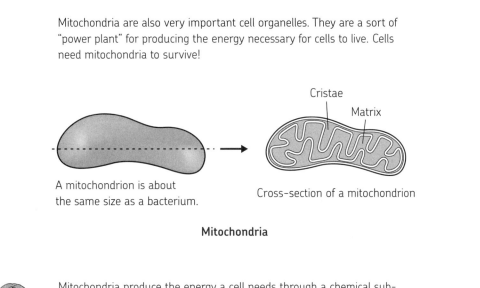

A mitochondrion is about the same size as a bacterium.

Cross-section of a mitochondrion

Mitochondria

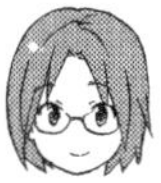

Mitochondria produce the energy a cell needs through a chemical substance called *ATP*. It creates ATP using oxygen taken into our body through aspiration (breathing) and pyruvic acid, a product made from the breakdown of glucose (sugar).

Oxygen
Glucose
Pyruvic acid
Degradation
ATP
ATP
ATP
Produces energy
Mitochondrion

Power plant

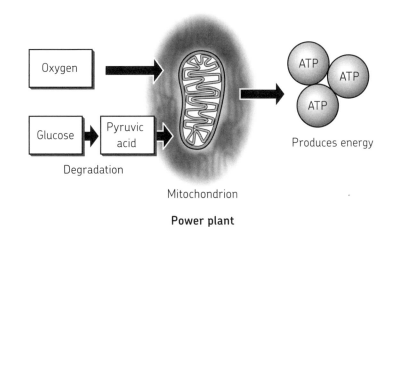

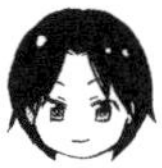

Well, I think I can remember Golgi apparatus and mitochondria, but ribosome, lysosome, and peroxisome sound similar and are difficult to remember.

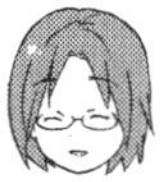

Yes, people often get these names mixed up. But if you know a little bit about Greek, you might remember their names better. *Some* means body, like a bag. *Lyso* means to break down. So it makes sense that a lysosome is a bag-like organelle that breaks down molecules. In the same way, *peroxi* means oxygen. And if you can remember that you use hydrogen peroxide to clean a cut, you can remember that a *peroxisome* is an organelle that uses oxygen to kill bacteria and other dangerous things.

THE NUCLEUS: A LITTLE BRAIN

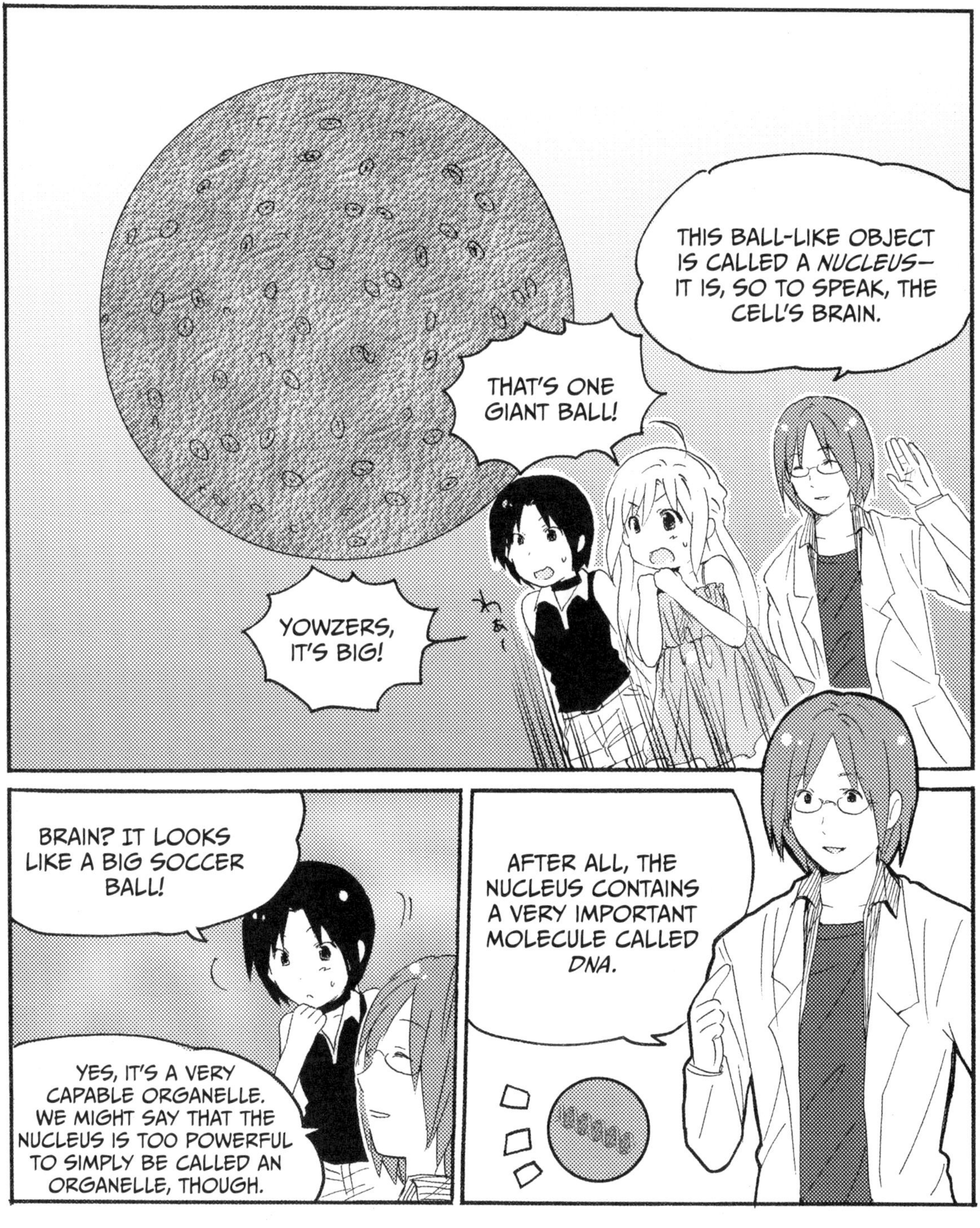
THIS BALL-LIKE OBJECT IS CALLED A *NUCLEUS*— IT IS, SO TO SPEAK, THE CELL'S BRAIN.
THAT'S ONE GIANT BALL!
YOWZERS, IT'S BIG!
BRAIN? IT LOOKS LIKE A BIG SOCCER BALL!
YES, IT'S A VERY CAPABLE ORGANELLE. WE MIGHT SAY THAT THE NUCLEUS IS TOO POWERFUL TO SIMPLY BE CALLED AN ORGANELLE, THOUGH.
AFTER ALL, THE NUCLEUS CONTAINS A VERY IMPORTANT MOLECULE CALLED *DNA*.

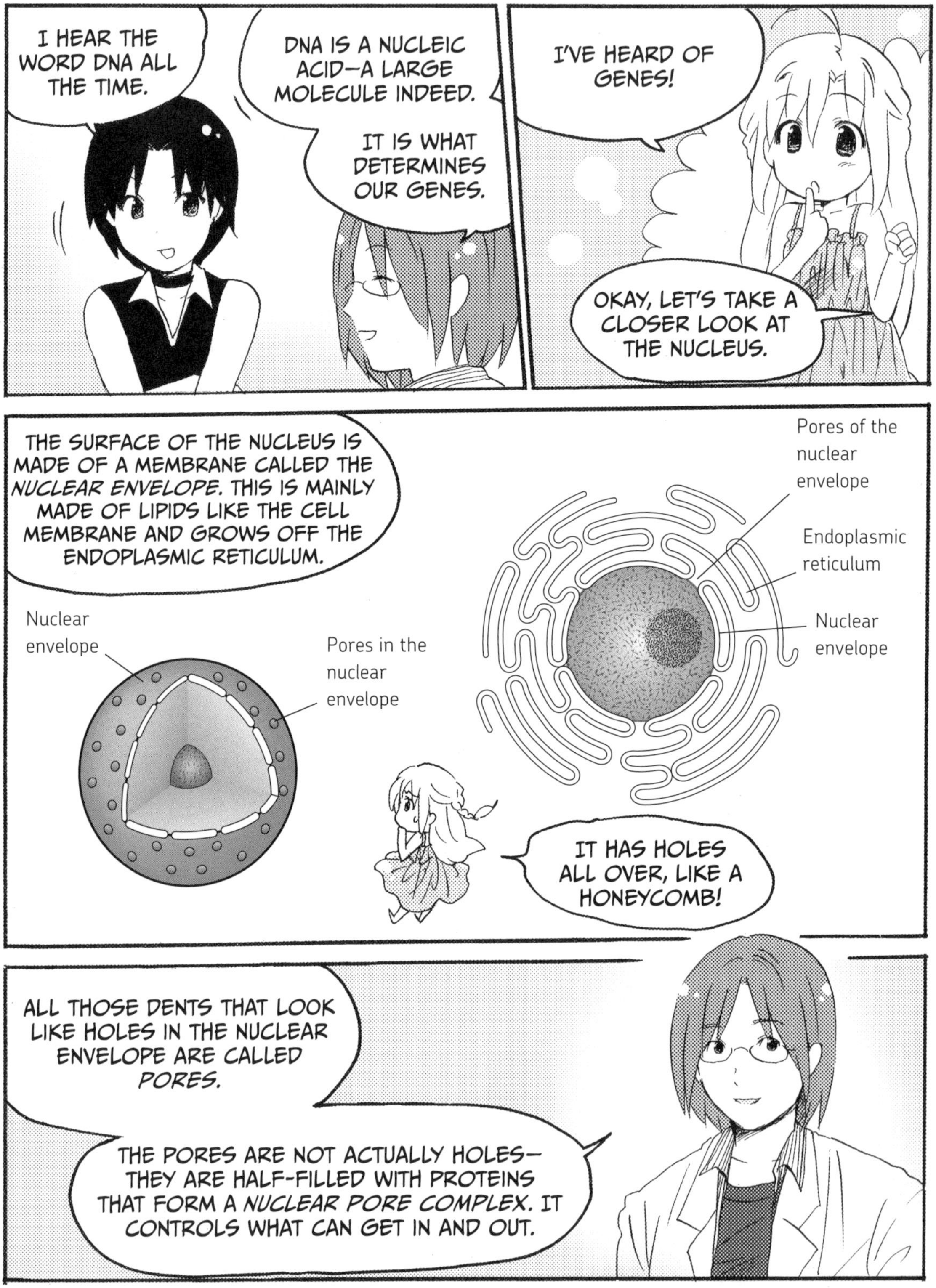
I HEAR THE WORD DNA ALL THE TIME.
DNA IS A NUCLEIC ACID—A LARGE MOLECULE INDEED.
IT IS WHAT DETERMINES OUR GENES.
I'VE HEARD OF GENES!
OKAY, LET'S TAKE A CLOSER LOOK AT THE NUCLEUS.
THE SURFACE OF THE NUCLEUS IS MADE OF A MEMBRANE CALLED THE *NUCLEAR ENVELOPE*. THIS IS MAINLY MADE OF LIPIDS LIKE THE CELL MEMBRANE AND GROWS OFF THE ENDOPLASMIC RETICULUM.
Nuclear envelope
Pores in the nuclear envelope
Pores of the nuclear envelope
Endoplasmic reticulum
Nuclear envelope
IT HAS HOLES ALL OVER, LIKE A HONEYCOMB!
ALL THOSE DENTS THAT LOOK LIKE HOLES IN THE NUCLEAR ENVELOPE ARE CALLED *PORES*.
THE PORES ARE NOT ACTUALLY HOLES—THEY ARE HALF-FILLED WITH PROTEINS THAT FORM A *NUCLEAR PORE COMPLEX*. IT CONTROLS WHAT CAN GET IN AND OUT.

WHAT'S INSIDE THE NUCLEUS?

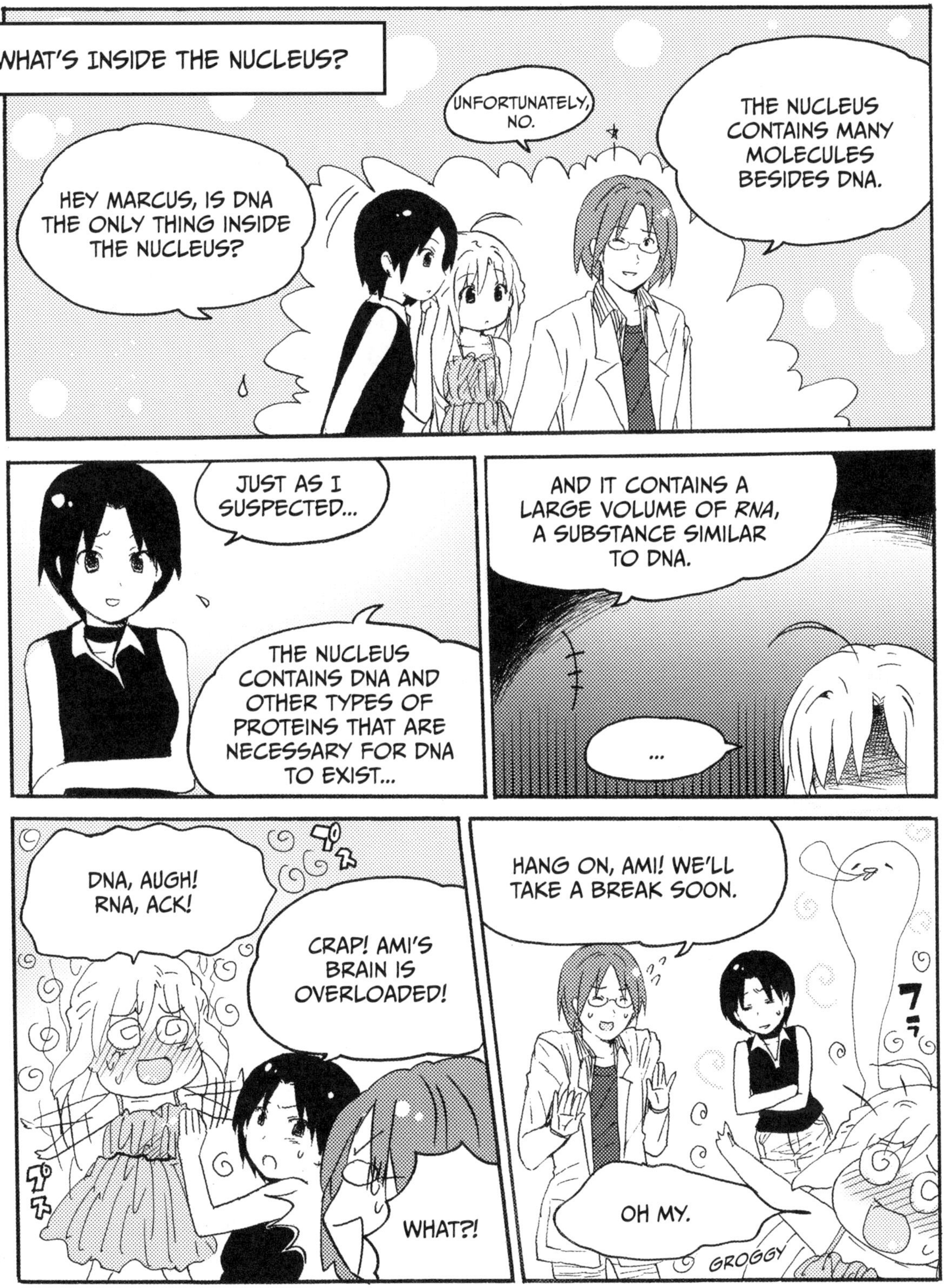
HEY MARCUS, IS DNA THE ONLY THING INSIDE THE NUCLEUS?
UNFORTUNATELY, NO.
THE NUCLEUS CONTAINS MANY MOLECULES BESIDES DNA.
JUST AS I SUSPECTED...
THE NUCLEUS CONTAINS DNA AND OTHER TYPES OF PROTEINS THAT ARE NECESSARY FOR DNA TO EXIST...
AND IT CONTAINS A LARGE VOLUME OF *RNA*, A SUBSTANCE SIMILAR TO DNA.
...
DNA, AUGH! RNA, ACK!
CRAP! AMI'S BRAIN IS OVERLOADED!
WHAT?!
HANG ON, AMI! WE'LL TAKE A BREAK SOON.
OH MY.
GROGGY

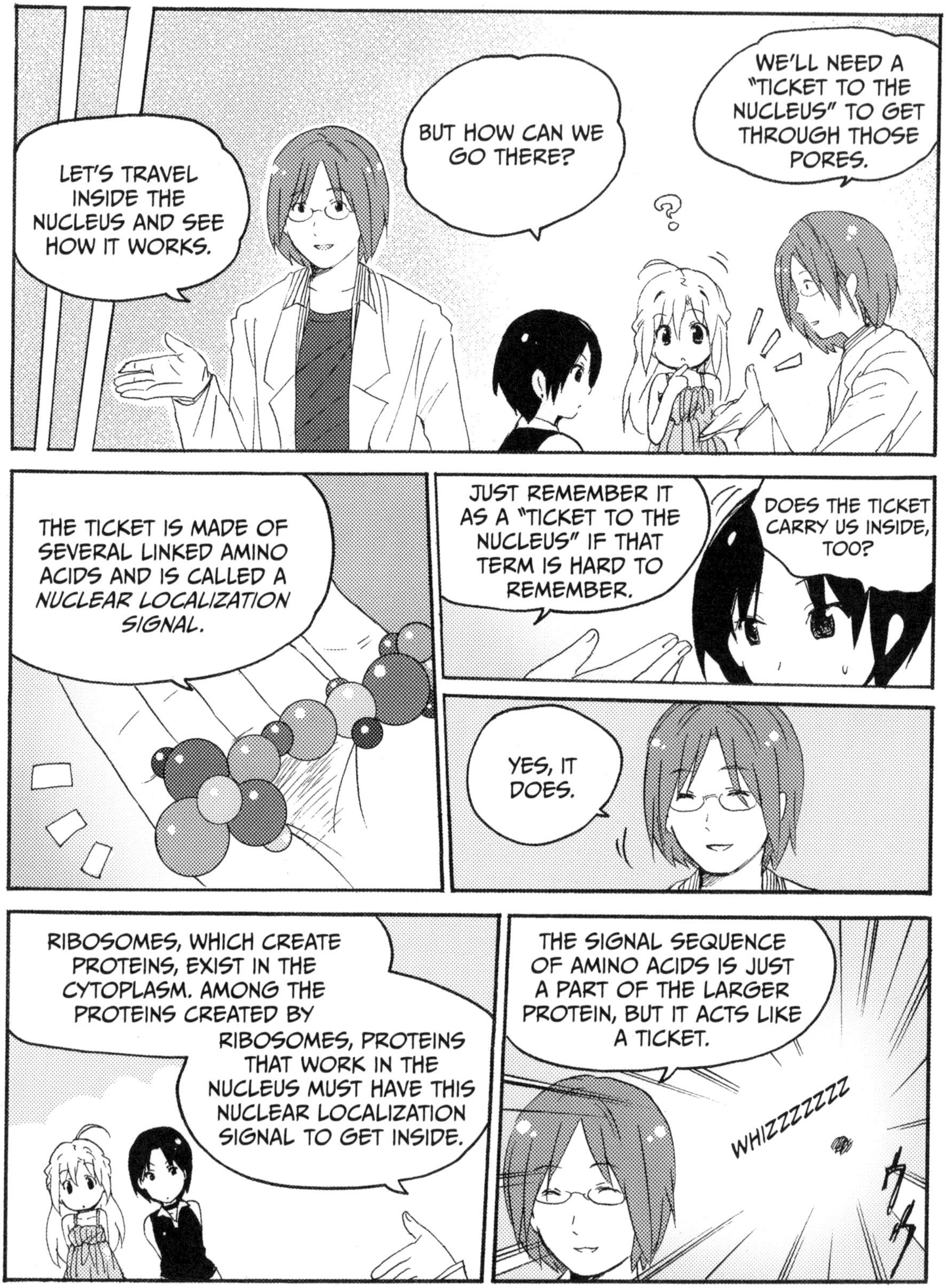
LET'S TRAVEL INSIDE THE NUCLEUS AND SEE HOW IT WORKS.
BUT HOW CAN WE GO THERE?
WE'LL NEED A "TICKET TO THE NUCLEUS" TO GET THROUGH THOSE PORES.
THE TICKET IS MADE OF SEVERAL LINKED AMINO ACIDS AND IS CALLED A *NUCLEAR LOCALIZATION SIGNAL*.
JUST REMEMBER IT AS A "TICKET TO THE NUCLEUS" IF THAT TERM IS HARD TO REMEMBER.
DOES THE TICKET CARRY US INSIDE, TOO?
YES, IT DOES.
RIBOSOMES, WHICH CREATE PROTEINS, EXIST IN THE CYTOPLASM. AMONG THE PROTEINS CREATED BY RIBOSOMES, PROTEINS THAT WORK IN THE NUCLEUS MUST HAVE THIS NUCLEAR LOCALIZATION SIGNAL TO GET INSIDE.
THE SIGNAL SEQUENCE OF AMINO ACIDS IS JUST A PART OF THE LARGER PROTEIN, BUT IT ACTS LIKE A TICKET.
WHIZZZZZZZ

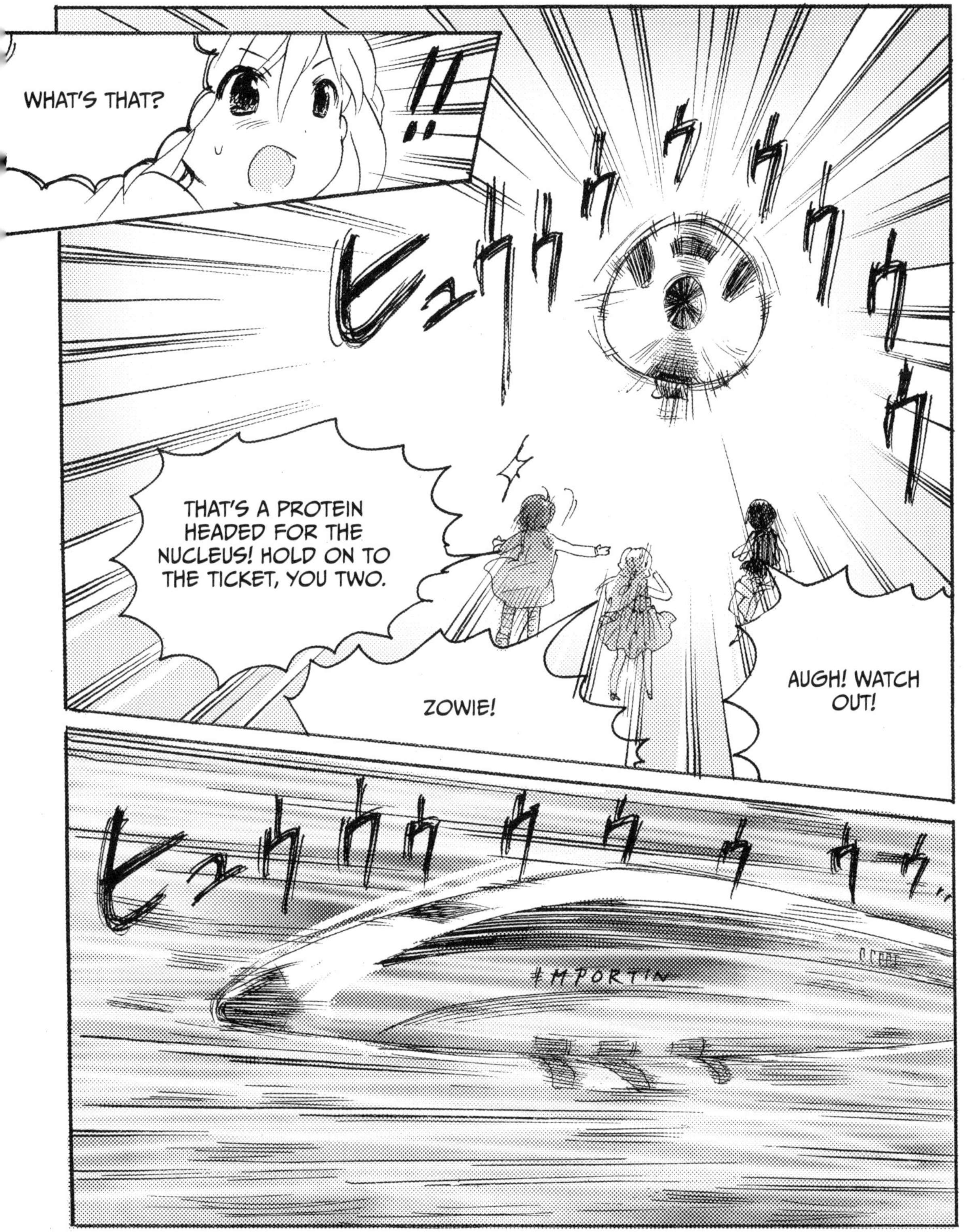
WHAT'S THAT?
!!
THAT'S A PROTEIN HEADED FOR THE NUCLEUS! HOLD ON TO THE TICKET, YOU TWO.
ZOWIE!
AUGH! WATCH OUT!
IMPORTIN

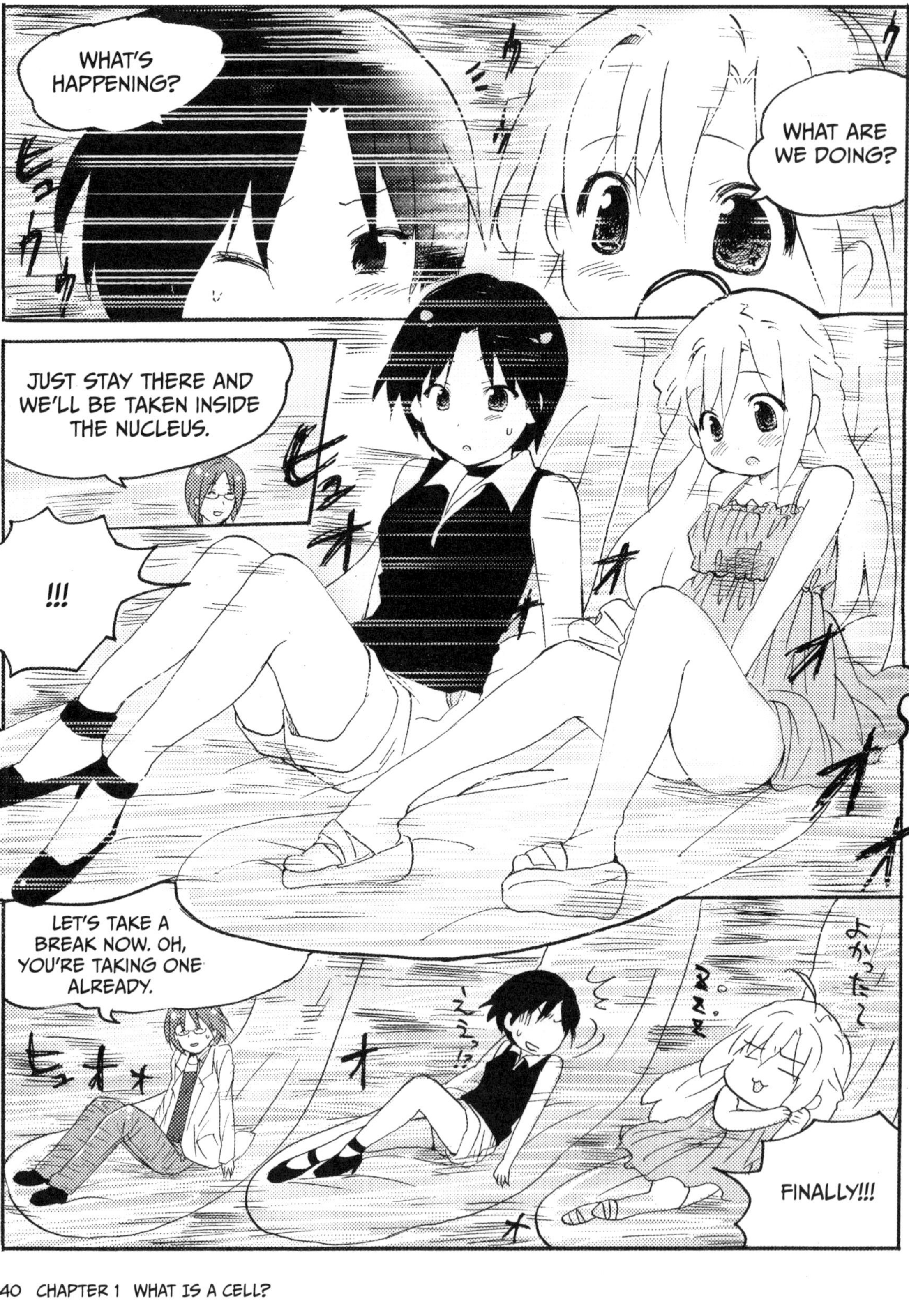
WHAT'S HAPPENING?
WHAT ARE WE DOING?
JUST STAY THERE AND WE'LL BE TAKEN INSIDE THE NUCLEUS.
!!!
LET'S TAKE A BREAK NOW. OH, YOU'RE TAKING ONE ALREADY.
FINALLY!!!

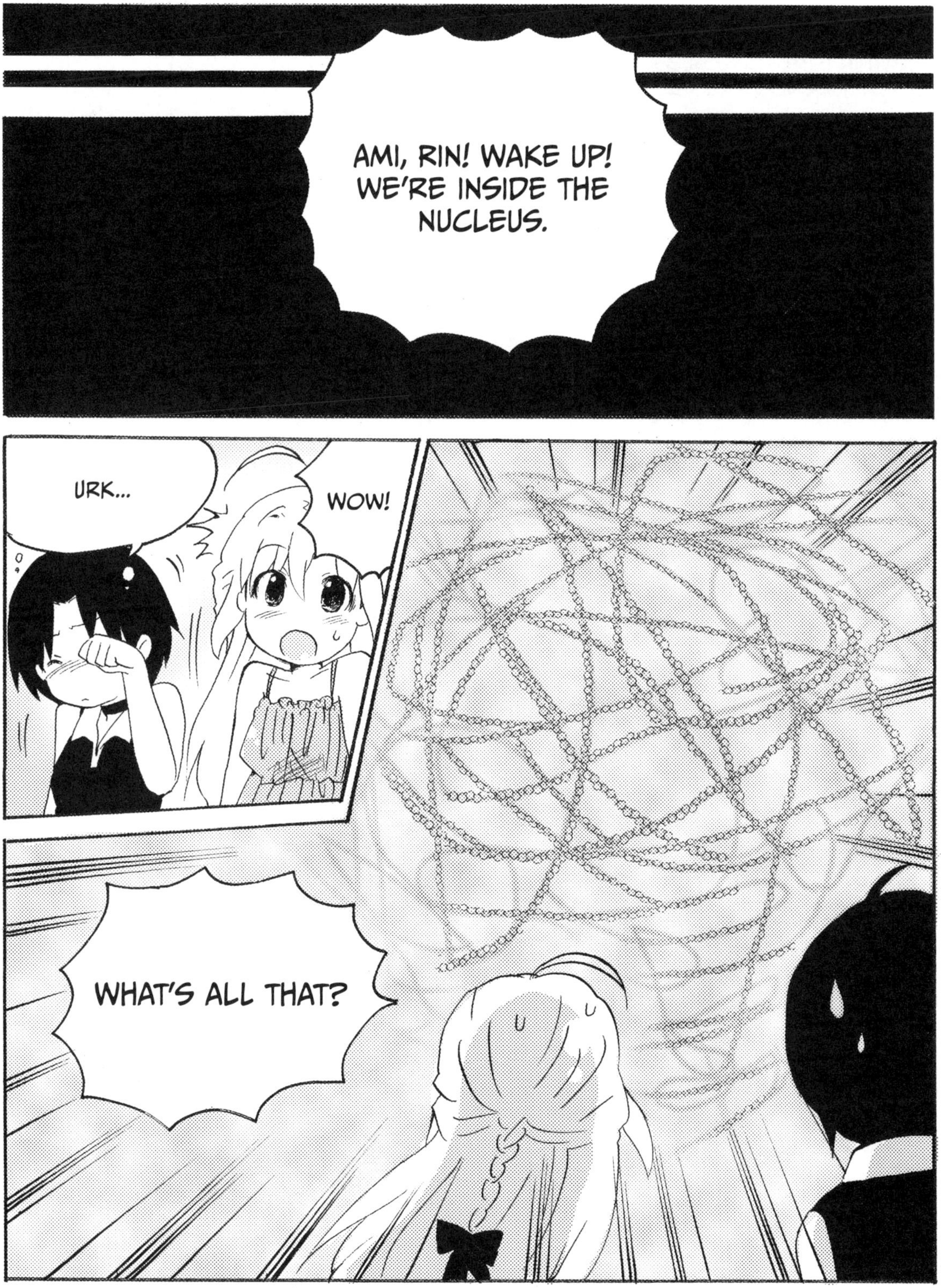
AMI, RIN! WAKE UP! WE'RE INSIDE THE NUCLEUS.
URK...
WOW!
WHAT'S ALL THAT?

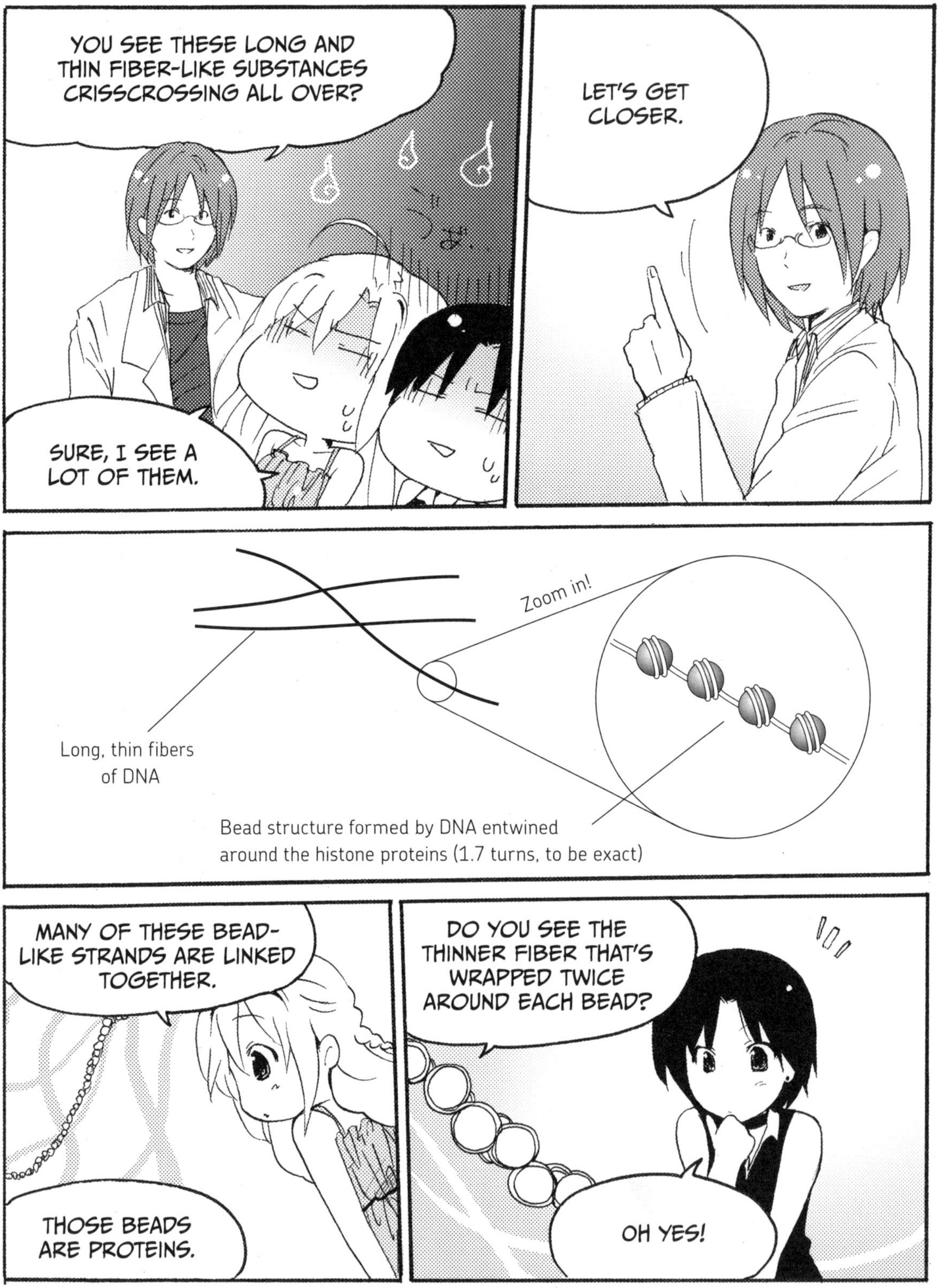
YOU SEE THESE LONG AND THIN FIBER-LIKE SUBSTANCES CRISSCROSSING ALL OVER?
SURE, I SEE A LOT OF THEM.
LET'S GET CLOSER.
Zoom in!
Long, thin fibers of DNA
Bead structure formed by DNA entwined around the histone proteins (1.7 turns, to be exact)
MANY OF THESE BEAD-LIKE STRANDS ARE LINKED TOGETHER.
THOSE BEADS ARE PROTEINS.
DO YOU SEE THE THINNER FIBER THAT'S WRAPPED TWICE AROUND EACH BEAD?
OH YES!

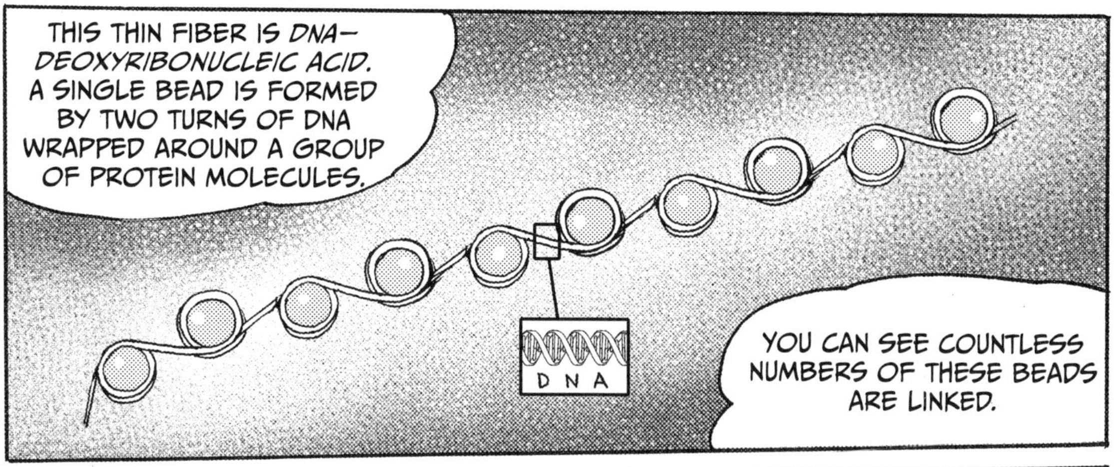
THIS THIN FIBER IS *DNA— DEOXYRIBONUCLEIC ACID.* A SINGLE BEAD IS FORMED BY TWO TURNS OF DNA WRAPPED AROUND A GROUP OF PROTEIN MOLECULES.
DNA
YOU CAN SEE COUNTLESS NUMBERS OF THESE BEADS ARE LINKED.

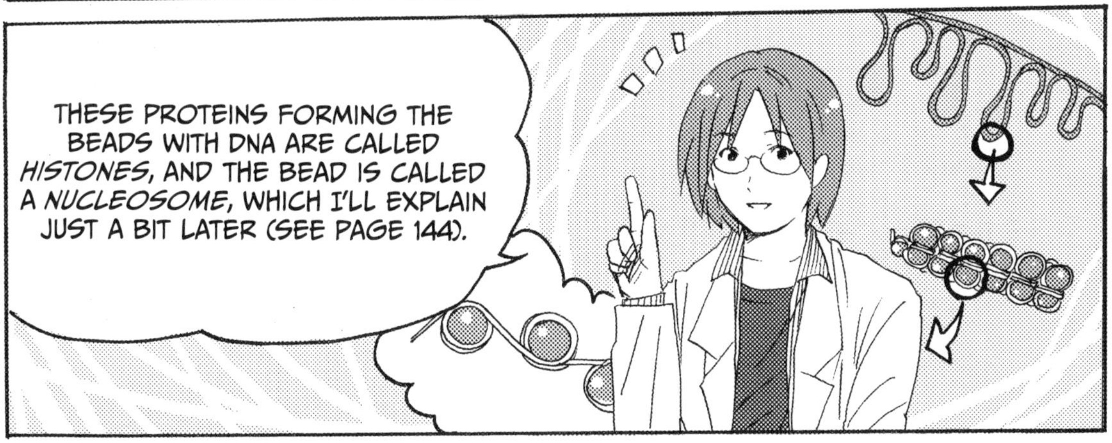
THESE PROTEINS FORMING THE BEADS WITH DNA ARE CALLED *HISTONES*, AND THE BEAD IS CALLED A *NUCLEOSOME*, WHICH I'LL EXPLAIN JUST A BIT LATER (SEE PAGE 144).

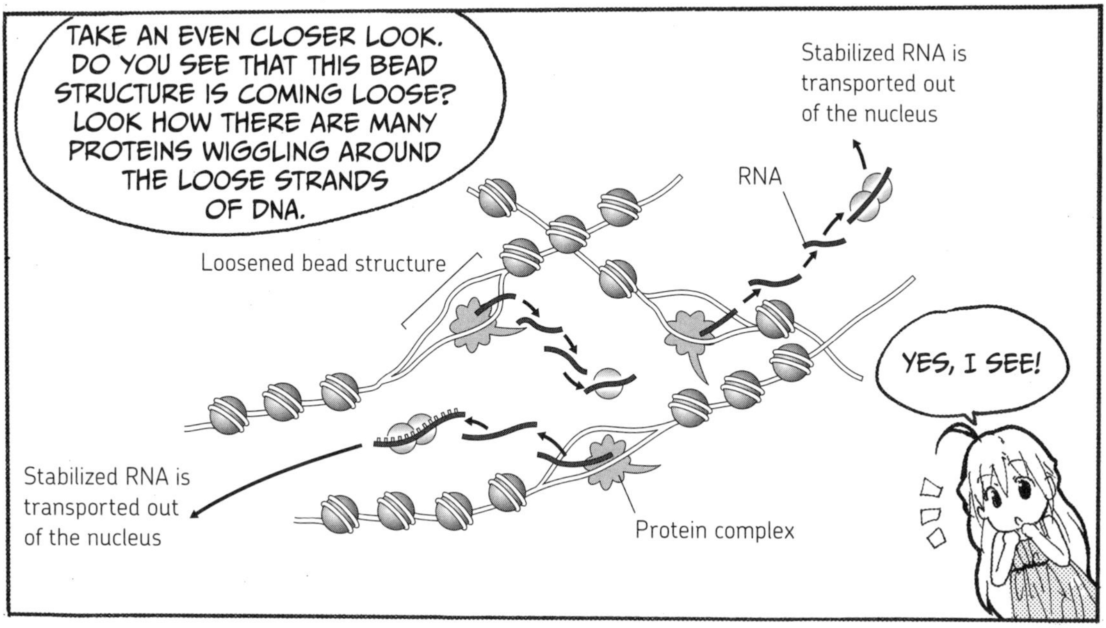
TAKE AN EVEN CLOSER LOOK. DO YOU SEE THAT THIS BEAD STRUCTURE IS COMING LOOSE? LOOK HOW THERE ARE MANY PROTEINS WIGGLING AROUND THE LOOSE STRANDS OF DNA.
Stabilized RNA is transported out of the nucleus
RNA
Loosened bead structure
YES, I SEE!
Stabilized RNA is transported out of the nucleus
Protein complex

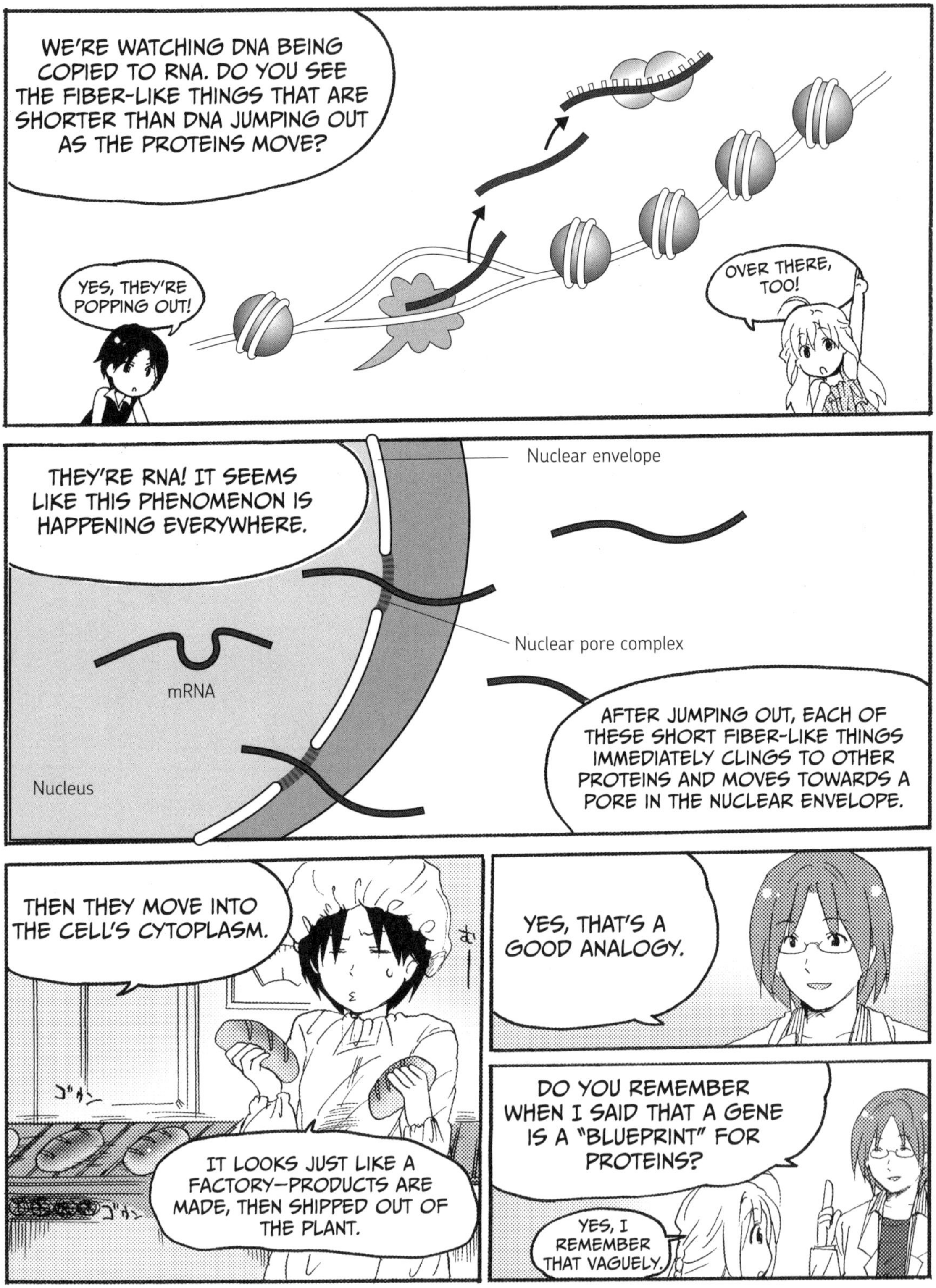
WE'RE WATCHING DNA BEING COPIED TO RNA. DO YOU SEE THE FIBER-LIKE THINGS THAT ARE SHORTER THAN DNA JUMPING OUT AS THE PROTEINS MOVE?
YES, THEY'RE POPPING OUT!
OVER THERE, TOO!
THEY'RE RNA! IT SEEMS LIKE THIS PHENOMENON IS HAPPENING EVERYWHERE.
Nuclear envelope
Nuclear pore complex
mRNA
Nucleus
AFTER JUMPING OUT, EACH OF THESE SHORT FIBER-LIKE THINGS IMMEDIATELY CLINGS TO OTHER PROTEINS AND MOVES TOWARDS A PORE IN THE NUCLEAR ENVELOPE.
THEN THEY MOVE INTO THE CELL'S CYTOPLASM.
IT LOOKS JUST LIKE A FACTORY—PRODUCTS ARE MADE, THEN SHIPPED OUT OF THE PLANT.
YES, THAT'S A GOOD ANALOGY.
DO YOU REMEMBER WHEN I SAID THAT A GENE IS A "BLUEPRINT" FOR PROTEINS?
YES, I REMEMBER THAT VAGUELY.

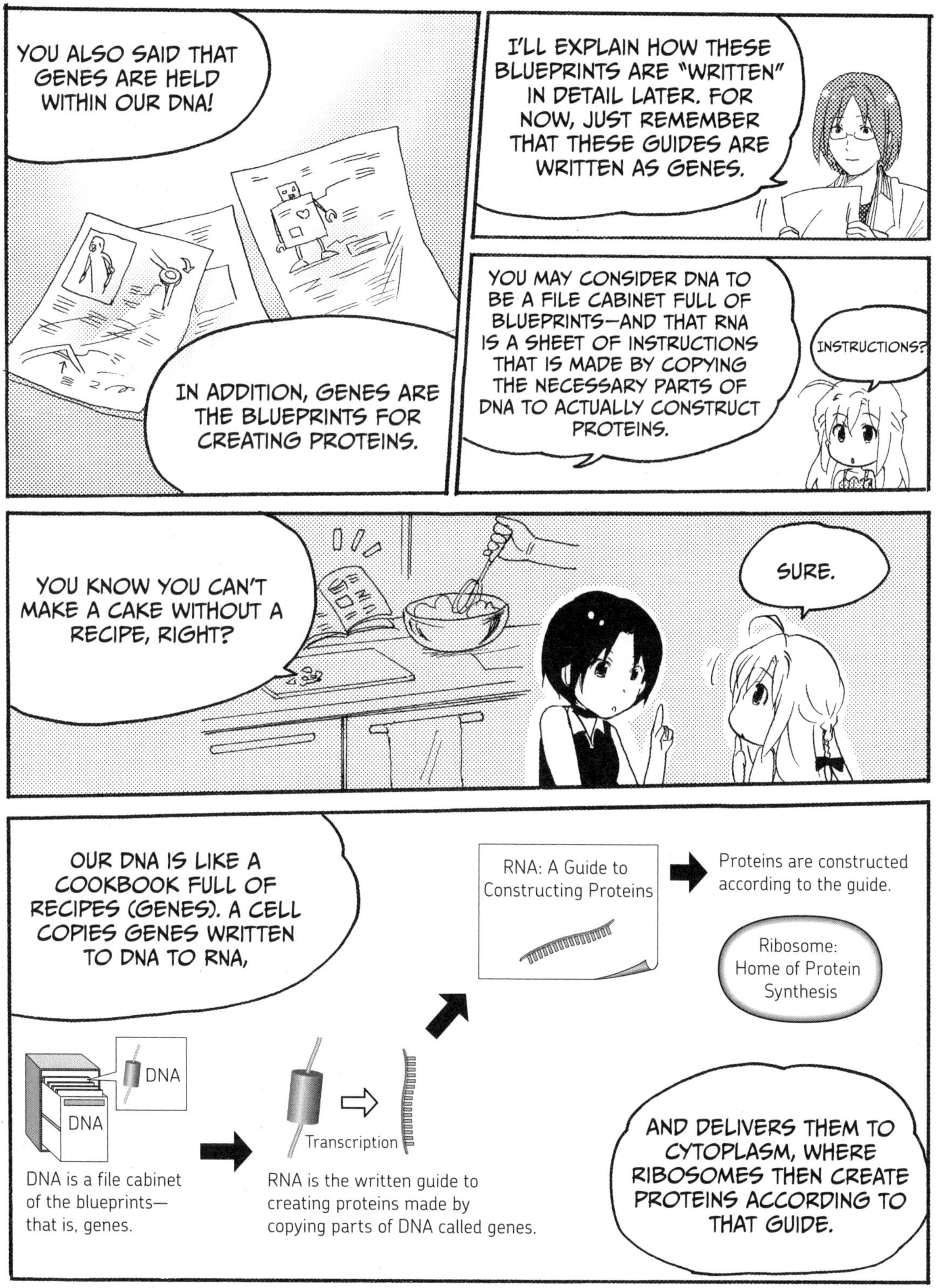
YOU ALSO SAID THAT GENES ARE HELD WITHIN OUR DNA!
IN ADDITION, GENES ARE THE BLUEPRINTS FOR CREATING PROTEINS.
I'LL EXPLAIN HOW THESE BLUEPRINTS ARE "WRITTEN" IN DETAIL LATER. FOR NOW, JUST REMEMBER THAT THESE GUIDES ARE WRITTEN AS GENES.
YOU MAY CONSIDER DNA TO BE A FILE CABINET FULL OF BLUEPRINTS—AND THAT RNA IS A SHEET OF INSTRUCTIONS THAT IS MADE BY COPYING THE NECESSARY PARTS OF DNA TO ACTUALLY CONSTRUCT PROTEINS.
INSTRUCTIONS?
YOU KNOW YOU CAN'T MAKE A CAKE WITHOUT A RECIPE, RIGHT?
SURE.
OUR DNA IS LIKE A COOKBOOK FULL OF RECIPES (GENES). A CELL COPIES GENES WRITTEN TO DNA TO RNA,
DNA
DNA
DNA is a file cabinet of the blueprints—that is, genes.
Transcription
RNA is the written guide to creating proteins made by copying parts of DNA called genes.
RNA: A Guide to Constructing Proteins
Proteins are constructed according to the guide.
Ribosome: Home of Protein Synthesis
AND DELIVERS THEM TO CYTOPLASM, WHERE RIBOSOMES THEN CREATE PROTEINS ACCORDING TO THAT GUIDE.

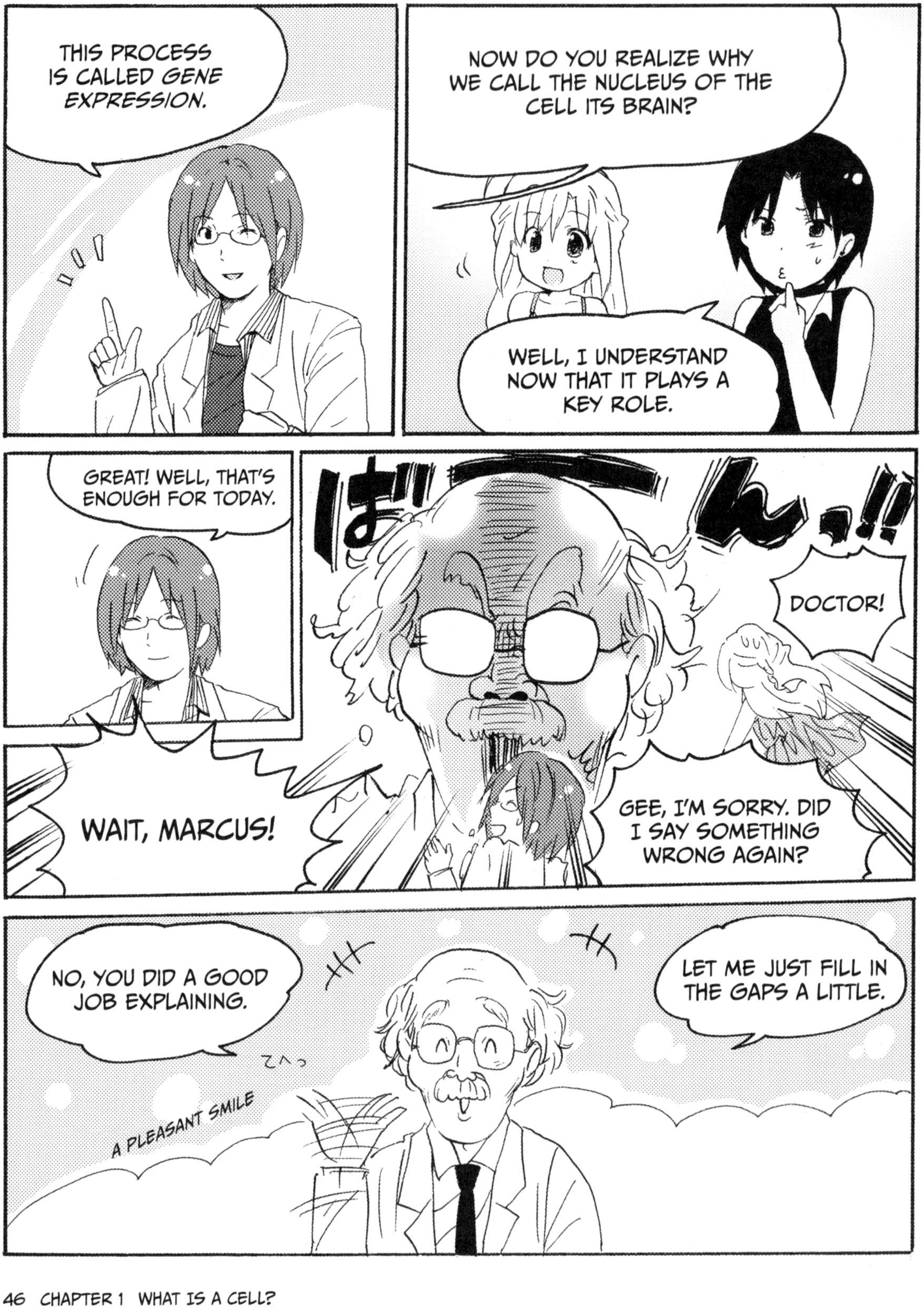
THIS PROCESS IS CALLED GENE EXPRESSION.
NOW DO YOU REALIZE WHY WE CALL THE NUCLEUS OF THE CELL ITS BRAIN?
WELL, I UNDERSTAND NOW THAT IT PLAYS A KEY ROLE.
GREAT! WELL, THAT'S ENOUGH FOR TODAY.
ばーーん!!
DOCTOR!
WAIT, MARCUS!
GEE, I'M SORRY. DID I SAY SOMETHING WRONG AGAIN?
NO, YOU DID A GOOD JOB EXPLAINING.
LET ME JUST FILL IN THE GAPS A LITTLE.
てへっ
A PLEASANT SMILE

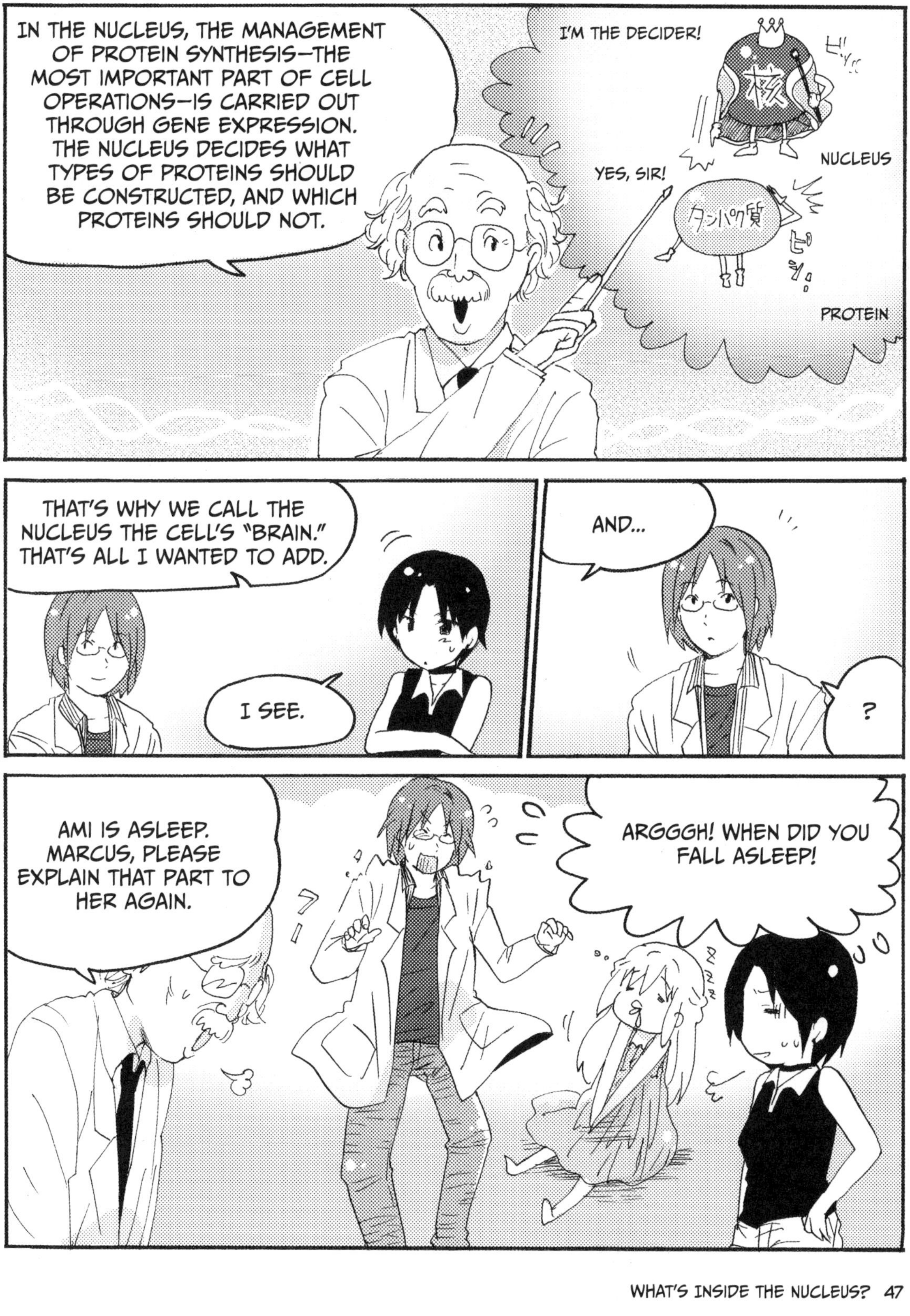
IN THE NUCLEUS, THE MANAGEMENT OF PROTEIN SYNTHESIS—THE MOST IMPORTANT PART OF CELL OPERATIONS—IS CARRIED OUT THROUGH GENE EXPRESSION. THE NUCLEUS DECIDES WHAT TYPES OF PROTEINS SHOULD BE CONSTRUCTED, AND WHICH PROTEINS SHOULD NOT.
I'M THE DECIDER!
核
NUCLEUS
YES, SIR!
タンパク質
PROTEIN
THAT'S WHY WE CALL THE NUCLEUS THE CELL'S "BRAIN." THAT'S ALL I WANTED TO ADD.
I SEE.
AND...
?
AMI IS ASLEEP. MARCUS, PLEASE EXPLAIN THAT PART TO HER AGAIN.
ARGGGH! WHEN DID YOU FALL ASLEEP!

SINGLE-CELLED AND MULTICELLULAR ORGANISMS

 I used the term *single cell* at the beginning of this chapter—the term can refer to one organism or to a "society" of living organisms.

A single-celled organism is a living creature made of only one cell. Most single-celled organisms are invisible to the human eye. You may think they are far away, but they're actually wriggling all around us.

In fact, we have trillions of single-celled organisms living inside our bodies. One example is intestinal bacteria, which live in our large intestines. Intestinal bacteria survive by grabbing nutrients from the debris of our digested food, but at the same time they prevent the propagation of harmful disease-causing bacteria. Our bodies and intestinal bacterial both benefit from each other in a "give and take" relationship. A mutually beneficial relationship between two organisms like this is called a *symbiotic* relationship.

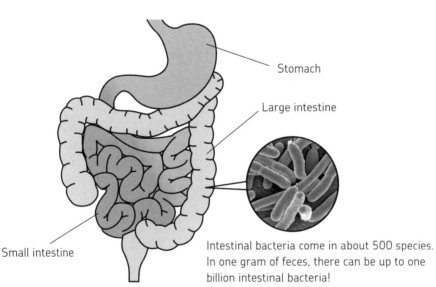

Stomach

Large intestine

Small intestine

Intestinal bacteria come in about 500 species. In one gram of feces, there can be up to one billion intestinal bacteria!

 Bacteria are just one kind of single-celled organism. There are also protozoa—such as paramecia and amoebas.

Organisms with more than one cell are called . . . you guessed it, *multicellular*. Almost all living organisms visible to us (and many that aren't!) are multicellular: human beings, cherry trees, moss, dogs, fleas, and elephants are multicellular.

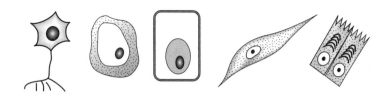

Various types of cells

 Cells that make up different organs—nerves, stomach, and skin, for example—have different shapes and different functions. Cells with the same shape and the same function that gather together are called *tissue*. Bodies of animals, including human beings, are primarily made up of four kinds of tissue: epithelial tissue, connective tissue, muscular tissue, and nervous tissue.

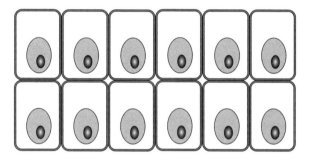

Cells get together to form tissue.

Epithelial tissue (or the epithelium)   This tissue forms the skin of organisms and the surface of internal organs, like digestive canals. There are several types, including flat epithelium, columnar epithelium, and sensory epithelium.

Connective tissue   This tissue performs diverse roles throughout the body. It works to connect different tissues, cells, and organs to one another. Fibrous connective tissue, like ligaments and tendons, helps connect muscles to bone. Connective tissue is abundant in the protein collagen. Blood, bone, and adipose tissue (which you may know as fat!) are also types of connective tissue.

Muscular tissue   As the name suggests, this tissue forms muscles. Skeletal muscle, cardiac muscle, and visceral muscle are included in this category.

Nervous tissue   This tissue forms the nerves that make up the nervous system. Nerve cells use electrical signals to transmit messages back and forth from the brain to the rest of the body.

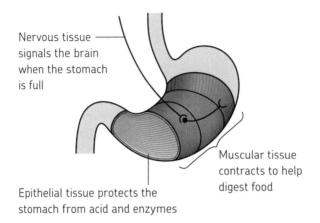

Nervous tissue signals the brain when the stomach is full

Muscular tissue contracts to help digest food

Epithelial tissue protects the stomach from acid and enzymes

The tissues in your stomach: The connective tissue in the stomach is in a layer between the epithelium and muscular layer—it holds the whole stomach together.

A collection of tissues that are gathered together for a specific purpose is called an *organ*. Each tissue is joined to another in a specific way to form respective organs.

As you've seen, the stomach is made of a collection of four types of tissues.

Other organs are formed in a similar way, and carry out their specific duties. These organs in turn combine to form systems in our body.

Within multicellular organisms, cells gather together to form organs, and organs function together to form systems. In this way, a tiny cell can carry out very complicated tasks, indeed!

Trachea
Blood vessel
Heart
Lung
Stomach
Liver
Large intestine
Small intestine

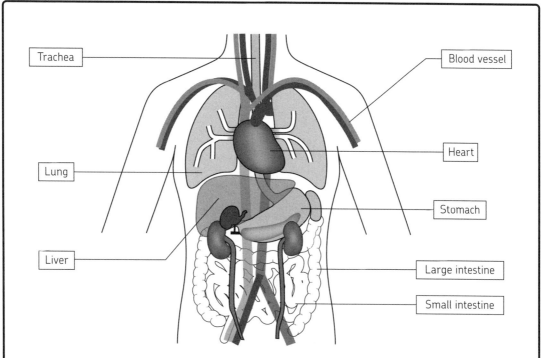

Organs in the digestive system: mouth, pharynx, esophagus, stomach, small intestine, large intestine, anus, rectum, liver, gallbladder, pancreas
Organs in the circulatory system: heart, aorta, arteries, veins, lymphatic vessels
Organs in the respiratory system: nares, pharynx, larynx, trachea, bronchi, lungs

PROKARYOTIC ORGANISMS AND EUKARYOTIC ORGANISMS

As you have learned before, the concepts of a single-celled organism and a multicellular organism are used for broadly classifying the world of living organisms into two groups. However, there is another way to classify the world of living organisms into two groups. This approach depends on the presence or absence of a nucleus in a cell. If this approach is employed, living organisms are broadly divided into prokaryotic organisms (those that don't have a nucleus) and eukaryotic organisms (those that do).

Presence or absence of a nucleus? Isn't the nucleus the "brain of the cell?" It's indispensable, isn't it?

You might think so, but hold on for a moment. Try to remember why the nucleus is called the "brain." It's because the nucleus stores DNA and controls the expression of genes written in DNA, right? But as long as the DNA is in the cell and the genes are appropriately expressed, does the structure called the nucleus matter? When the genetic material of an organism is not held within a nucleus, this organism is called a *prokaryotic organism*. Bacteria are the only prokaryotic organisms on the earth. In prokaryotic organisms, all important jobs of the cell are done within the cytoplasm, instead of within membrane-bound organelles.

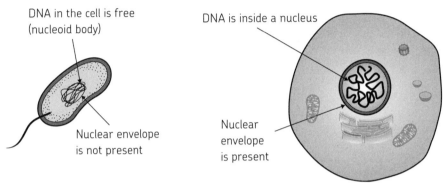

DNA in the cell is free
(nucleoid body)

Nuclear envelope
is not present

Prokaryotic organism

DNA is inside a nucleus

Nuclear
envelope
is present

Eukaryotic organism

When the nucleus of an organism has a membrane and a distinct shape, this organism is called a *eukaryotic organism*. Every single-celled organism (excluding bacteria) is included in the eukaryotic category. Protozoa such as paramecia are eukaryotic organisms.

You might think the terms *nonnucleated organism* and *nucleated organism* would be sufficient for classifying organisms, depending on the presence or absence of a nucleus. But that is not correct. Although we say that prokaryotic organisms do not have nuclei, they actually do have their own structure that functions the same as a nucleus. The area where DNA is stored (it's contained in a nucleoid body) is separate from the surrounding cytosol. Thus, the term *prokaryon* is used to indicate the existence of a nucleuslike structure (like the nucleoid body), more primitive than a true nucleus. This prokaryon is not surrounded by a nuclear envelope like a true nucleus.

This term *true nucleus* is used for organisms with a membrane surrounding and protecting their genetic material. The nuclear envelope is just a membrane, but it's an important one!

2
PROTEINS AND DNA: DECIPHERING THE GENETIC CODE

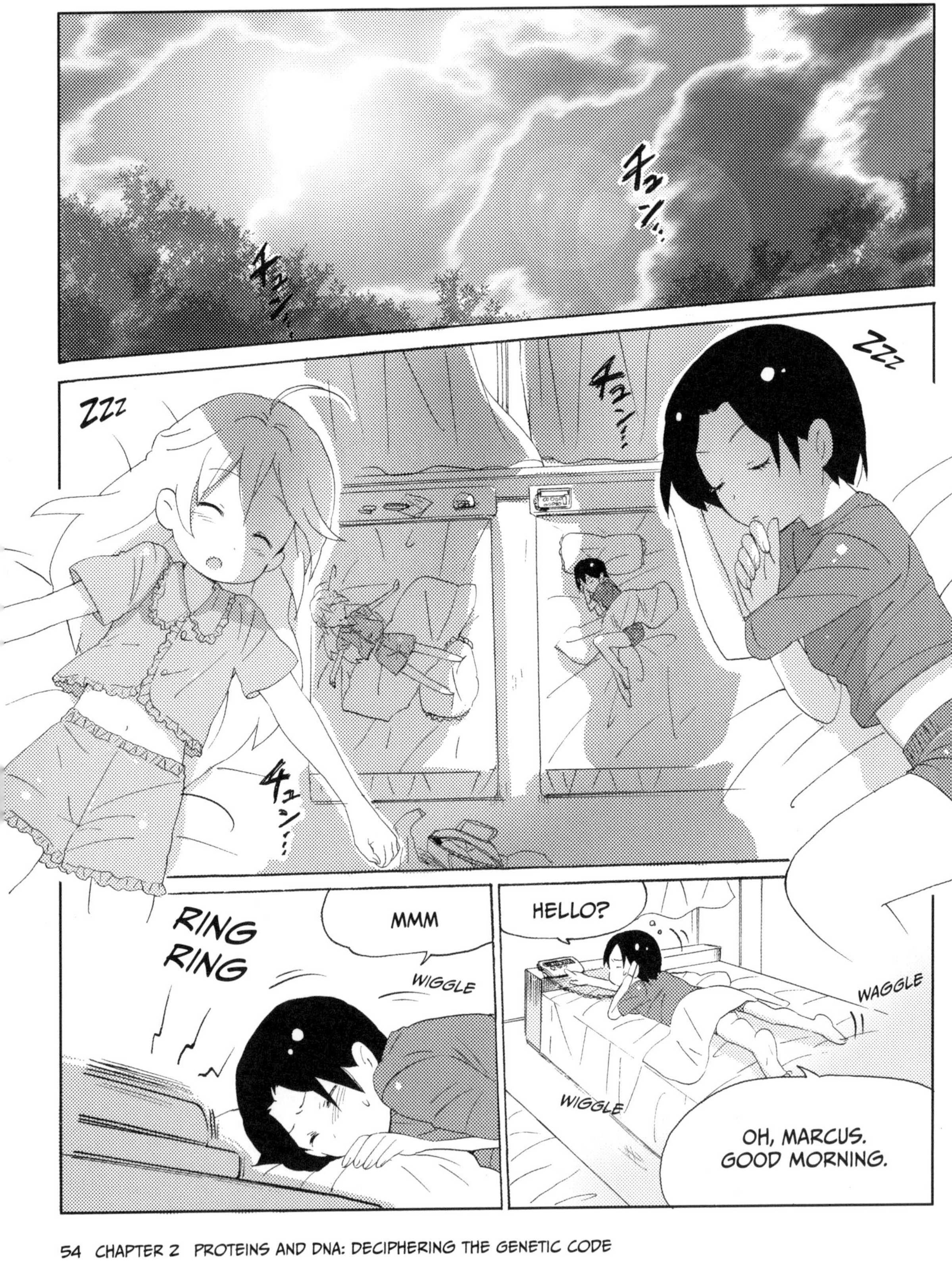
チュン…
チュン…
ZZZ
チュン…
ZZZ
チュン…
RING RING
MMM
WIGGLE
HELLO?
WAGGLE
WIGGLE
OH, MARCUS. GOOD MORNING.

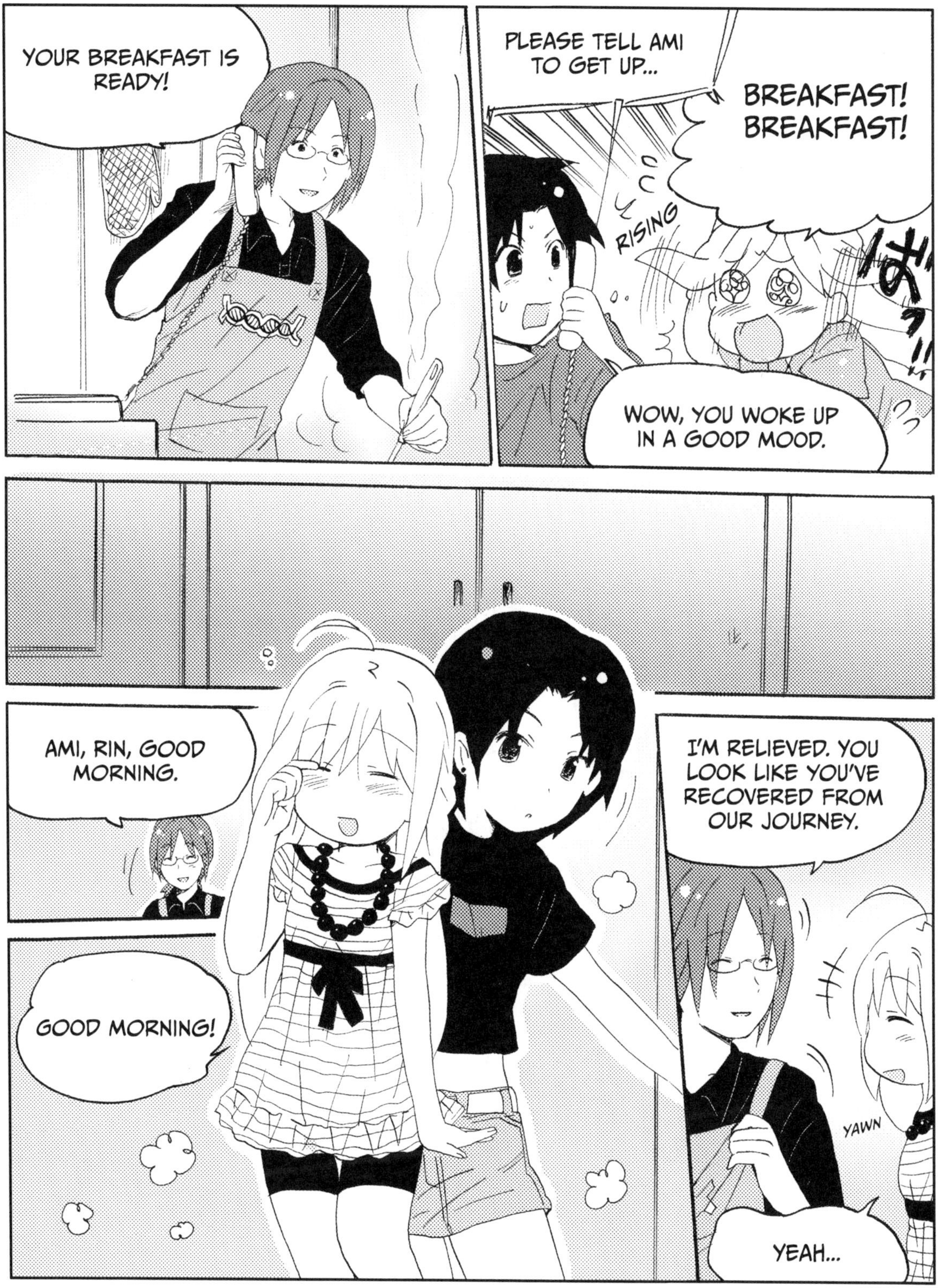
YOUR BREAKFAST IS READY!
PLEASE TELL AMI TO GET UP...
BREAKFAST! BREAKFAST!
RISING
WOW, YOU WOKE UP IN A GOOD MOOD.
AMI, RIN, GOOD MORNING.
GOOD MORNING!
I'M RELIEVED. YOU LOOK LIKE YOU'VE RECOVERED FROM OUR JOURNEY.
YAWN
YEAH...

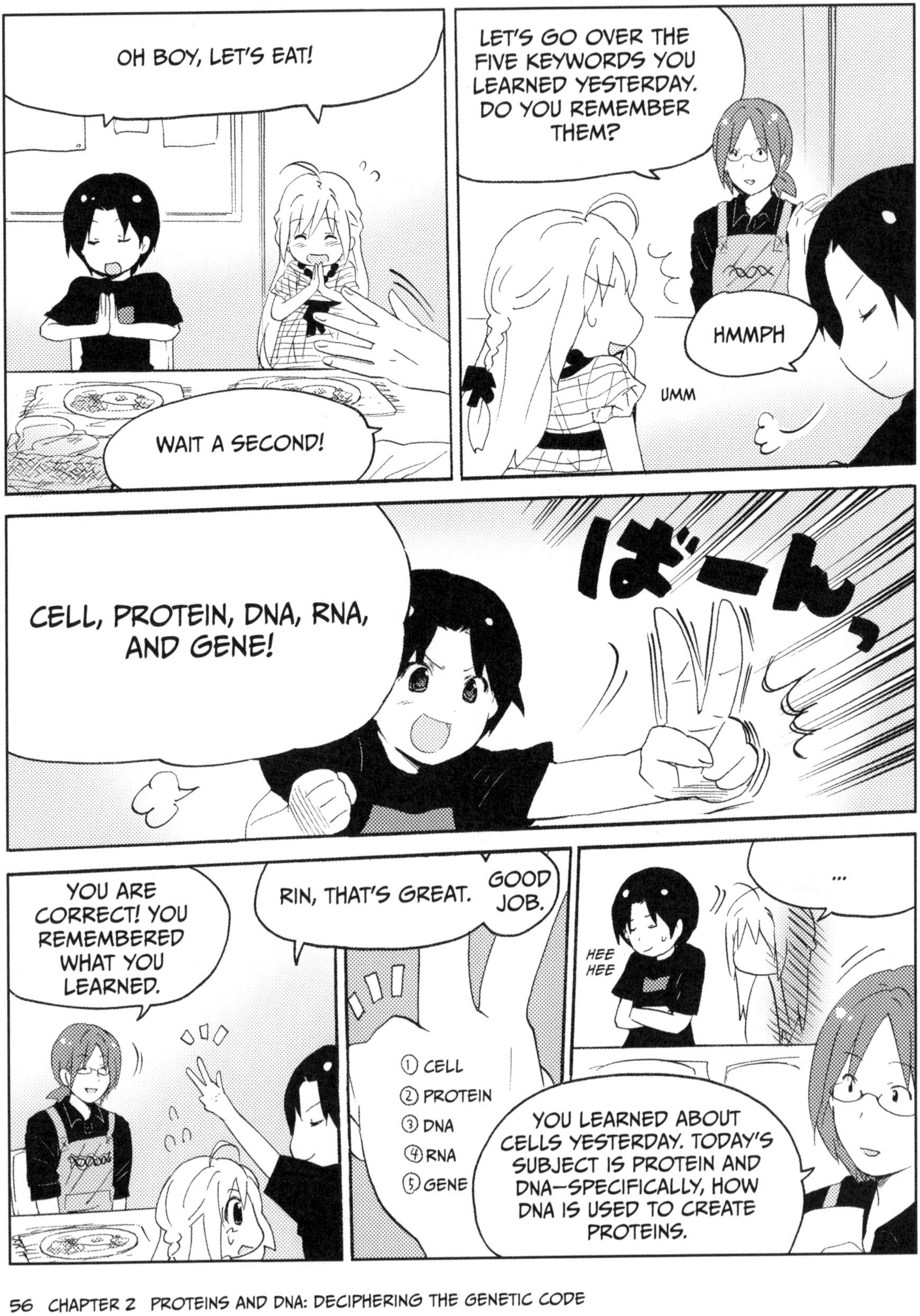
OH BOY, LET'S EAT!
WAIT A SECOND!
LET'S GO OVER THE FIVE KEYWORDS YOU LEARNED YESTERDAY. DO YOU REMEMBER THEM?
HMMPH
UMM
CELL, PROTEIN, DNA, RNA, AND GENE!
ばーん
YOU ARE CORRECT! YOU REMEMBERED WHAT YOU LEARNED.
RIN, THAT'S GREAT.
GOOD JOB.
① CELL
② PROTEIN
③ DNA
④ RNA
⑤ GENE
...
HEE HEE
YOU LEARNED ABOUT CELLS YESTERDAY. TODAY'S SUBJECT IS PROTEIN AND DNA—SPECIFICALLY, HOW DNA IS USED TO CREATE PROTEINS.

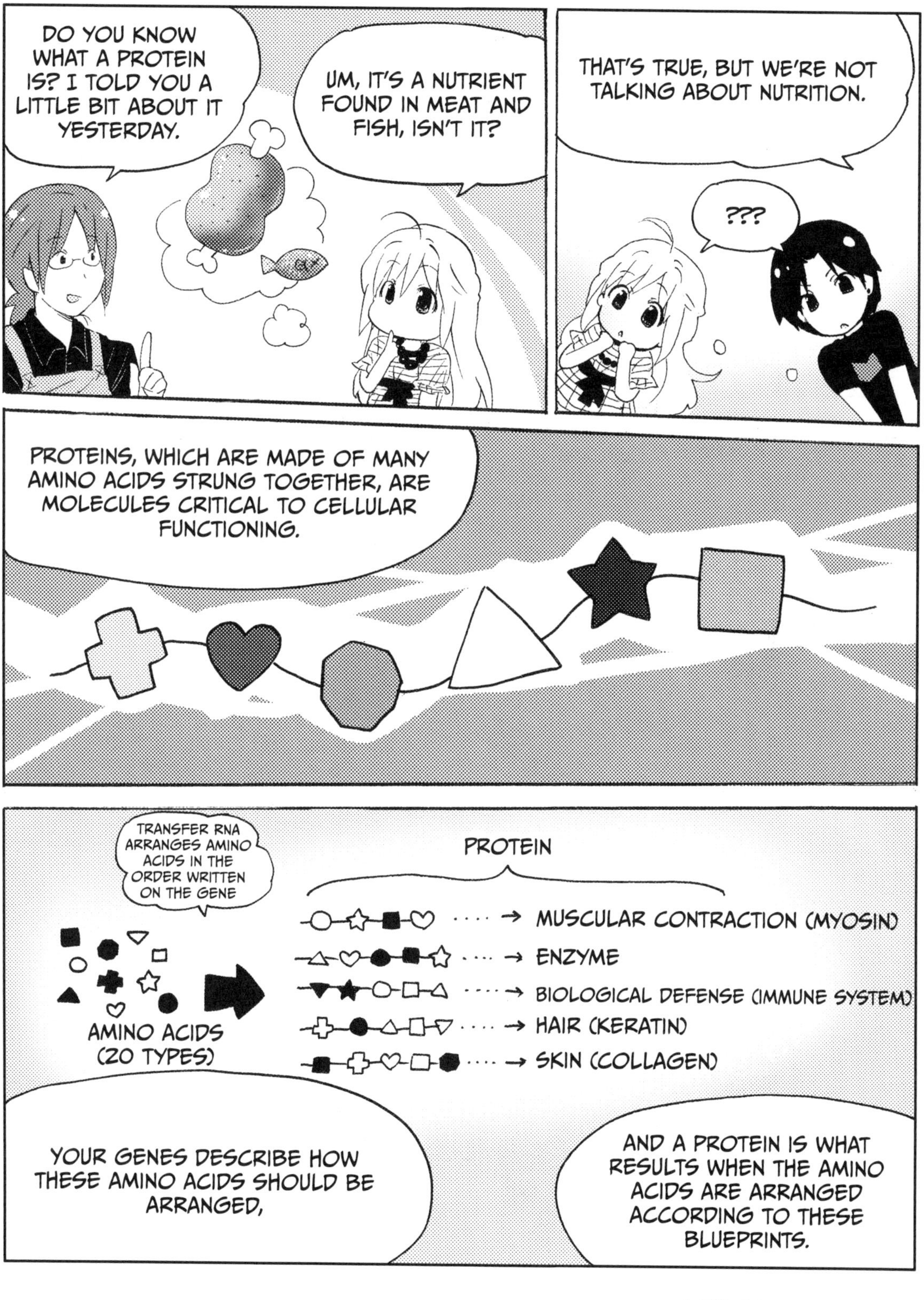
DO YOU KNOW WHAT A PROTEIN IS? I TOLD YOU A LITTLE BIT ABOUT IT YESTERDAY.
UM, IT'S A NUTRIENT FOUND IN MEAT AND FISH, ISN'T IT?
THAT'S TRUE, BUT WE'RE NOT TALKING ABOUT NUTRITION.
???
PROTEINS, WHICH ARE MADE OF MANY AMINO ACIDS STRUNG TOGETHER, ARE MOLECULES CRITICAL TO CELLULAR FUNCTIONING.
TRANSFER RNA ARRANGES AMINO ACIDS IN THE ORDER WRITTEN ON THE GENE
PROTEIN
MUSCULAR CONTRACTION (MYOSIN)
ENZYME
BIOLOGICAL DEFENSE (IMMUNE SYSTEM)
HAIR (KERATIN)
SKIN (COLLAGEN)
AMINO ACIDS (20 TYPES)
YOUR GENES DESCRIBE HOW THESE AMINO ACIDS SHOULD BE ARRANGED,
AND A PROTEIN IS WHAT RESULTS WHEN THE AMINO ACIDS ARE ARRANGED ACCORDING TO THESE BLUEPRINTS.

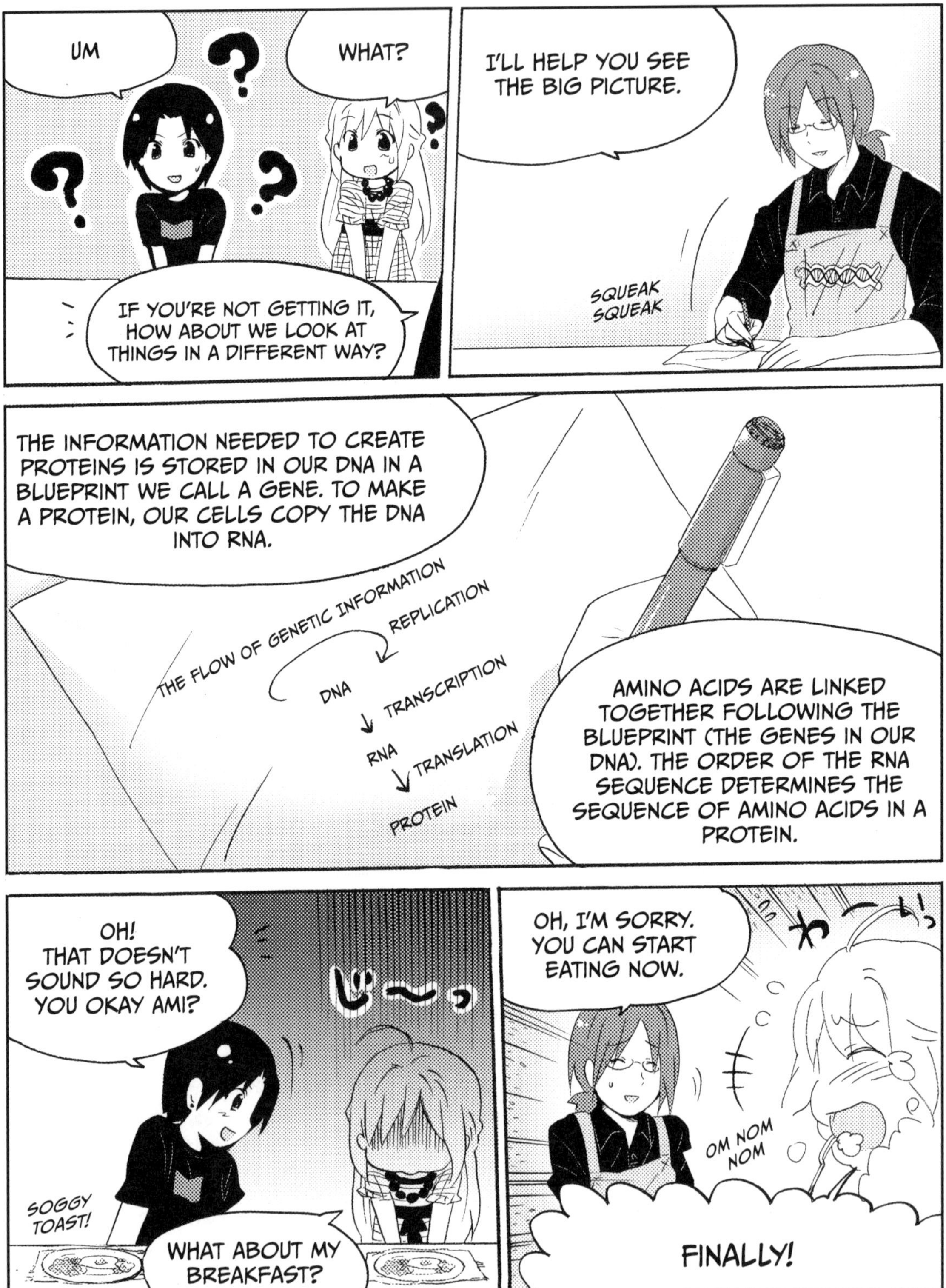
UM
WHAT?
IF YOU'RE NOT GETTING IT, HOW ABOUT WE LOOK AT THINGS IN A DIFFERENT WAY?
I'LL HELP YOU SEE THE BIG PICTURE.
SQUEAK SQUEAK
THE INFORMATION NEEDED TO CREATE PROTEINS IS STORED IN OUR DNA IN A BLUEPRINT WE CALL A GENE. TO MAKE A PROTEIN, OUR CELLS COPY THE DNA INTO RNA.
THE FLOW OF GENETIC INFORMATION
REPLICATION
DNA
TRANSCRIPTION
RNA
TRANSLATION
PROTEIN
AMINO ACIDS ARE LINKED TOGETHER FOLLOWING THE BLUEPRINT (THE GENES IN OUR DNA). THE ORDER OF THE RNA SEQUENCE DETERMINES THE SEQUENCE OF AMINO ACIDS IN A PROTEIN.
OH! THAT DOESN'T SOUND SO HARD. YOU OKAY AMI?
じ〜っ
SOGGY TOAST!
WHAT ABOUT MY BREAKFAST?
OH, I'M SORRY. YOU CAN START EATING NOW.
OM NOM NOM
FINALLY!

PROTEINS DRIVE CELLULAR ACTIVITY

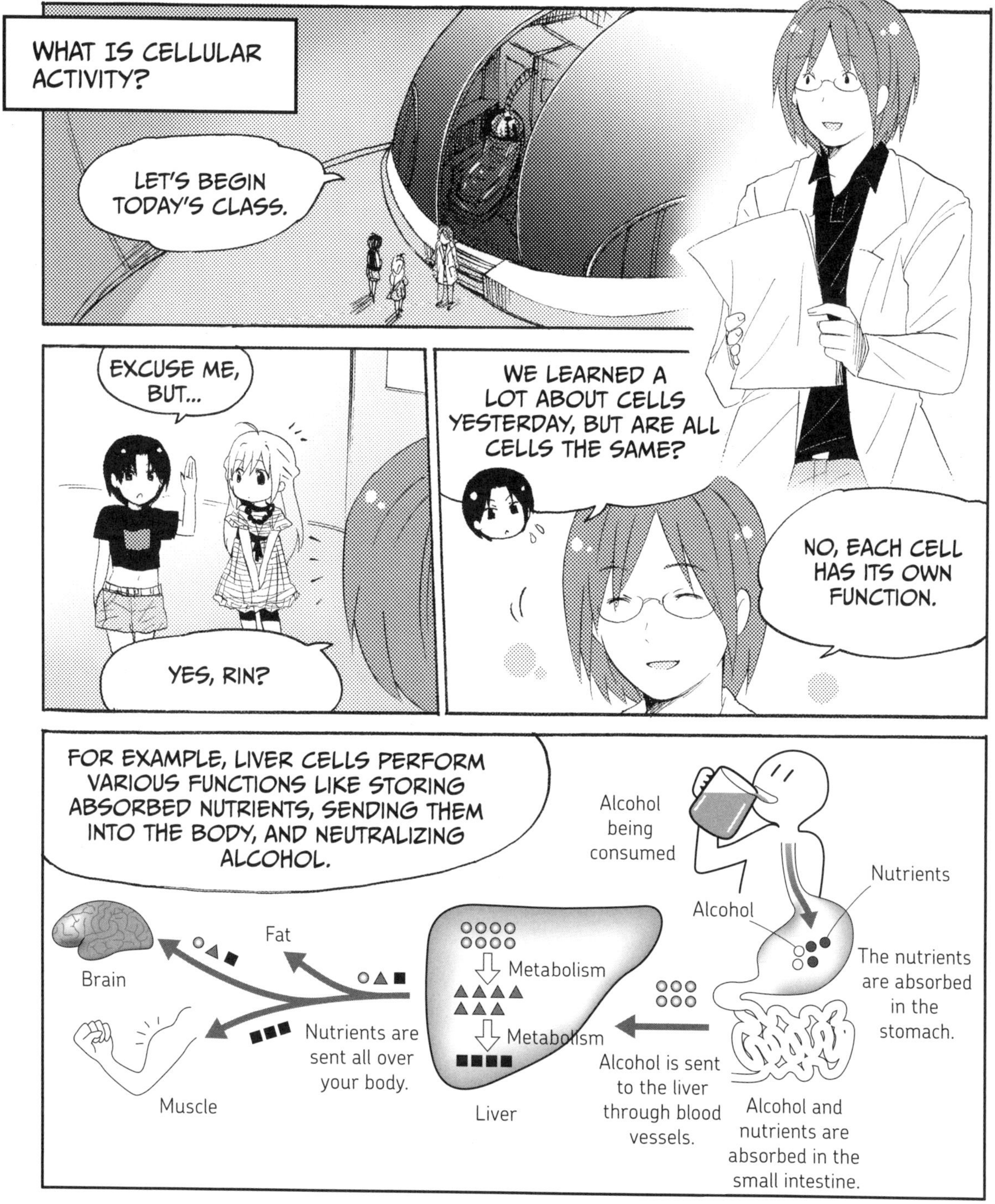
WHAT IS CELLULAR ACTIVITY?
LET'S BEGIN TODAY'S CLASS.
EXCUSE ME, BUT...
YES, RIN?
WE LEARNED A LOT ABOUT CELLS YESTERDAY, BUT ARE ALL CELLS THE SAME?
NO, EACH CELL HAS ITS OWN FUNCTION.
FOR EXAMPLE, LIVER CELLS PERFORM VARIOUS FUNCTIONS LIKE STORING ABSORBED NUTRIENTS, SENDING THEM INTO THE BODY, AND NEUTRALIZING ALCOHOL.
Alcohol being consumed
Nutrients
Alcohol
The nutrients are absorbed in the stomach.
Alcohol and nutrients are absorbed in the small intestine.
Alcohol is sent to the liver through blood vessels.
Metabolism
Metabolism
Liver
Fat
Brain
Nutrients are sent all over your body.
Muscle

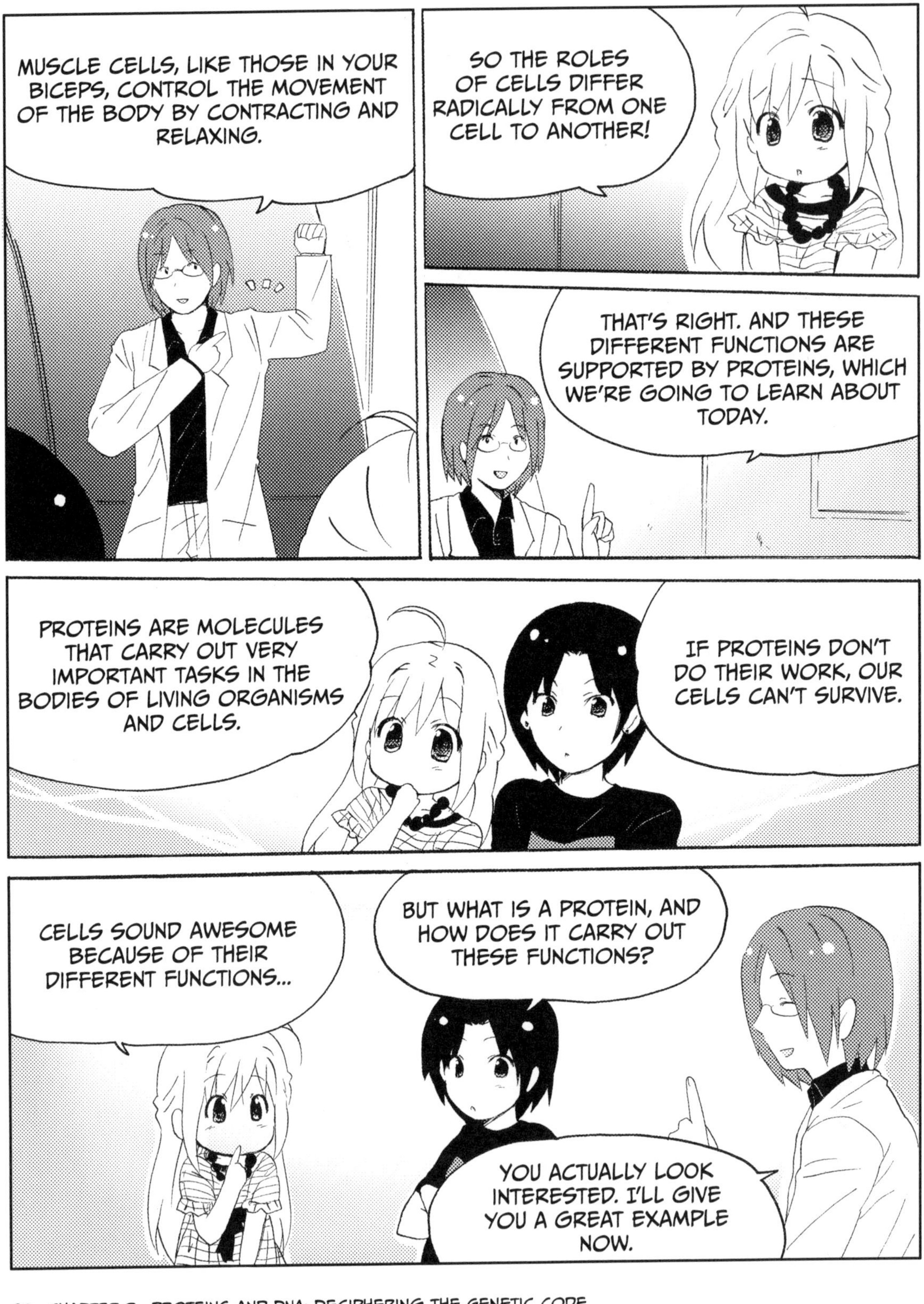
MUSCLE CELLS, LIKE THOSE IN YOUR BICEPS, CONTROL THE MOVEMENT OF THE BODY BY CONTRACTING AND RELAXING.
SO THE ROLES OF CELLS DIFFER RADICALLY FROM ONE CELL TO ANOTHER!
THAT'S RIGHT. AND THESE DIFFERENT FUNCTIONS ARE SUPPORTED BY PROTEINS, WHICH WE'RE GOING TO LEARN ABOUT TODAY.
PROTEINS ARE MOLECULES THAT CARRY OUT VERY IMPORTANT TASKS IN THE BODIES OF LIVING ORGANISMS AND CELLS.
IF PROTEINS DON'T DO THEIR WORK, OUR CELLS CAN'T SURVIVE.
CELLS SOUND AWESOME BECAUSE OF THEIR DIFFERENT FUNCTIONS...
BUT WHAT IS A PROTEIN, AND HOW DOES IT CARRY OUT THESE FUNCTIONS?
YOU ACTUALLY LOOK INTERESTED. I'LL GIVE YOU A GREAT EXAMPLE NOW.

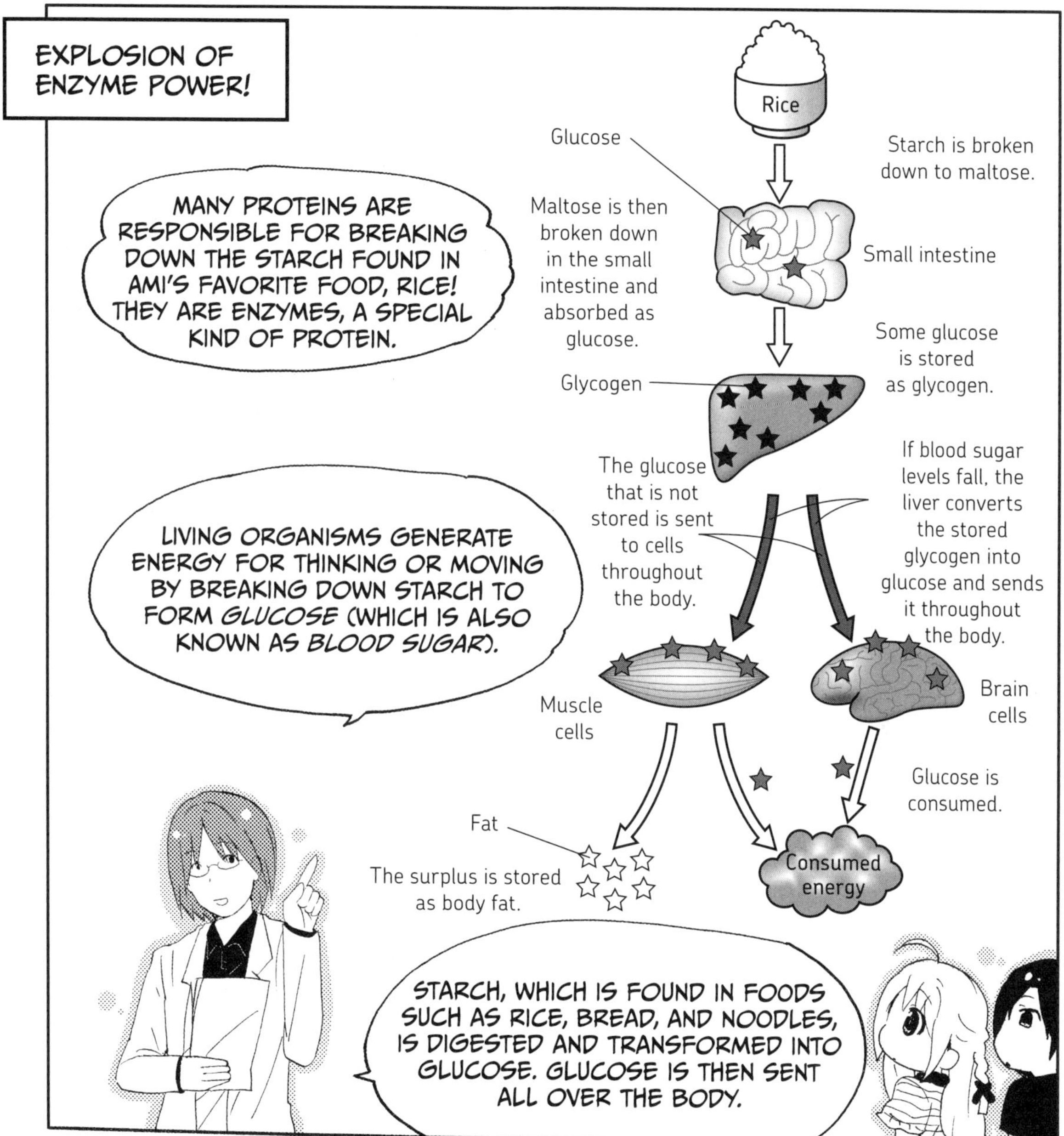
EXPLOSION OF ENZYME POWER!
MANY PROTEINS ARE RESPONSIBLE FOR BREAKING DOWN THE STARCH FOUND IN AMI'S FAVORITE FOOD, RICE! THEY ARE ENZYMES, A SPECIAL KIND OF PROTEIN.
Rice
Glucose
Starch is broken down to maltose.
Maltose is then broken down in the small intestine and absorbed as glucose.
Small intestine
Some glucose is stored as glycogen.
Glycogen
The glucose that is not stored is sent to cells throughout the body.
If blood sugar levels fall, the liver converts the stored glycogen into glucose and sends it throughout the body.
Muscle cells
Brain cells
Glucose is consumed.
Fat
The surplus is stored as body fat.
Consumed energy
LIVING ORGANISMS GENERATE ENERGY FOR THINKING OR MOVING BY BREAKING DOWN STARCH TO FORM *GLUCOSE* (WHICH IS ALSO KNOWN AS *BLOOD SUGAR*).
STARCH, WHICH IS FOUND IN FOODS SUCH AS RICE, BREAD, AND NOODLES, IS DIGESTED AND TRANSFORMED INTO GLUCOSE. GLUCOSE IS THEN SENT ALL OVER THE BODY.
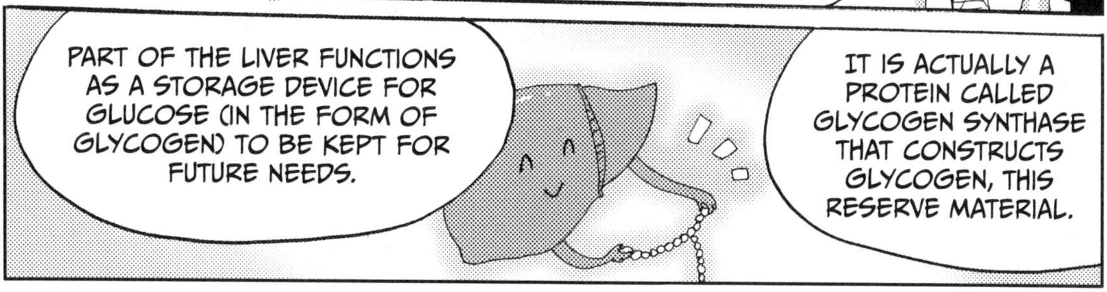
PART OF THE LIVER FUNCTIONS AS A STORAGE DEVICE FOR GLUCOSE (IN THE FORM OF GLYCOGEN) TO BE KEPT FOR FUTURE NEEDS.
IT IS ACTUALLY A PROTEIN CALLED GLYCOGEN SYNTHASE THAT CONSTRUCTS GLYCOGEN, THIS RESERVE MATERIAL.

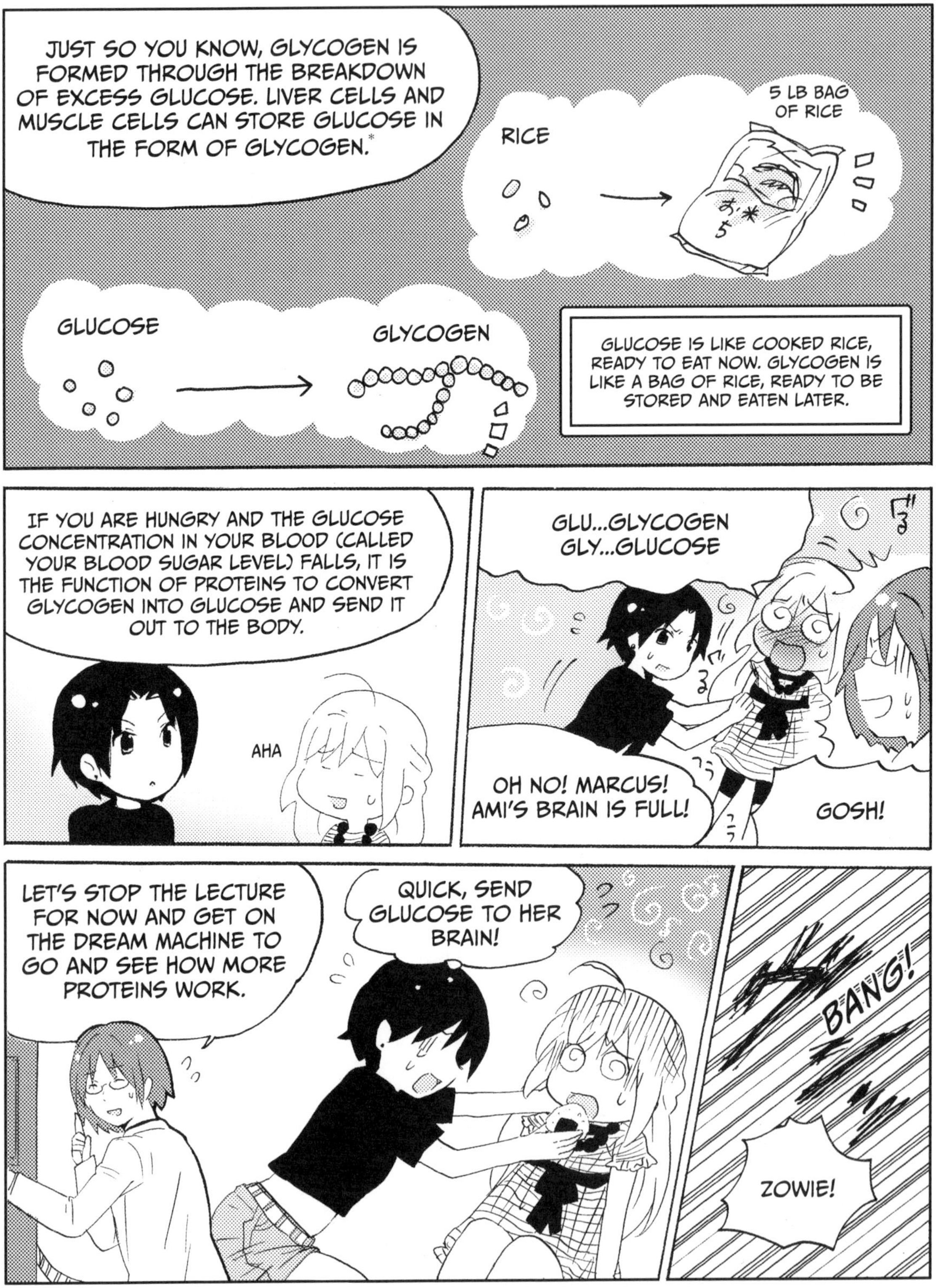
JUST SO YOU KNOW, GLYCOGEN IS FORMED THROUGH THE BREAKDOWN OF EXCESS GLUCOSE. LIVER CELLS AND MUSCLE CELLS CAN STORE GLUCOSE IN THE FORM OF GLYCOGEN.*
RICE
5 LB BAG OF RICE
GLUCOSE
GLYCOGEN
GLUCOSE IS LIKE COOKED RICE, READY TO EAT NOW. GLYCOGEN IS LIKE A BAG OF RICE, READY TO BE STORED AND EATEN LATER.
IF YOU ARE HUNGRY AND THE GLUCOSE CONCENTRATION IN YOUR BLOOD (CALLED YOUR BLOOD SUGAR LEVEL) FALLS, IT IS THE FUNCTION OF PROTEINS TO CONVERT GLYCOGEN INTO GLUCOSE AND SEND IT OUT TO THE BODY.
AHA
GLU...GLYCOGEN GLY...GLUCOSE
OH NO! MARCUS! AMI'S BRAIN IS FULL!
GOSH!
LET'S STOP THE LECTURE FOR NOW AND GET ON THE DREAM MACHINE TO GO AND SEE HOW MORE PROTEINS WORK.
QUICK, SEND GLUCOSE TO HER BRAIN!
BANG!
ZOWIE!

* ALTHOUGH ALMOST ALL CELLS OF LIVING ORGANISMS STORE GLYCOGEN, MOST OF IT IS STORED IN THE LIVER AND MUSCLES.

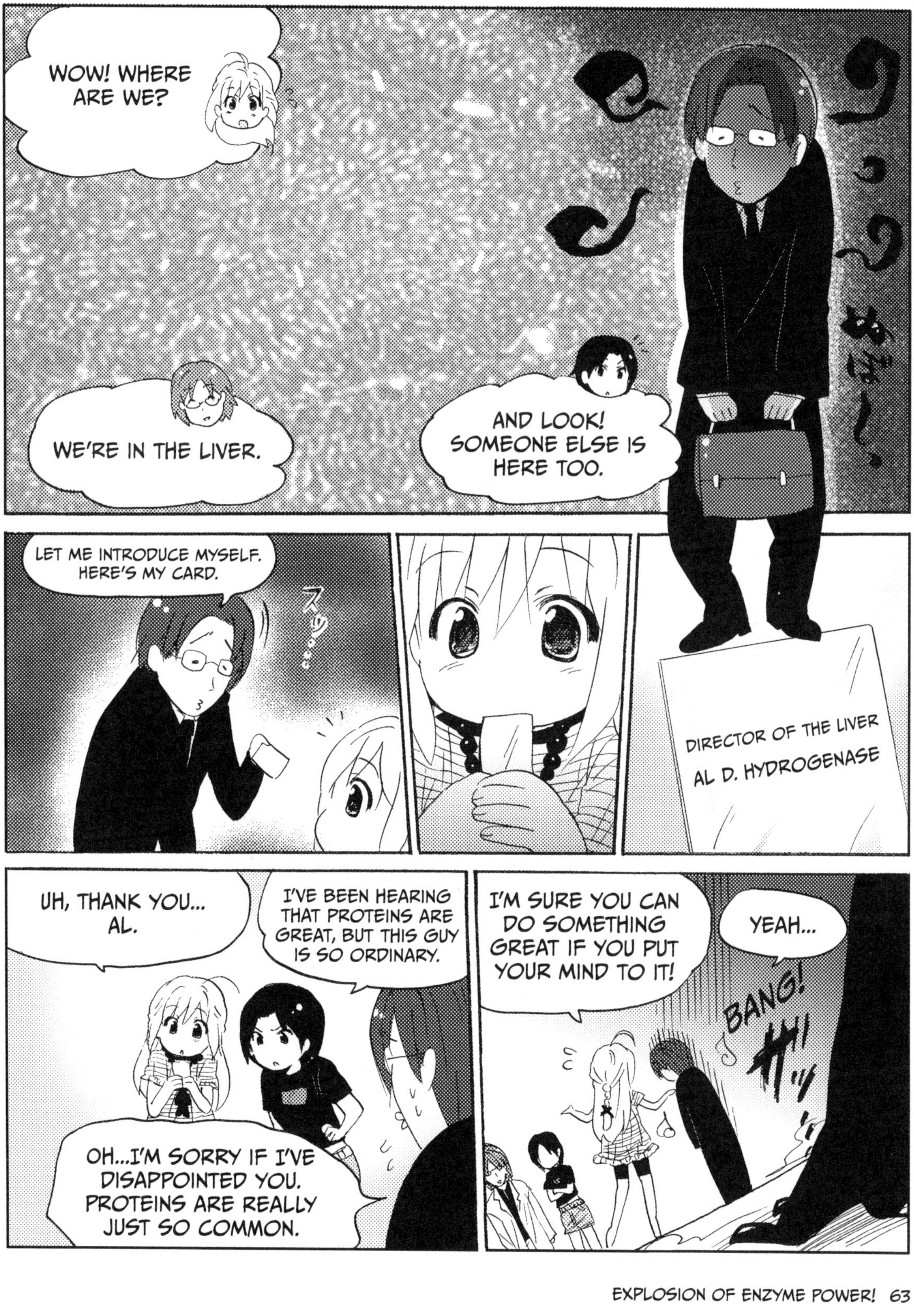
WOW! WHERE ARE WE?
WE'RE IN THE LIVER.
AND LOOK! SOMEONE ELSE IS HERE TOO.
LET ME INTRODUCE MYSELF. HERE'S MY CARD.
DIRECTOR OF THE LIVER
AL D. HYDROGENASE
UH, THANK YOU... AL.
I'VE BEEN HEARING THAT PROTEINS ARE GREAT, BUT THIS GUY IS SO ORDINARY.
OH...I'M SORRY IF I'VE DISAPPOINTED YOU. PROTEINS ARE REALLY JUST SO COMMON.
I'M SURE YOU CAN DO SOMETHING GREAT IF YOU PUT YOUR MIND TO IT!
YEAH...
BANG!

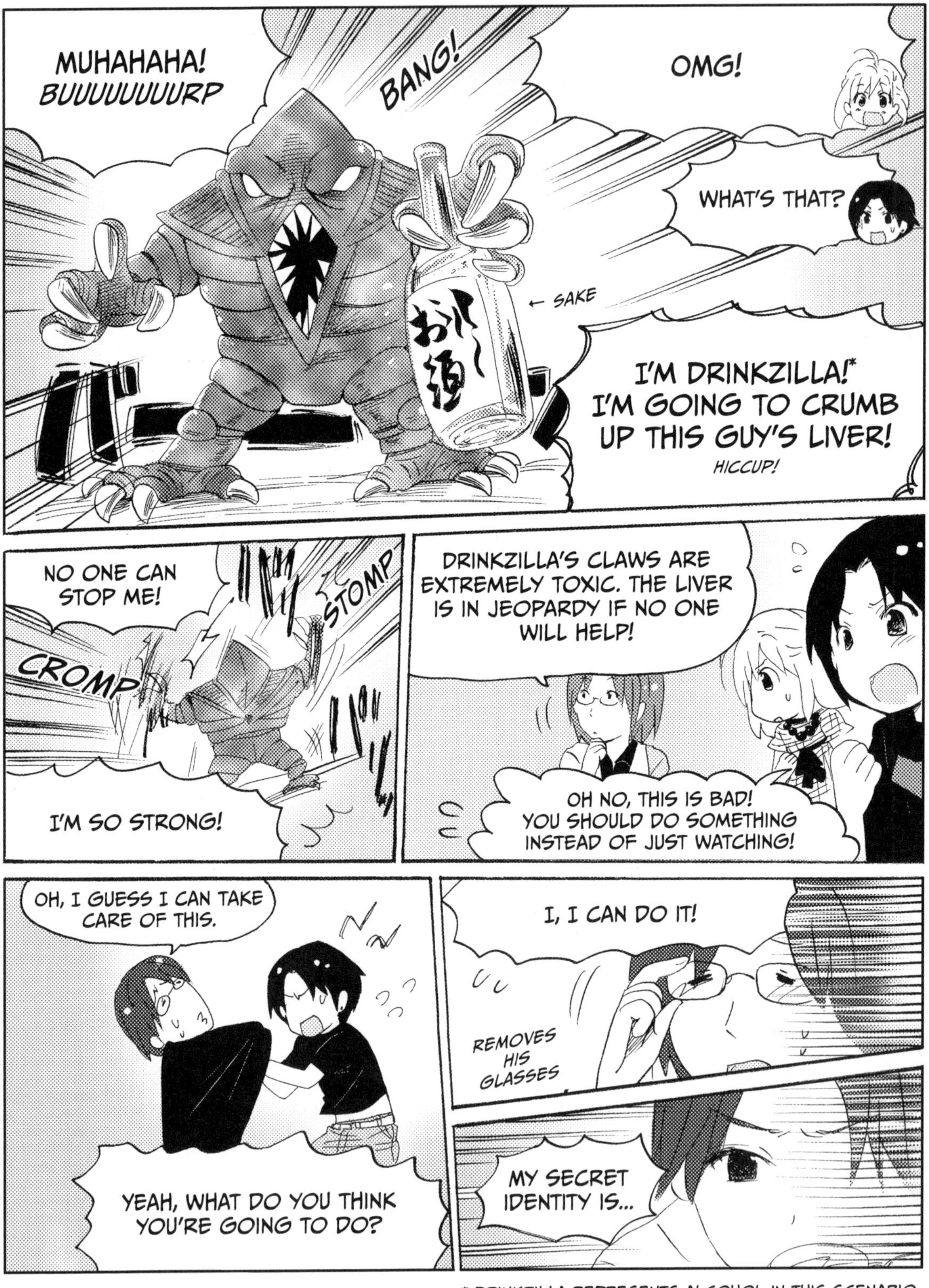
MUHAHAHA! BUUUUUUUURP
BANG!
OMG!
WHAT'S THAT?
← SAKE
I'M DRINKZILLA!* I'M GOING TO CRUMB UP THIS GUY'S LIVER!
HICCUP!
NO ONE CAN STOP ME!
STOMP
CROMP
I'M SO STRONG!
DRINKZILLA'S CLAWS ARE EXTREMELY TOXIC. THE LIVER IS IN JEOPARDY IF NO ONE WILL HELP!
OH NO, THIS IS BAD! YOU SHOULD DO SOMETHING INSTEAD OF JUST WATCHING!
OH, I GUESS I CAN TAKE CARE OF THIS.
YEAH, WHAT DO YOU THINK YOU'RE GOING TO DO?
I, I CAN DO IT!
REMOVES HIS GLASSES
MY SECRET IDENTITY IS...

* DRINKZILLA REPRESENTS ALCOHOL IN THIS SCENARIO.

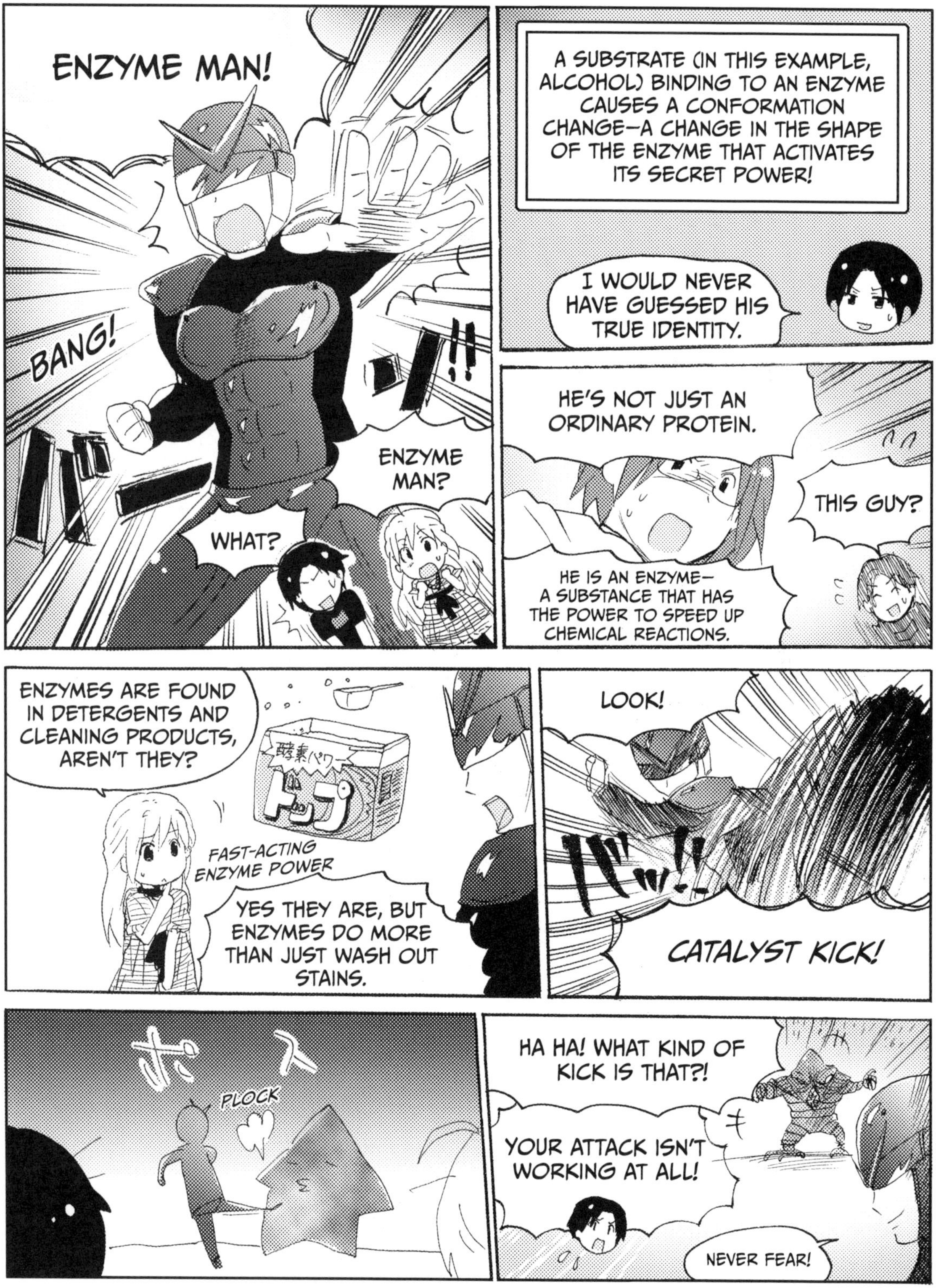
ENZYME MAN!
BANG!
WHAT?
ENZYME MAN?
A SUBSTRATE (IN THIS EXAMPLE, ALCOHOL) BINDING TO AN ENZYME CAUSES A CONFORMATION CHANGE—A CHANGE IN THE SHAPE OF THE ENZYME THAT ACTIVATES ITS SECRET POWER!
I WOULD NEVER HAVE GUESSED HIS TRUE IDENTITY.
HE'S NOT JUST AN ORDINARY PROTEIN.
THIS GUY?
HE IS AN ENZYME—A SUBSTANCE THAT HAS THE POWER TO SPEED UP CHEMICAL REACTIONS.
ENZYMES ARE FOUND IN DETERGENTS AND CLEANING PRODUCTS, AREN'T THEY?
酵素パワー
FAST-ACTING ENZYME POWER
YES THEY ARE, BUT ENZYMES DO MORE THAN JUST WASH OUT STAINS.
LOOK!
CATALYST KICK!
PLOCK
HA HA! WHAT KIND OF KICK IS THAT?!
YOUR ATTACK ISN'T WORKING AT ALL!
NEVER FEAR!

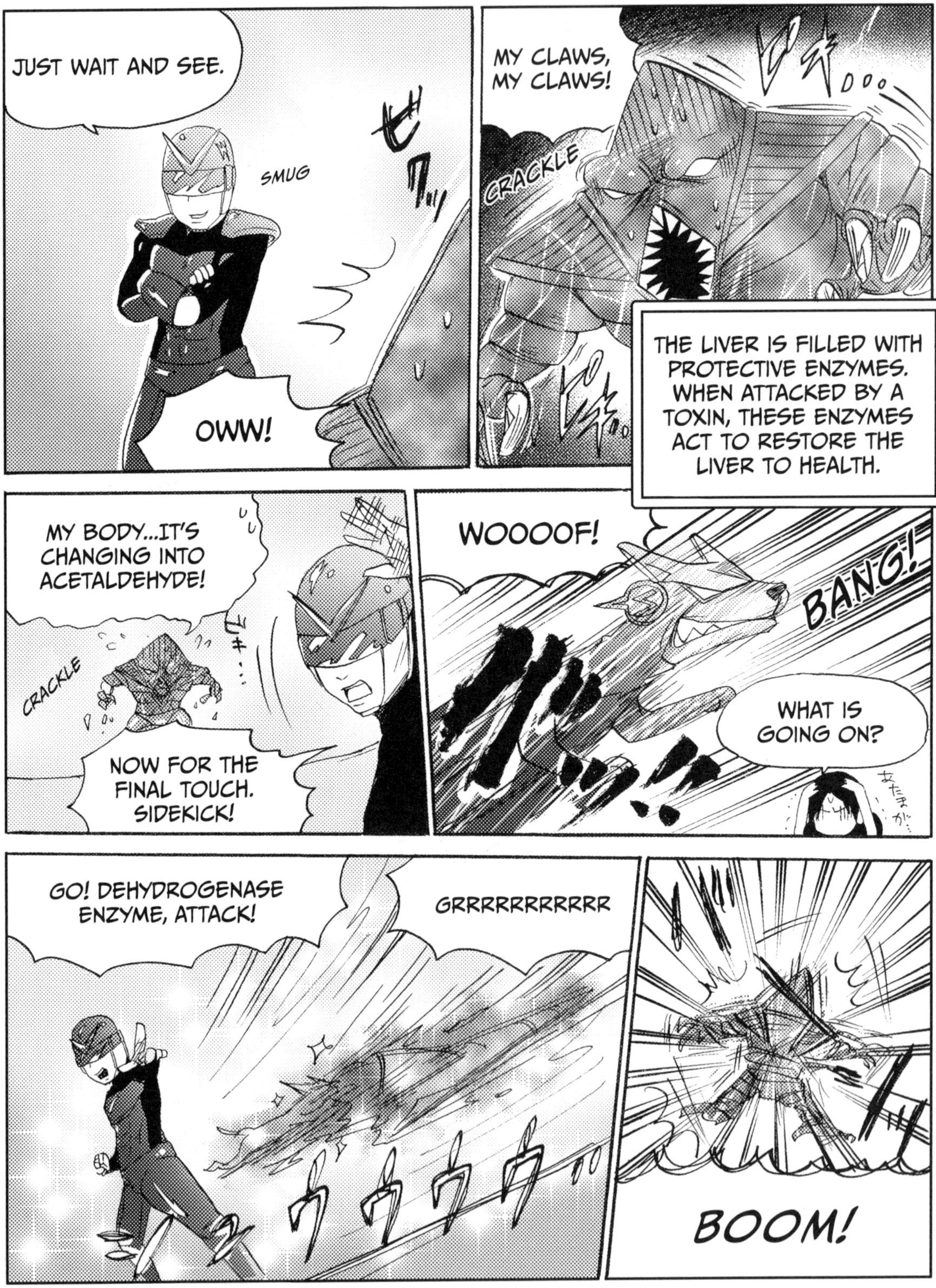
JUST WAIT AND SEE.
SMUG
OWW!
MY CLAWS, MY CLAWS!
CRACKLE
THE LIVER IS FILLED WITH PROTECTIVE ENZYMES. WHEN ATTACKED BY A TOXIN, THESE ENZYMES ACT TO RESTORE THE LIVER TO HEALTH.
MY BODY...IT'S CHANGING INTO ACETALDEHYDE!
CRACKLE
NOW FOR THE FINAL TOUCH. SIDEKICK!
WOOOOF!
BANG!
WHAT IS GOING ON?
GO! DEHYDROGENASE ENZYME, ATTACK!
GRRRRRRRRRR
BOOM!

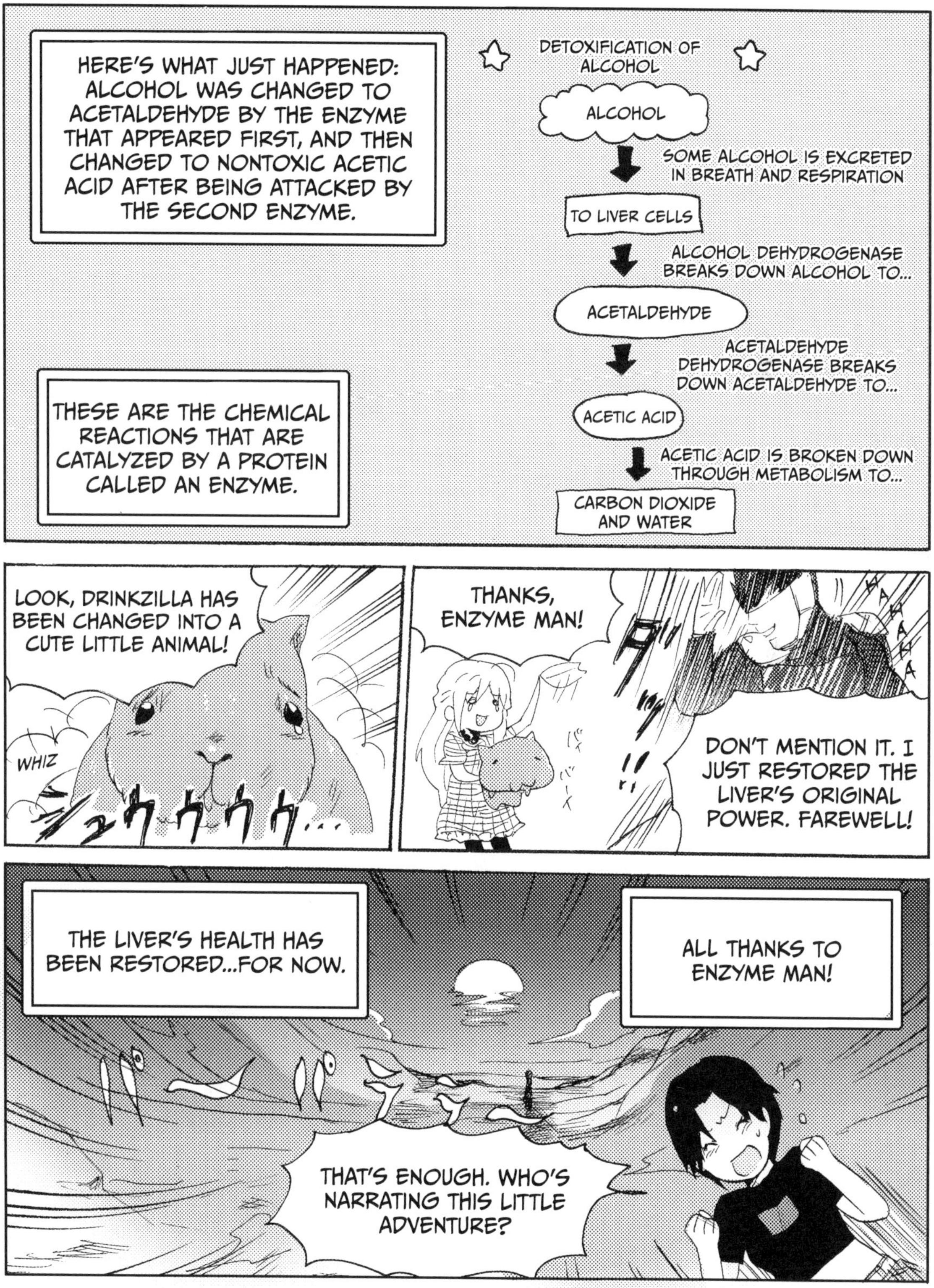
HERE'S WHAT JUST HAPPENED: ALCOHOL WAS CHANGED TO ACETALDEHYDE BY THE ENZYME THAT APPEARED FIRST, AND THEN CHANGED TO NONTOXIC ACETIC ACID AFTER BEING ATTACKED BY THE SECOND ENZYME.
DETOXIFICATION OF ALCOHOL
ALCOHOL
SOME ALCOHOL IS EXCRETED IN BREATH AND RESPIRATION
TO LIVER CELLS
ALCOHOL DEHYDROGENASE BREAKS DOWN ALCOHOL TO...
ACETALDEHYDE
ACETALDEHYDE DEHYDROGENASE BREAKS DOWN ACETALDEHYDE TO...
ACETIC ACID
ACETIC ACID IS BROKEN DOWN THROUGH METABOLISM TO...
CARBON DIOXIDE AND WATER
THESE ARE THE CHEMICAL REACTIONS THAT ARE CATALYZED BY A PROTEIN CALLED AN ENZYME.
LOOK, DRINKZILLA HAS BEEN CHANGED INTO A CUTE LITTLE ANIMAL!
WHIZ
THANKS, ENZYME MAN!
DON'T MENTION IT. I JUST RESTORED THE LIVER'S ORIGINAL POWER. FAREWELL!
THE LIVER'S HEALTH HAS BEEN RESTORED...FOR NOW.
ALL THANKS TO ENZYME MAN!
THAT'S ENOUGH. WHO'S NARRATING THIS LITTLE ADVENTURE?

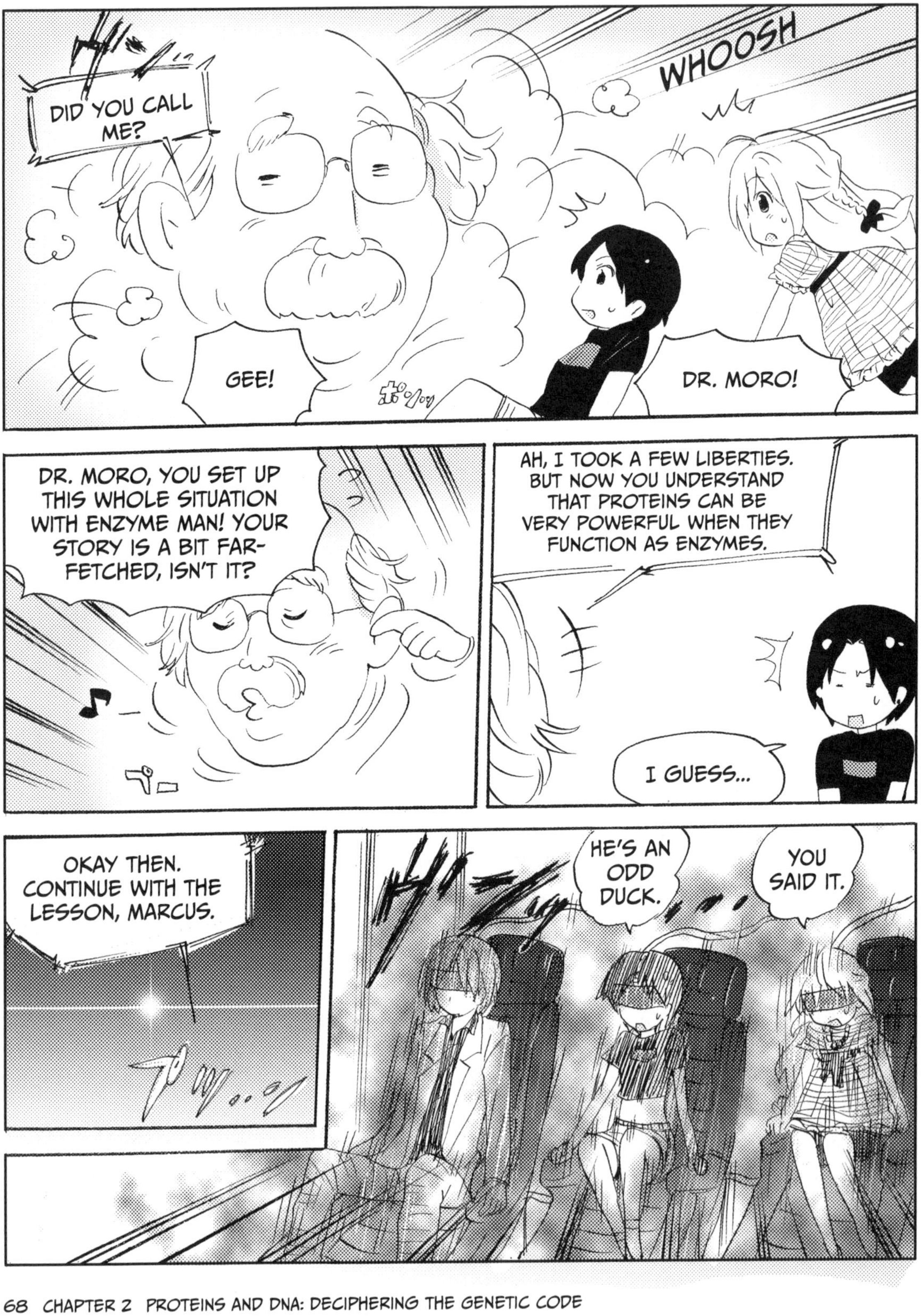
DID YOU CALL ME?
WHOOSH
GEE!
DR. MORO!
DR. MORO, YOU SET UP THIS WHOLE SITUATION WITH ENZYME MAN! YOUR STORY IS A BIT FAR-FETCHED, ISN'T IT?
AH, I TOOK A FEW LIBERTIES. BUT NOW YOU UNDERSTAND THAT PROTEINS CAN BE VERY POWERFUL WHEN THEY FUNCTION AS ENZYMES.
I GUESS...
OKAY THEN. CONTINUE WITH THE LESSON, MARCUS.
HE'S AN ODD DUCK.
YOU SAID IT.

PROTEINS ACTING AS ENZYMES

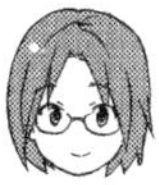

Enzymes are a special type of protein that have the power to accelerate chemical reactions. There are tens of thousands of types of proteins in our bodies, but not all of them are enzymes. You'll soon learn that some proteins, such as those that fight bacteria as part of our immune system, are not enzymes.

Various processes that happen within our bodies, such as digestion, absorption of nutrients, and replication of DNA, occur due to chemical reactions. A different protein is responsible for each chemical reaction. A unique enzyme starts almost all the chemical reactions carried out by living organisms. All enzymes are *catalysts* (or *biocatalysts*). Catalysts help start chemical reactions and make them easier to move forward, but they do not actually react with any other molecules in the reaction.

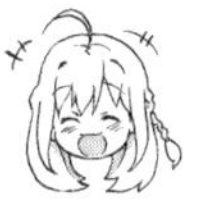

That's why Enzyme Man called his attack a *Catalyst Kick*!

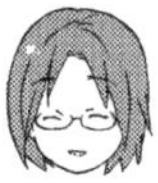

By the way, the proteins we saw that stored glucose and detoxified alcohol are all enzymes. The protein that stores glucose as glycogen is called *glycogen synthase*, and the protein that degrades alcohol to the harmless acetaldehyde is *alcohol dehydrogenase*.

Alcohol dehydrogenase (ADH), etc.
Acetaldehyde dehydrogenase (ALDH1, ALDH2), etc.
Alcohol
Acetaldehyde
Acetic acid
Carbon dioxide and water

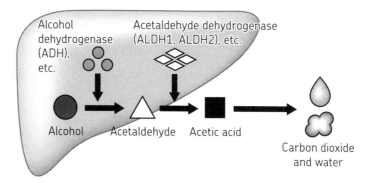

As we saw earlier, alcohol is broken down in the liver. One enzyme breaks down alcohol to acetaldehyde, and yet another enzyme further breaks down acetaldehyde into acetic acid.

Do you remember? It was Enzyme Man, or alcohol dehydrogenase, that broke down alcohol, and his sidekick dog, acetaldehyde dehydrogenase, that broke down acetaldehyde to acetic acid. A different enzyme is responsible for each chemical reaction.

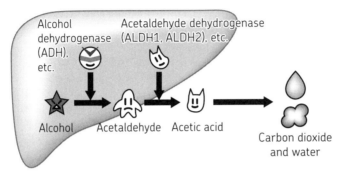

Alcohol dehydrogenase (ADH), etc.

Acetaldehyde dehydrogenase (ALDH1, ALDH2), etc.

Alcohol → Acetaldehyde → Acetic acid → Carbon dioxide and water

PROTEINS' ROLE IN CELL DIVISION

Proteins also run the process of cell division. Division is how a cell reproduces, that is, by splitting in two. Before a cell undergoes division, the DNA in the nucleus is copied so that one copy may be given to the two new cells formed after division. An enzyme is responsible for starting this process of copying DNA (called *DNA replication*).

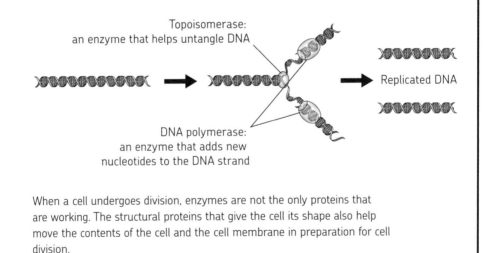

Topoisomerase: an enzyme that helps untangle DNA

Replicated DNA

DNA polymerase: an enzyme that adds new nucleotides to the DNA strand

When a cell undergoes division, enzymes are not the only proteins that are working. The structural proteins that give the cell its shape also help move the contents of the cell and the cell membrane in preparation for cell division.

PROTEINS AND MUSCLE CONTRACTION

Marcus, are there proteins in muscle cells, like the biceps you talked about at breakfast?

Good question. Muscles like biceps are formed by a group of bundles of muscle cells. Each cell is called a *muscle fiber*.

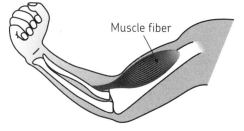

Muscle fiber

Muscle cells are made up of two long, fine fibers called *actin filament* and *myosin filament*. These fibers are made up of two types of proteins called *actin* and *myosin*, respectively. A muscle contracts when the fibers slide against each other.

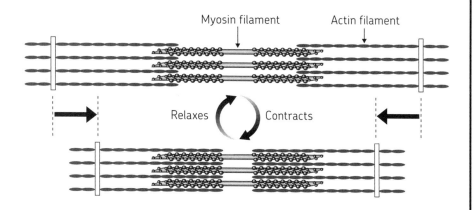

Myosin filament   Actin filament

Relaxes   Contracts

Proteins help muscles maintain their shape, and they move muscles themselves. They're not simply catalysts for a chemical reaction.

 I guess Enzyme Man had a bit part after all.

 Rin! That's not very nice. I feel sorry for Enzyme Man.

 Oh, whatever. He definitely didn't do as much as a protein in a muscle does.

SUMMARY

Proteins help cells perform various functions. Over 100,000 types of proteins exist in the human body. Each protein is responsible for carrying out specific work.
These are some of the main functions of proteins:

- Controlling chemical reactions (enzymes)
- Contracting muscles (actin and myosin)
- Transporting oxygen and nutrients (hemoglobin)
- Maintaining the homeostasis (hormones like insulin)
- Defending the body from viruses and harmful bacteria (immunoglobin)
- Maintaining the structure of cells (collagen and keratin)

 Proteins perform vital functions in the body.

 I get it, proteins are great.

 Yes, we can carry on living thanks to the ongoing work of proteins.

 Oooh, proteins like Enzyme Man! He's so dreamy.

 What's with you and Enzyme Man?!

PROTEINS ARE MADE OF AMINO ACIDS

 We learned yesterday that living organisms are made up of cells. Groups of cells form tissues, and groups of tissues form organs, which ultimately form the body of living organisms. It's like stacking up bricks to build a house.

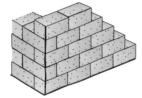 

Stacked bricks make a wall. **Cells join together to form tissue.**

Proteins also consist of smaller components—specifically, chains of molecules called *amino acids*.

Proteins are made of chains of amino acids.

DNA is also made up of chains—the substances that link together to form DNA are called *nucleotides*.

DNA is made of a chain of nucleotides.

 Do amino acids form proteins the way cells form tissues?

 No! There is a big difference between cells and proteins. Cells pile up in three dimensions to form body tissues, but amino acids link horizontally to form proteins.

 So if cells are like bricks, amino acids are like beads on a necklace, right?

Exactly. Of course, these strings take unique jumbled and tangled shapes as well. They certainly aren't just in a straight line.

These amino acids link together to form proteins and their sequence determines their function.

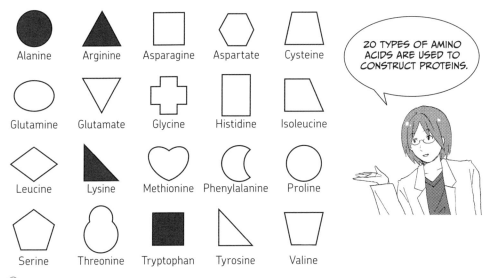

Alanine   Arginine   Asparagine   Aspartate   Cysteine

Glutamine   Glutamate   Glycine   Histidine   Isoleucine

Leucine   Lysine   Methionine   Phenylalanine   Proline

Serine   Threonine   Tryptophan   Tyrosine   Valine

20 TYPES OF AMINO ACIDS ARE USED TO CONSTRUCT PROTEINS.

MSG is made up of an amino acid, right? It makes my food taste so good!

Yes, that's right. MSG is made of glutamic acid, one type of amino acid. There are 20 types of amino acids, and arranging them in a predetermined order creates a specific protein. (There are actually more than 20 types of amino acids, but this is how many are used for creating proteins.) Let's look at the structure of amino acids. Each of the 20 amino acids used for creating proteins has a portion common to the other 19 amino acids, as well as a unique portion not found in the others.

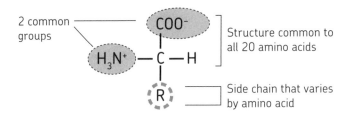

2 common groups

Structure common to all 20 amino acids

$H_3N^+ — C — H$

COO^-

R

Side chain that varies by amino acid

In this figure, R represents a side chain: This portion varies in each amino acid. So there are 20 types of side chains, ranging from simple ones consisting of a single hydrogen atom to complex ones made up of several linked benzene rings.

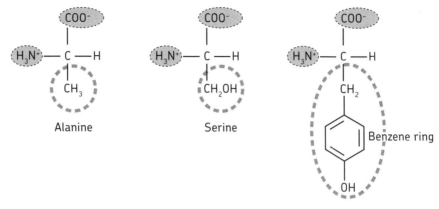

COO-
H3N+
C
H
CH3
Alanine
COO-
H3N+
C
H
CH2OH
Serine
COO-
H3N+
C
H
CH2
Benzene ring
OH
Tyrosine

Each of these 20 amino acids shares a common structure of H_3N^+ and COO^-.

REPLACING ONE AMINO ACID WITH ANOTHER IS A BIG DEAL!

This may seem unrelated, but do you know why human blood is red?

Because human blood is red with the flame of justice—just like Enzyme Man burns with passion for doing good!

Uh, that can't be right.

No, indeed, it's not. The correct answer is because of the red pigment called hemoglobin found in blood. Red blood cells carry out the important function of bringing oxygen molecules to cells throughout the body by attaching these oxygen molecules to hemoglobin. Once they have oxygen, the cells can generate energy. Hemoglobin also carries CO_2. Hemoglobin is bright red because of the high quantity of iron contained within it.

But Marcus, why are you suddenly talking about blood? What about proteins?

Well, the main component of hemoglobin is protein! Hemoglobin is made up of two types of proteins called globins: α and β. Each of these proteins is called a *subunit*. There are two α subunits and two β subunits. Because they are proteins, each of the two types of subunits is made up of a long sequence of 20 types of amino acids linked together in a particular order.

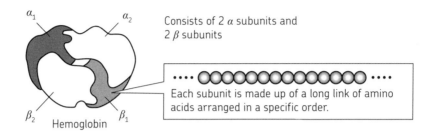

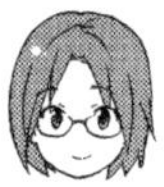

α1
α2
Consists of 2 α subunits and 2 β subunits
Each subunit is made up of a long link of amino acids arranged in a specific order.
β2
Hemoglobin
β1

Changing just a single amino acid in the chain can cause serious trouble. If, for instance, the sixth amino acid contained in the β subunit, glutamic acid, is replaced with valine, an abnormal deformation occurs in hemoglobin. This deformation prevents the hemoglobin from carrying oxygen properly and causes anemia. Switching amino acids can also cause red blood cells to deform into a sickle-like shape, causing sickle-cell disease.

Glutamic acid
1 2 3 4 5 6
Normal hemoglobin
Normal red blood cell
Abnormal hemoglobin
Sickle-shaped red blood cell
Valine

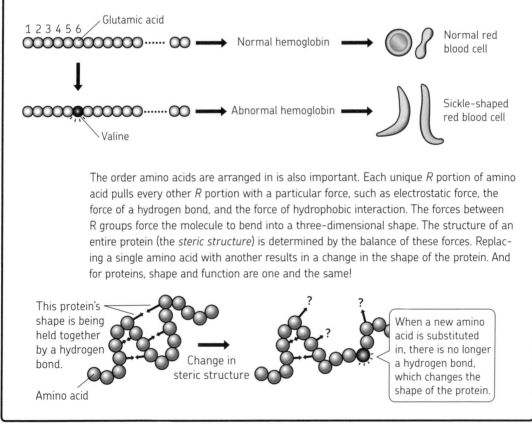

The order amino acids are arranged in is also important. Each unique *R* portion of amino acid pulls every other *R* portion with a particular force, such as electrostatic force, the force of a hydrogen bond, and the force of hydrophobic interaction. The forces between R groups force the molecule to bend into a three-dimensional shape. The structure of an entire protein (the *steric structure*) is determined by the balance of these forces. Replacing a single amino acid with another results in a change in the shape of the protein. And for proteins, shape and function are one and the same!

This protein's shape is being held together by a hydrogen bond.
Amino acid
Change in steric structure
?
?
?
When a new amino acid is substituted in, there is no longer a hydrogen bond, which changes the shape of the protein.

GENES: THE BLUEPRINT FOR BUILDING PROTEINS

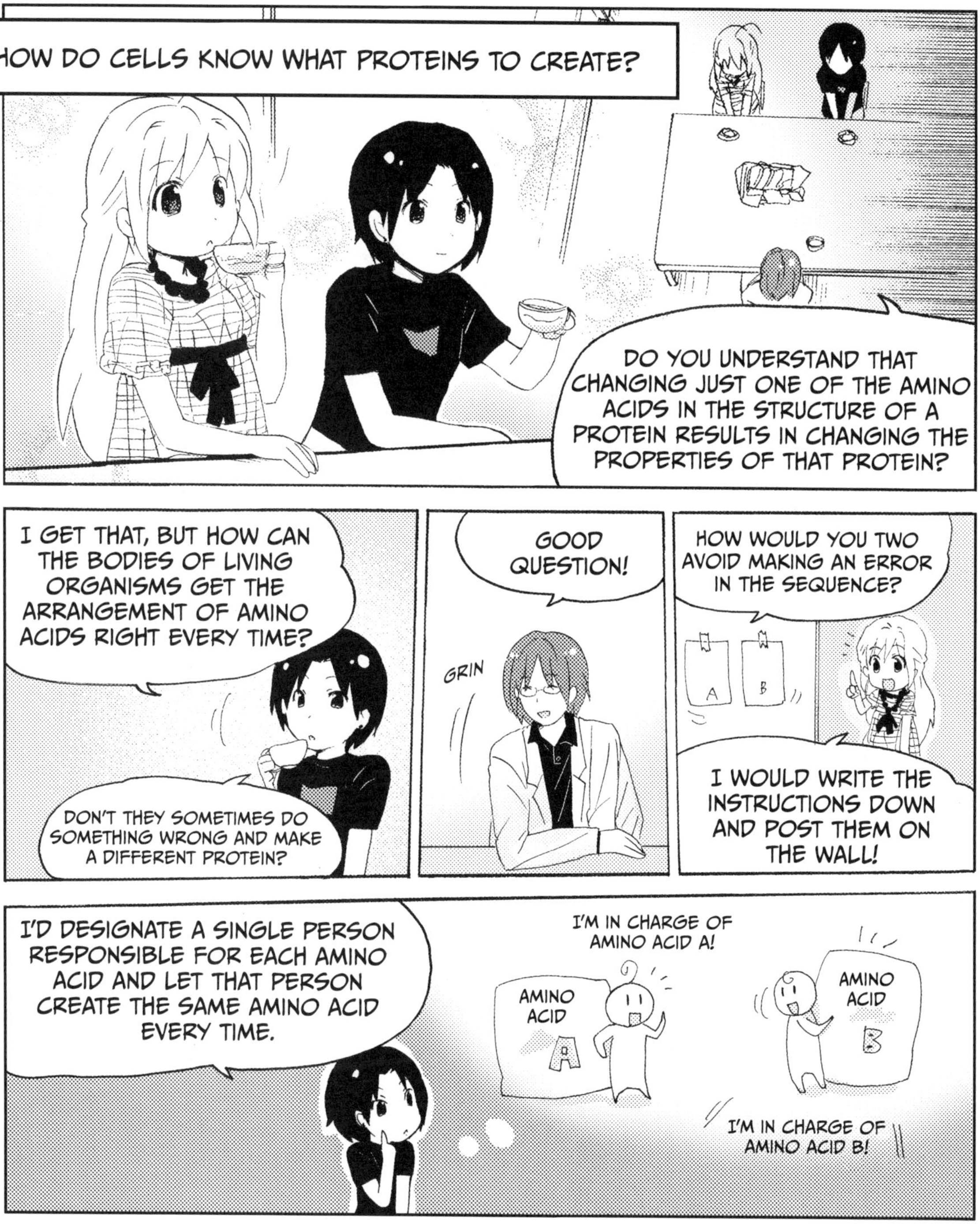
HOW DO CELLS KNOW WHAT PROTEINS TO CREATE?
DO YOU UNDERSTAND THAT CHANGING JUST ONE OF THE AMINO ACIDS IN THE STRUCTURE OF A PROTEIN RESULTS IN CHANGING THE PROPERTIES OF THAT PROTEIN?
I GET THAT, BUT HOW CAN THE BODIES OF LIVING ORGANISMS GET THE ARRANGEMENT OF AMINO ACIDS RIGHT EVERY TIME?
DON'T THEY SOMETIMES DO SOMETHING WRONG AND MAKE A DIFFERENT PROTEIN?
GOOD QUESTION!
GRIN
HOW WOULD YOU TWO AVOID MAKING AN ERROR IN THE SEQUENCE?
A
B
I WOULD WRITE THE INSTRUCTIONS DOWN AND POST THEM ON THE WALL!
I'D DESIGNATE A SINGLE PERSON RESPONSIBLE FOR EACH AMINO ACID AND LET THAT PERSON CREATE THE SAME AMINO ACID EVERY TIME.
I'M IN CHARGE OF AMINO ACID A!
AMINO ACID A
AMINO ACID B
I'M IN CHARGE OF AMINO ACID B!

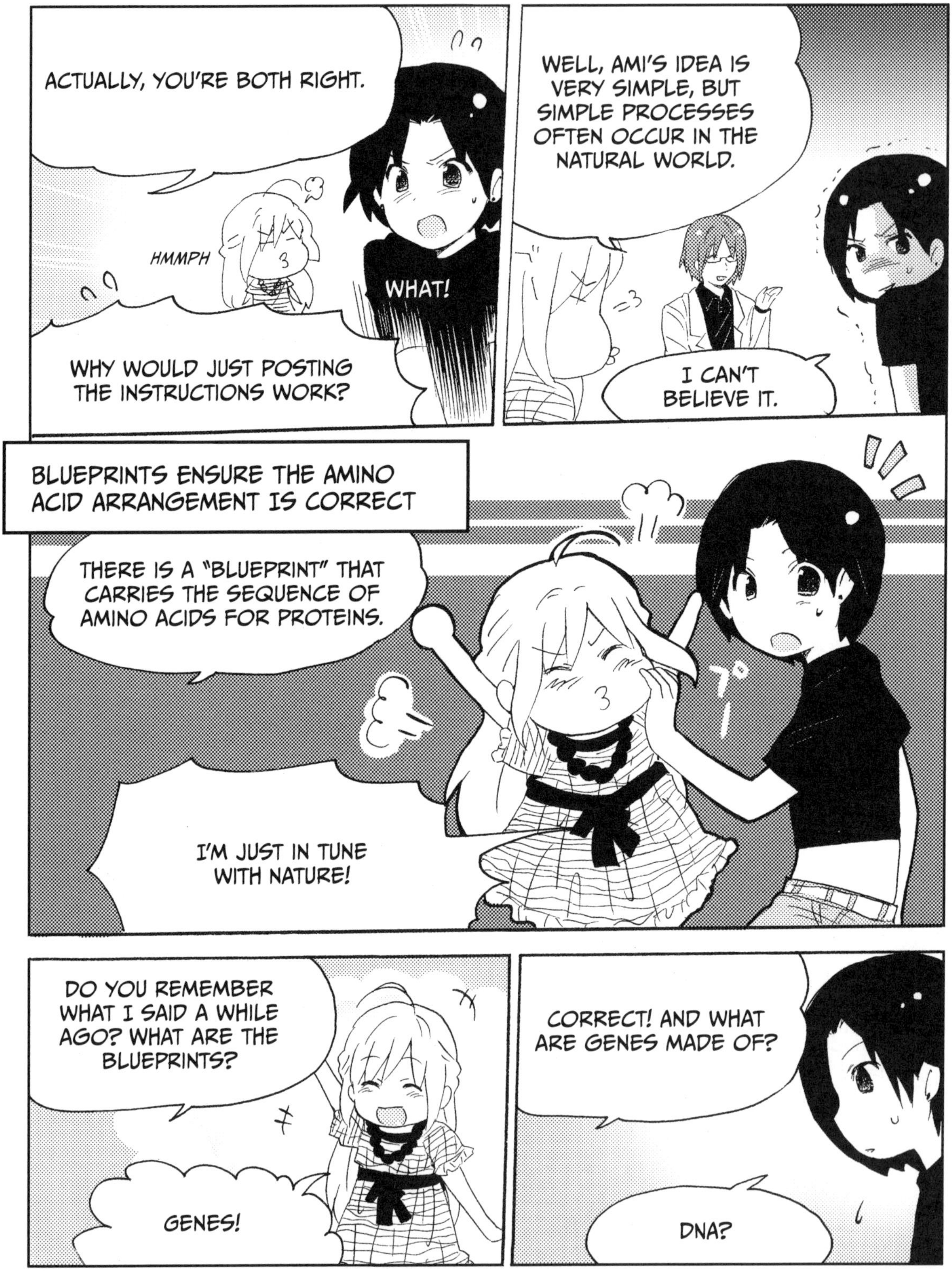
ACTUALLY, YOU'RE BOTH RIGHT.
HMMPH
WHAT!
WHY WOULD JUST POSTING THE INSTRUCTIONS WORK?
WELL, AMI'S IDEA IS VERY SIMPLE, BUT SIMPLE PROCESSES OFTEN OCCUR IN THE NATURAL WORLD.
I CAN'T BELIEVE IT.
BLUEPRINTS ENSURE THE AMINO ACID ARRANGEMENT IS CORRECT
THERE IS A "BLUEPRINT" THAT CARRIES THE SEQUENCE OF AMINO ACIDS FOR PROTEINS.
I'M JUST IN TUNE WITH NATURE!
DO YOU REMEMBER WHAT I SAID A WHILE AGO? WHAT ARE THE BLUEPRINTS?
GENES!
CORRECT! AND WHAT ARE GENES MADE OF?
DNA?

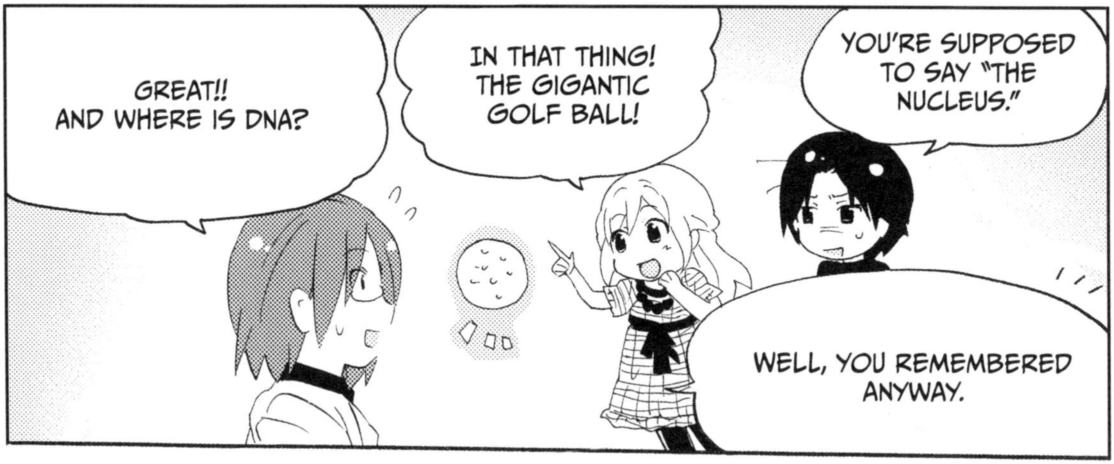
GREAT!!
AND WHERE IS DNA?
IN THAT THING!
THE GIGANTIC
GOLF BALL!
YOU'RE SUPPOSED
TO SAY "THE
NUCLEUS."
WELL, YOU REMEMBERED
ANYWAY.

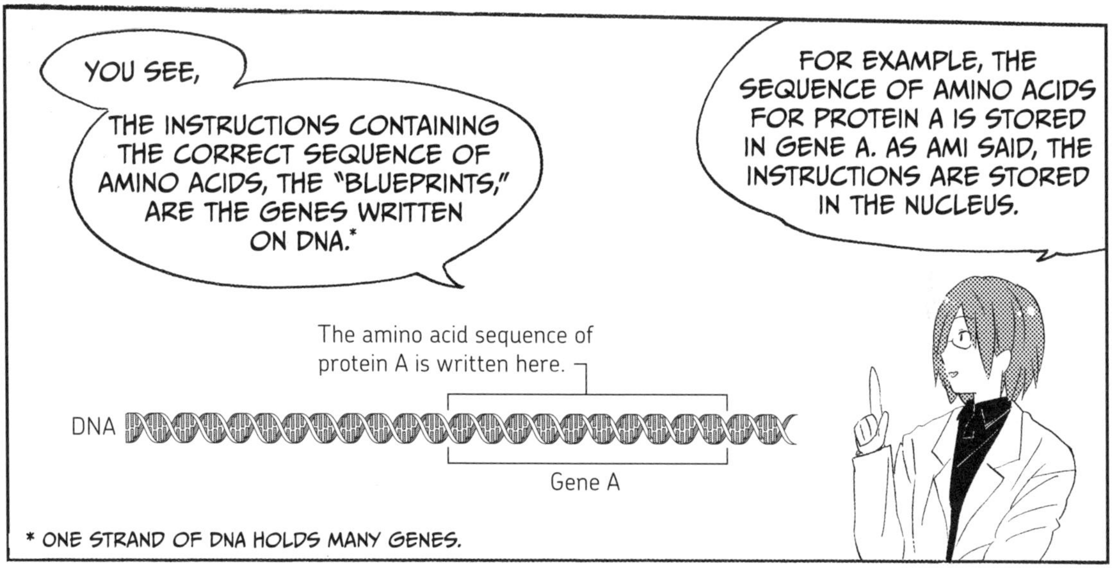
YOU SEE,
THE INSTRUCTIONS CONTAINING THE CORRECT SEQUENCE OF AMINO ACIDS, THE "BLUEPRINTS," ARE THE GENES WRITTEN ON DNA.*
FOR EXAMPLE, THE SEQUENCE OF AMINO ACIDS FOR PROTEIN A IS STORED IN GENE A. AS AMI SAID, THE INSTRUCTIONS ARE STORED IN THE NUCLEUS.
The amino acid sequence of protein A is written here.
DNA
Gene A
* ONE STRAND OF DNA HOLDS MANY GENES.

OUR GENES ARE WRITTEN IN CODE

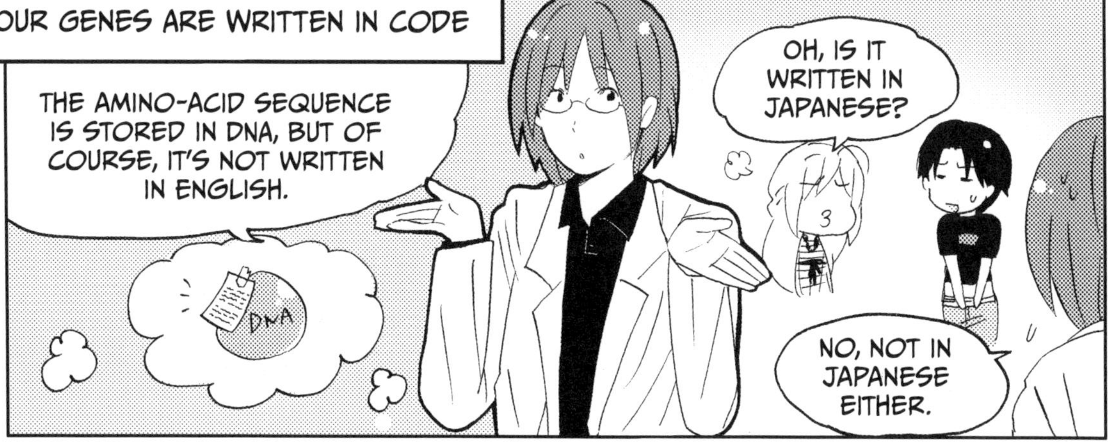
THE AMINO-ACID SEQUENCE IS STORED IN DNA, BUT OF COURSE, IT'S NOT WRITTEN IN ENGLISH.
DNA
OH, IS IT
WRITTEN IN
JAPANESE?
NO, NOT IN
JAPANESE
EITHER.

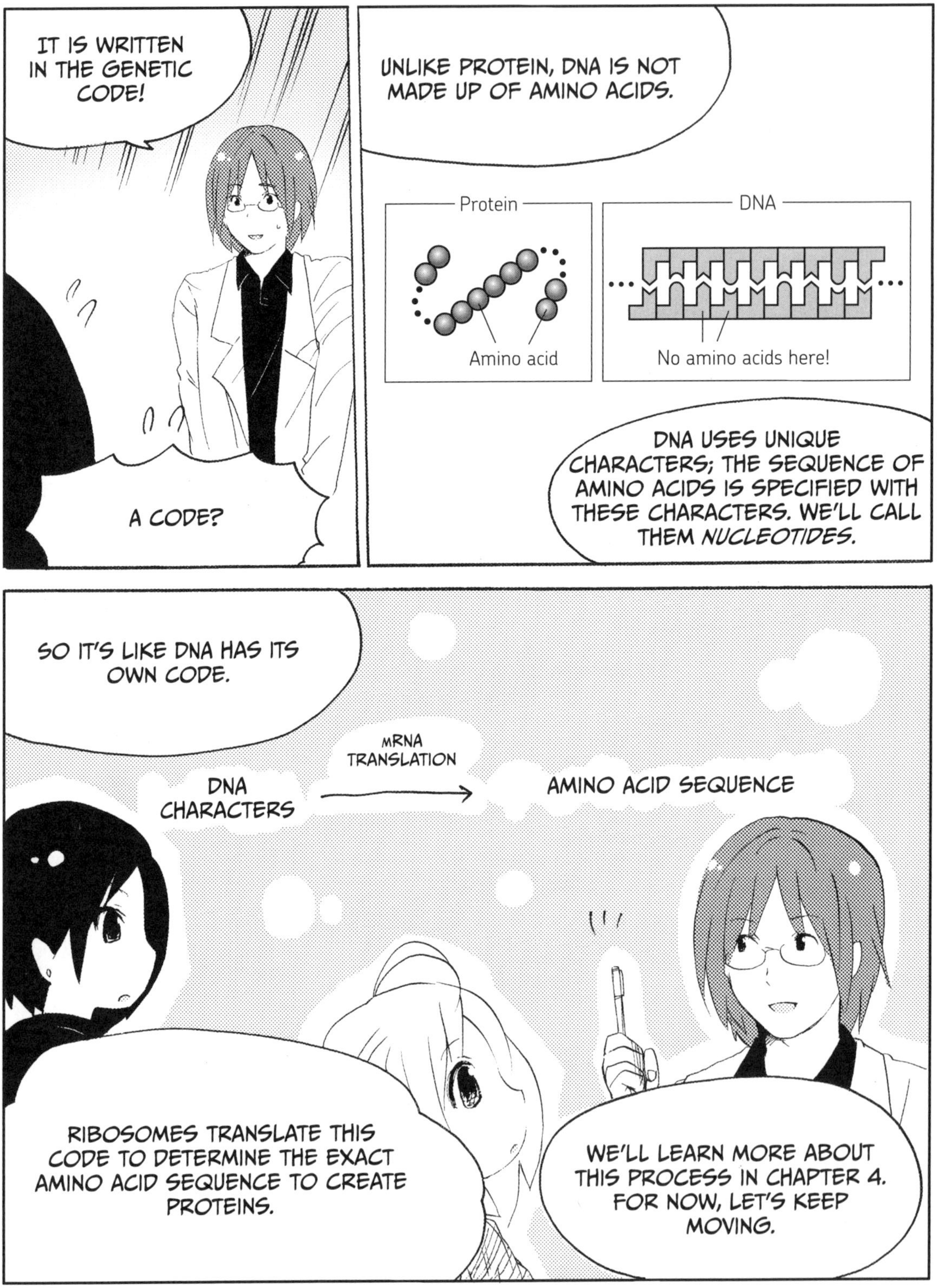
IT IS WRITTEN IN THE GENETIC CODE!
A CODE?
UNLIKE PROTEIN, DNA IS NOT MADE UP OF AMINO ACIDS.
Protein
DNA
Amino acid
No amino acids here!
DNA USES UNIQUE CHARACTERS; THE SEQUENCE OF AMINO ACIDS IS SPECIFIED WITH THESE CHARACTERS. WE'LL CALL THEM *NUCLEOTIDES*.
SO IT'S LIKE DNA HAS ITS OWN CODE.
DNA CHARACTERS
MRNA TRANSLATION
AMINO ACID SEQUENCE
RIBOSOMES TRANSLATE THIS CODE TO DETERMINE THE EXACT AMINO ACID SEQUENCE TO CREATE PROTEINS.
WE'LL LEARN MORE ABOUT THIS PROCESS IN CHAPTER 4. FOR NOW, LET'S KEEP MOVING.

DNA AND NUCLEOTIDES

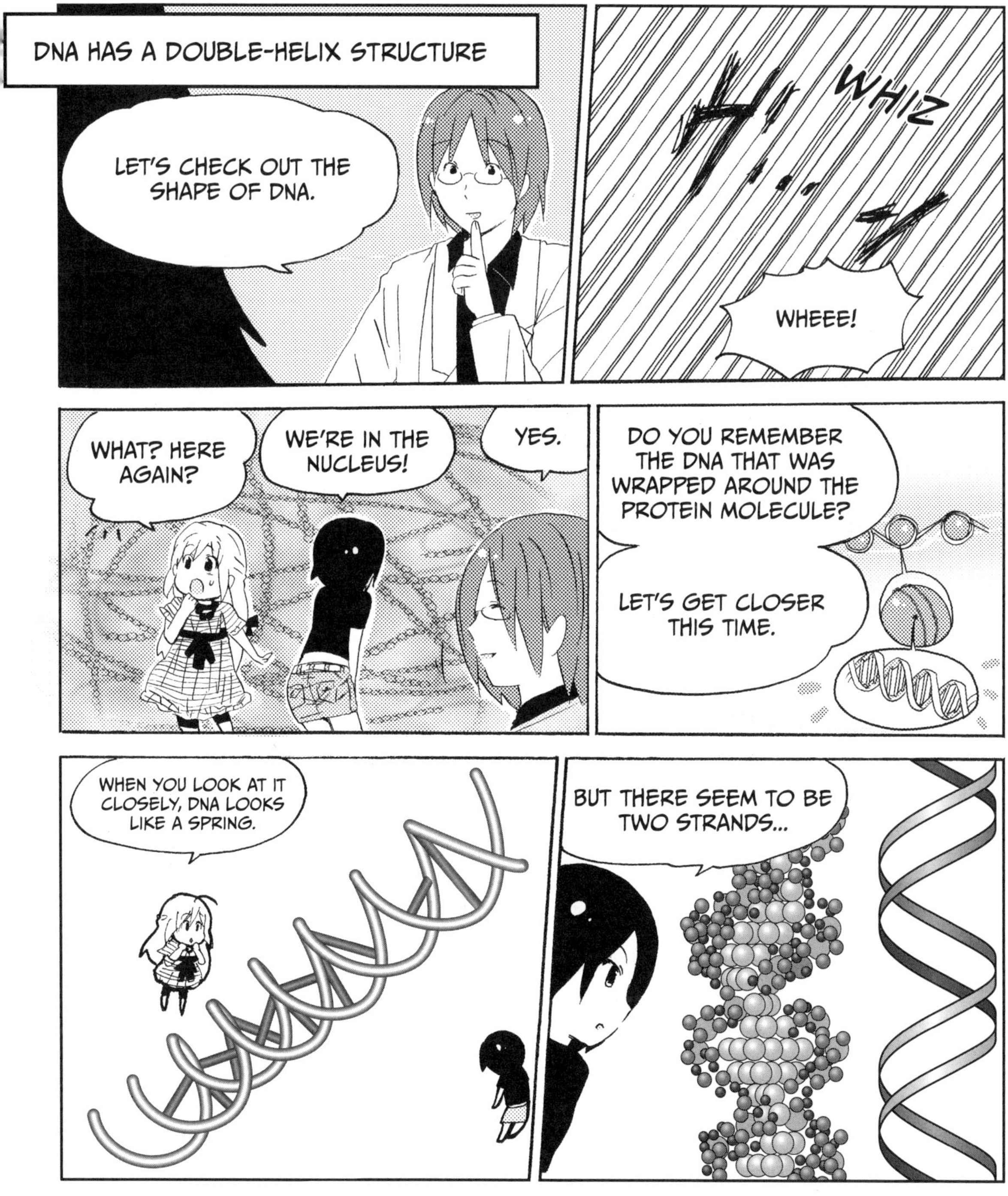
DNA HAS A DOUBLE-HELIX STRUCTURE
LET'S CHECK OUT THE SHAPE OF DNA.
WHIZ
WHEEE!
WHAT? HERE AGAIN?
WE'RE IN THE NUCLEUS!
YES.
DO YOU REMEMBER THE DNA THAT WAS WRAPPED AROUND THE PROTEIN MOLECULE?
LET'S GET CLOSER THIS TIME.
WHEN YOU LOOK AT IT CLOSELY, DNA LOOKS LIKE A SPRING.
BUT THERE SEEM TO BE TWO STRANDS...

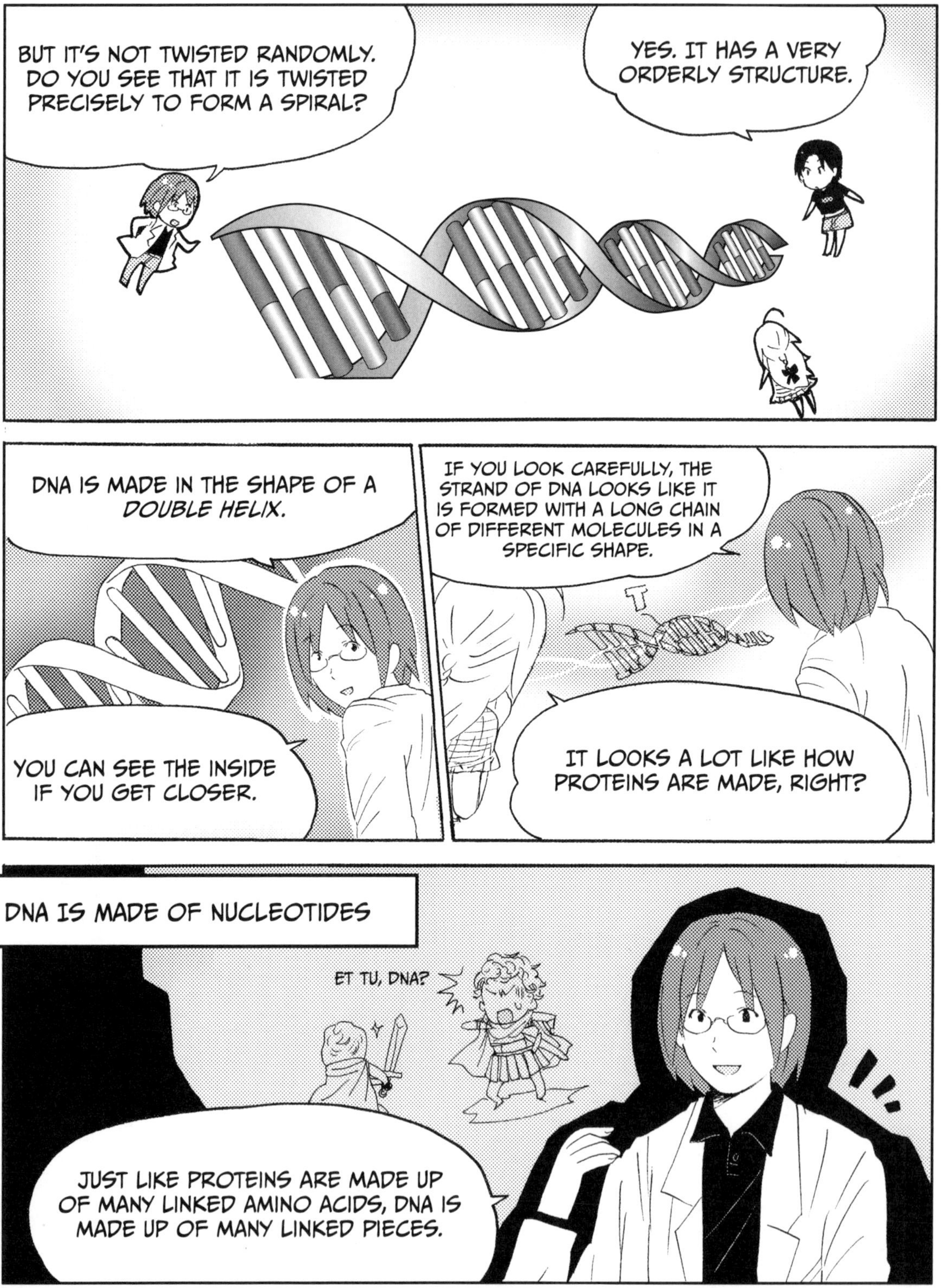
BUT IT'S NOT TWISTED RANDOMLY. DO YOU SEE THAT IT IS TWISTED PRECISELY TO FORM A SPIRAL?
YES. IT HAS A VERY ORDERLY STRUCTURE.
DNA IS MADE IN THE SHAPE OF A *DOUBLE HELIX*.
YOU CAN SEE THE INSIDE IF YOU GET CLOSER.
IF YOU LOOK CAREFULLY, THE STRAND OF DNA LOOKS LIKE IT IS FORMED WITH A LONG CHAIN OF DIFFERENT MOLECULES IN A SPECIFIC SHAPE.
T
IT LOOKS A LOT LIKE HOW PROTEINS ARE MADE, RIGHT?
DNA IS MADE OF NUCLEOTIDES
ET TU, DNA?
JUST LIKE PROTEINS ARE MADE UP OF MANY LINKED AMINO ACIDS, DNA IS MADE UP OF MANY LINKED PIECES.

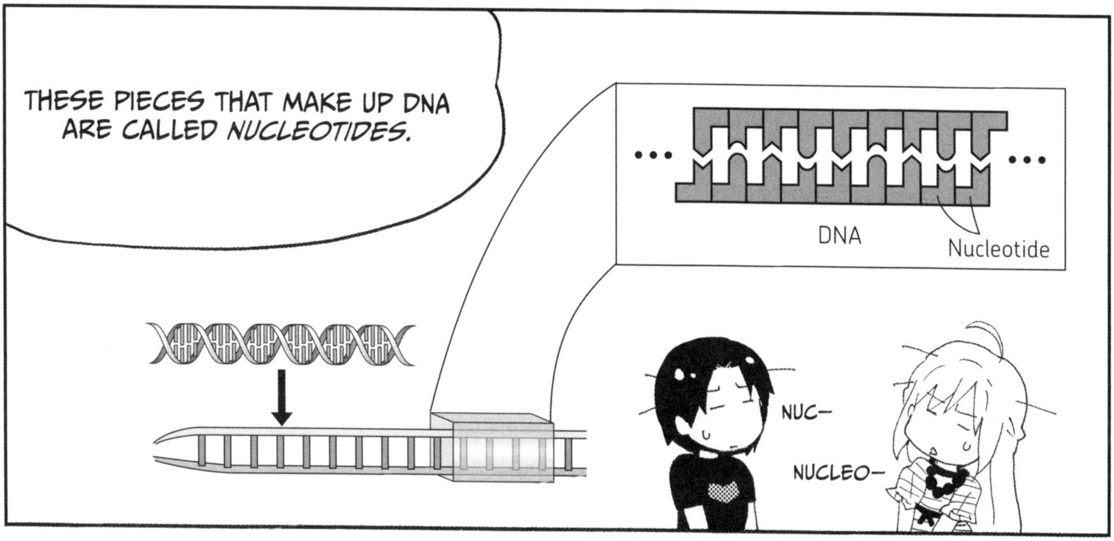
THESE PIECES THAT MAKE UP DNA ARE CALLED *NUCLEOTIDES.*
DNA
Nucleotide
NUC—
NUCLEO—

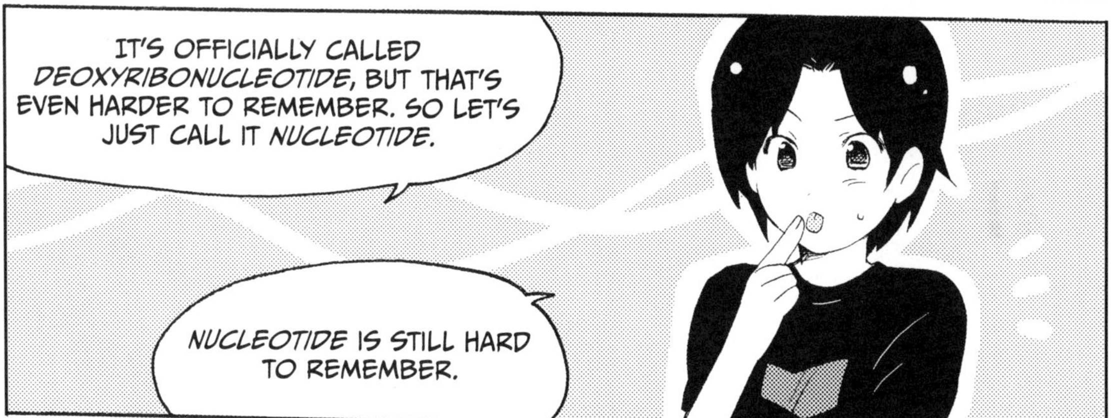
IT'S OFFICIALLY CALLED *DEOXYRIBONUCLEOTIDE*, BUT THAT'S EVEN HARDER TO REMEMBER. SO LET'S JUST CALL IT *NUCLEOTIDE.*
NUCLEOTIDE IS STILL HARD TO REMEMBER.

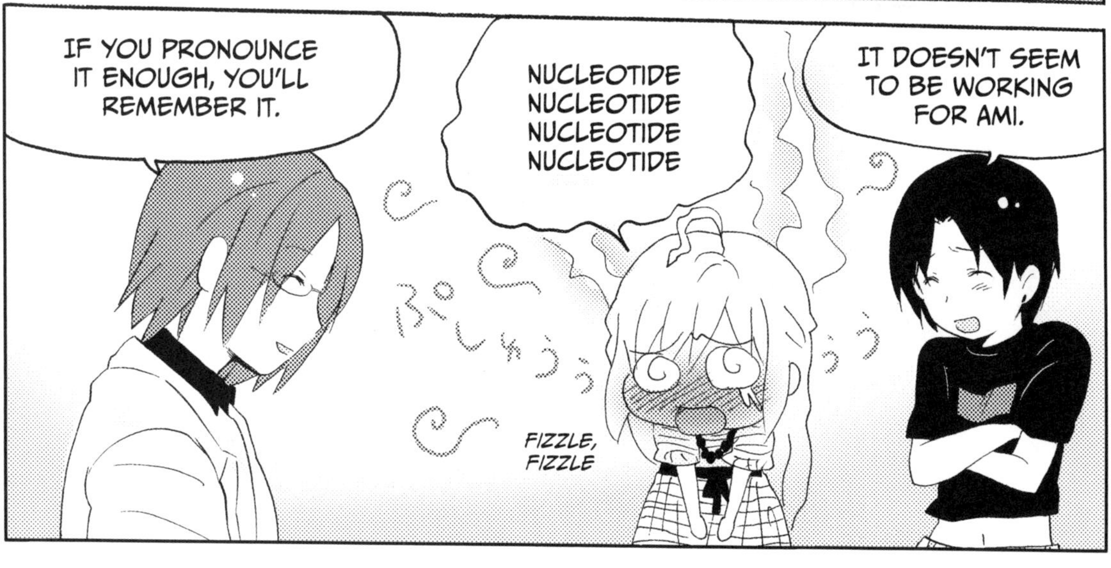
IF YOU PRONOUNCE IT ENOUGH, YOU'LL REMEMBER IT.
NUCLEOTIDE NUCLEOTIDE NUCLEOTIDE NUCLEOTIDE
IT DOESN'T SEEM TO BE WORKING FOR AMI.
FIZZLE, FIZZLE

NUCLEOTIDES ARE THE CHARACTERS IN THE "CODE"

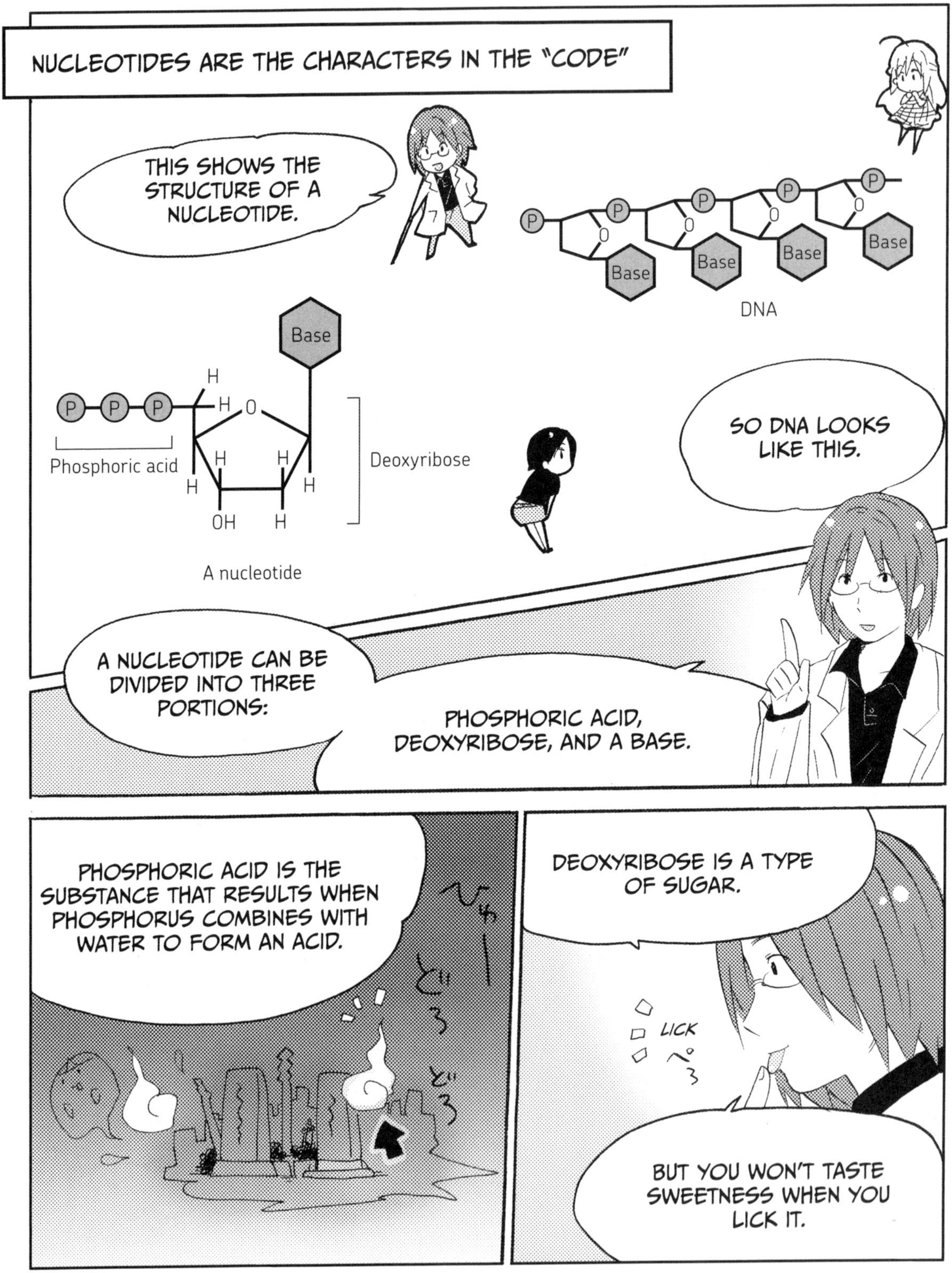
THIS SHOWS THE STRUCTURE OF A NUCLEOTIDE.
P
Base
O
DNA
Base
P
H
H
O
Phosphoric acid
H
H
H
H
OH
H
Deoxyribose
A nucleotide
SO DNA LOOKS LIKE THIS.
A NUCLEOTIDE CAN BE DIVIDED INTO THREE PORTIONS:
PHOSPHORIC ACID, DEOXYRIBOSE, AND A BASE.
PHOSPHORIC ACID IS THE SUBSTANCE THAT RESULTS WHEN PHOSPHORUS COMBINES WITH WATER TO FORM AN ACID.
どろ
どろ
DEOXYRIBOSE IS A TYPE OF SUGAR.
LICK
ぺろ
BUT YOU WON'T TASTE SWEETNESS WHEN YOU LICK IT.

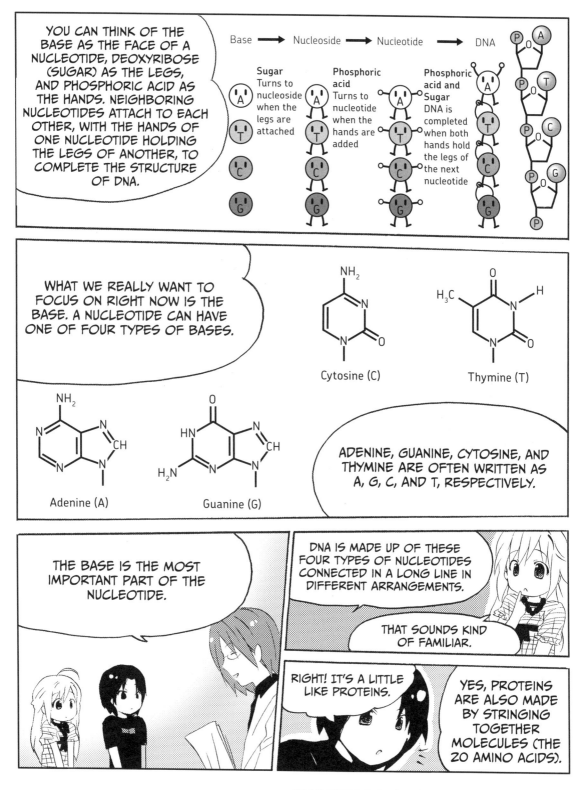

YOU CAN THINK OF THE BASE AS THE FACE OF A NUCLEOTIDE, DEOXYRIBOSE (SUGAR) AS THE LEGS, AND PHOSPHORIC ACID AS THE HANDS. NEIGHBORING NUCLEOTIDES ATTACH TO EACH OTHER, WITH THE HANDS OF ONE NUCLEOTIDE HOLDING THE LEGS OF ANOTHER, TO COMPLETE THE STRUCTURE OF DNA.

Base → Nucleoside → Nucleotide → DNA

Sugar
Turns to nucleoside when the legs are attached

Phosphoric acid
Turns to nucleotide when the hands are added

Phosphoric acid and Sugar
DNA is completed when both hands hold the legs of the next nucleotide

WHAT WE REALLY WANT TO FOCUS ON RIGHT NOW IS THE BASE. A NUCLEOTIDE CAN HAVE ONE OF FOUR TYPES OF BASES.

Cytosine (C)

Thymine (T)

Adenine (A)

Guanine (G)

ADENINE, GUANINE, CYTOSINE, AND THYMINE ARE OFTEN WRITTEN AS A, G, C, AND T, RESPECTIVELY.

THE BASE IS THE MOST IMPORTANT PART OF THE NUCLEOTIDE.

DNA IS MADE UP OF THESE FOUR TYPES OF NUCLEOTIDES CONNECTED IN A LONG LINE IN DIFFERENT ARRANGEMENTS.

THAT SOUNDS KIND OF FAMILIAR.

RIGHT! IT'S A LITTLE LIKE PROTEINS.

YES, PROTEINS ARE ALSO MADE BY STRINGING TOGETHER MOLECULES (THE 20 AMINO ACIDS).

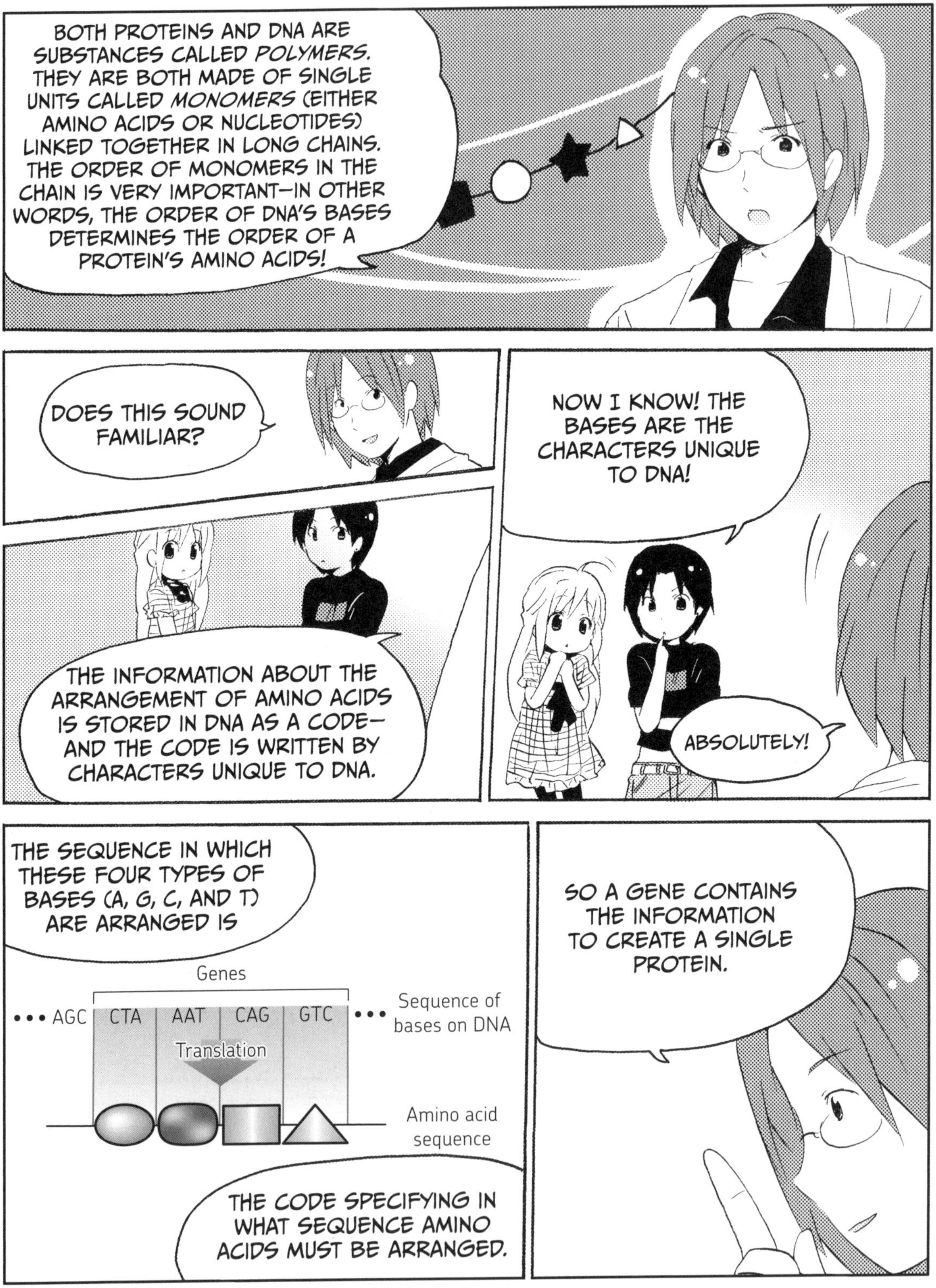
BOTH PROTEINS AND DNA ARE SUBSTANCES CALLED *POLYMERS*. THEY ARE BOTH MADE OF SINGLE UNITS CALLED *MONOMERS* (EITHER AMINO ACIDS OR NUCLEOTIDES) LINKED TOGETHER IN LONG CHAINS. THE ORDER OF MONOMERS IN THE CHAIN IS VERY IMPORTANT—IN OTHER WORDS, THE ORDER OF DNA'S BASES DETERMINES THE ORDER OF A PROTEIN'S AMINO ACIDS!
DOES THIS SOUND FAMILIAR?
THE INFORMATION ABOUT THE ARRANGEMENT OF AMINO ACIDS IS STORED IN DNA AS A CODE—AND THE CODE IS WRITTEN BY CHARACTERS UNIQUE TO DNA.
NOW I KNOW! THE BASES ARE THE CHARACTERS UNIQUE TO DNA!
ABSOLUTELY!
THE SEQUENCE IN WHICH THESE FOUR TYPES OF BASES (A, G, C, AND T) ARE ARRANGED IS
Genes
••• AGC
CTA
AAT
CAG
GTC •••
Sequence of bases on DNA
Translation
Amino acid sequence
THE CODE SPECIFYING IN WHAT SEQUENCE AMINO ACIDS MUST BE ARRANGED.
SO A GENE CONTAINS THE INFORMATION TO CREATE A SINGLE PROTEIN.

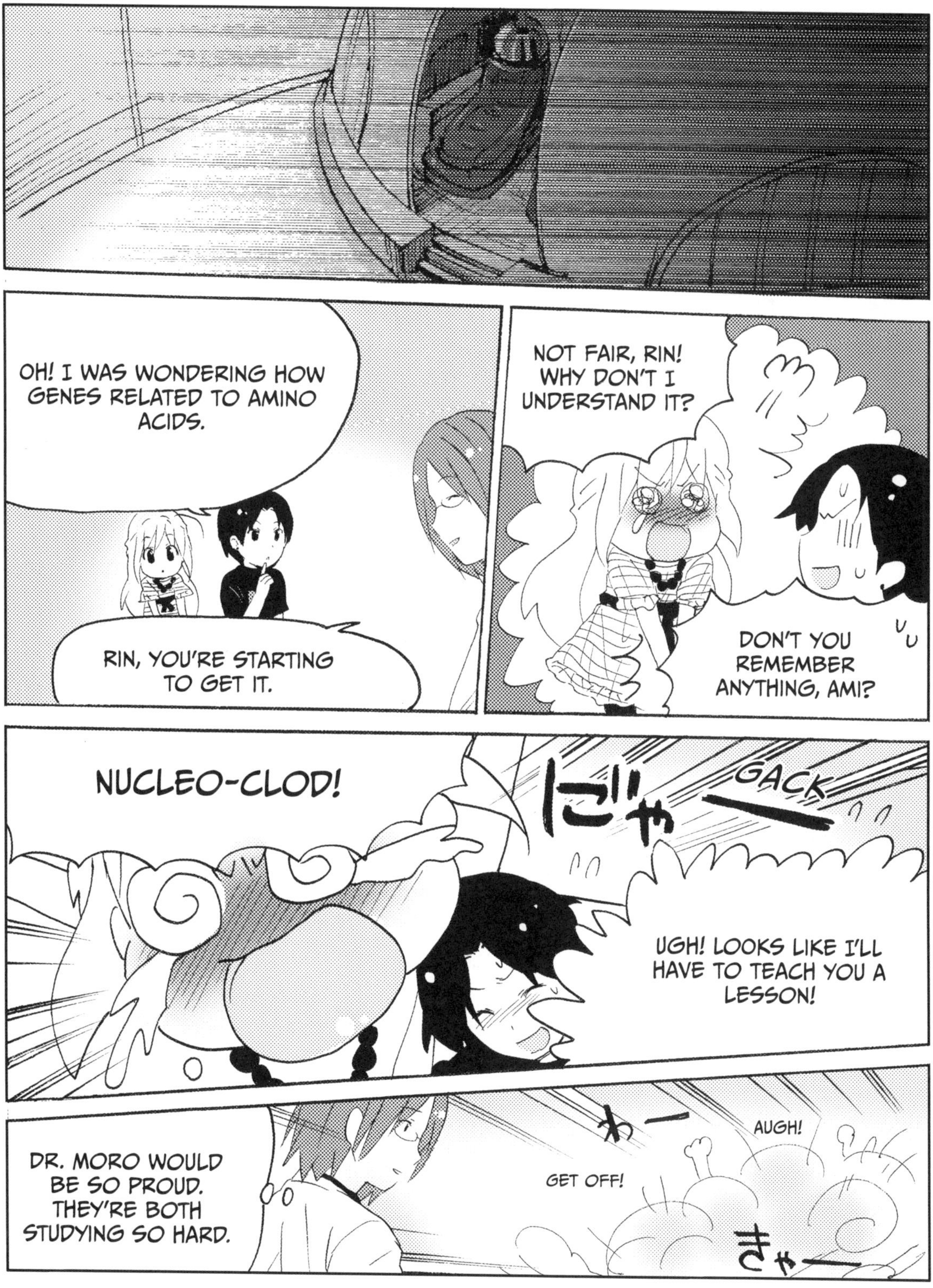
OH! I WAS WONDERING HOW GENES RELATED TO AMINO ACIDS.
RIN, YOU'RE STARTING TO GET IT.
NOT FAIR, RIN! WHY DON'T I UNDERSTAND IT?
DON'T YOU REMEMBER ANYTHING, AMI?
NUCLEO-CLOD!
GACK
UGH! LOOKS LIKE I'LL HAVE TO TEACH YOU A LESSON!
DR. MORO WOULD BE SO PROUD. THEY'RE BOTH STUDYING SO HARD.
わー
AUGH!
GET OFF!
きゃー

THE GENOME: A LIBRARY OF GENES

Knowing the sequence of every gene can be helpful in the fields of medicine and biology. In the future, people may be able to screen their DNA to see if they are predisposed for certain cancers or illnesses. A collection of the sequences of every strand of DNA in an organism is called a *genome*. Every living organism contains a genome in the nucleus of each cell.

The Human Genome Project was completed in 2003. This project found the sequence of bases in DNA and read every gene in the human body.

In human beings, this means looking at 30,000 genes, each with a unique long combination of the four bases A, G, C, and T.

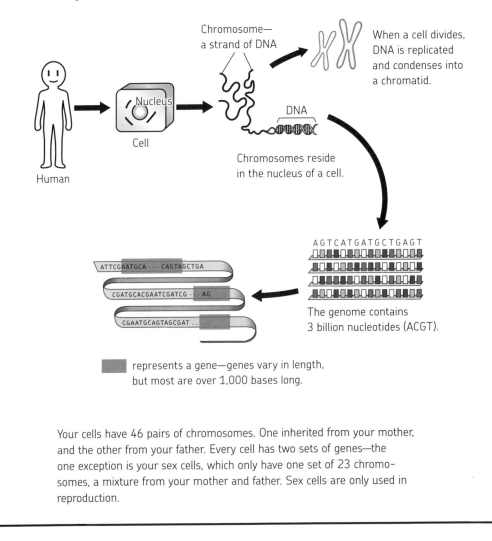

Chromosome—
a strand of DNA
When a cell divides, DNA is replicated and condenses into a chromatid.
Nucleus
Cell
Human
DNA
Chromosomes reside in the nucleus of a cell.
AGTCATGATGCTGAGT
ATTCGAATGCA · · · CAGTAGCTGA
CGATGCACGAATCGATCG · · · AG
CGAATGCAGTAGCGAT . . .
The genome contains 3 billion nucleotides (ACGT).
represents a gene—genes vary in length, but most are over 1,000 bases long.

Your cells have 46 pairs of chromosomes. One inherited from your mother, and the other from your father. Every cell has two sets of genes—the one exception is your sex cells, which only have one set of 23 chromosomes, a mixture from your mother and father. Sex cells are only used in reproduction.

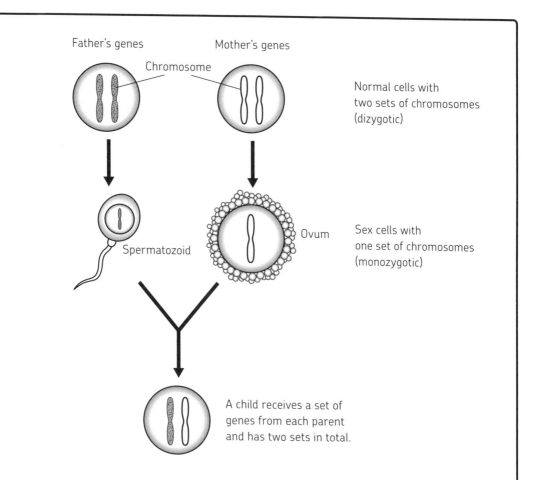

Father's genes

Mother's genes

Chromosome

Normal cells with
two sets of chromosomes
(dizygotic)

Spermatozoid

Ovum

Sex cells with
one set of chromosomes
(monozygotic)

A child receives a set of
genes from each parent
and has two sets in total.

We actually have 23 chromosomes derived from our father and 23 from
our mother (a total of 46 chromosomes made up of 23 pairs). A single pair
of chromosomes is shown above.

A genome can be compared to a library containing books of short stories.
Each book is a chromosome; each story is a gene about how to make one
protein.

But genomes contain more than just genes. There are base pairs in
between genes that do not code for genes. Research is currently going on
to learn more about these parts of the genome. Noncoding sections of
DNA can be important for functions such as regulating the expression of
genes.

If every base is equivalent to a letter, the genome would be over 100 million words long. That's equivalent to a library of 5,000 books, each 300 pages long; the entire library fits into a cell nucleus the size of a pinpoint. A complete copy of the library (all 5,000 volumes) is contained in almost every cell.

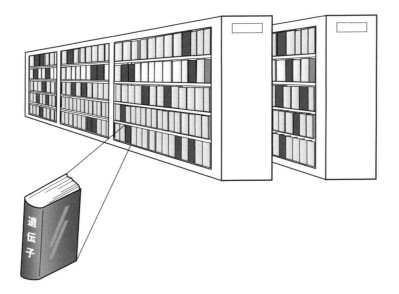

3

DNA REPLICATION AND CELL DIVISION

CELLS MULTIPLY THROUGH DIVISION

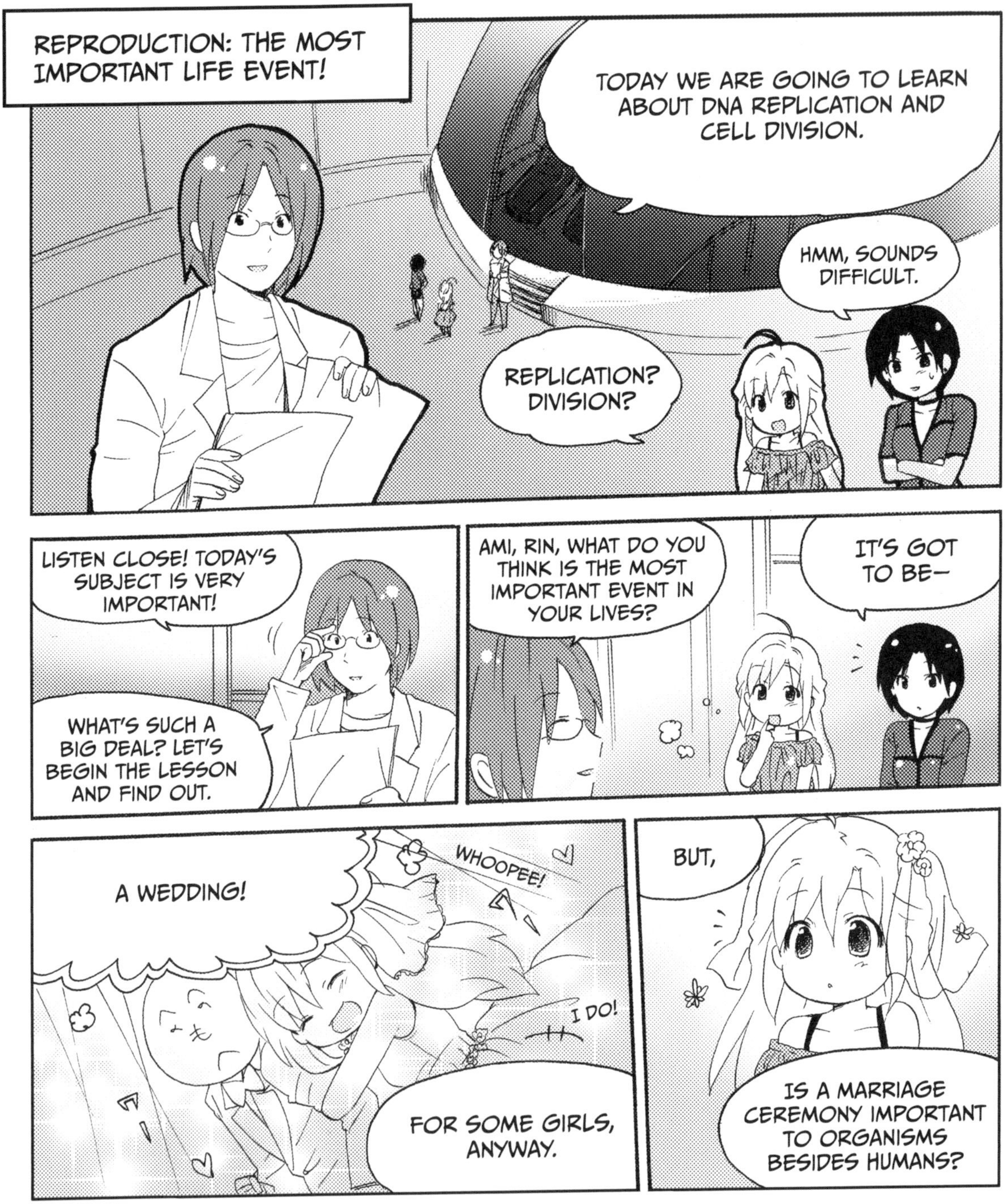
REPRODUCTION: THE MOST IMPORTANT LIFE EVENT!
TODAY WE ARE GOING TO LEARN ABOUT DNA REPLICATION AND CELL DIVISION.
HMM, SOUNDS DIFFICULT.
REPLICATION? DIVISION?
LISTEN CLOSE! TODAY'S SUBJECT IS VERY IMPORTANT!
WHAT'S SUCH A BIG DEAL? LET'S BEGIN THE LESSON AND FIND OUT.
AMI, RIN, WHAT DO YOU THINK IS THE MOST IMPORTANT EVENT IN YOUR LIVES?
IT'S GOT TO BE—
A WEDDING!
WHOOPEE!
I DO!
FOR SOME GIRLS, ANYWAY.
BUT,
IS A MARRIAGE CEREMONY IMPORTANT TO ORGANISMS BESIDES HUMANS?

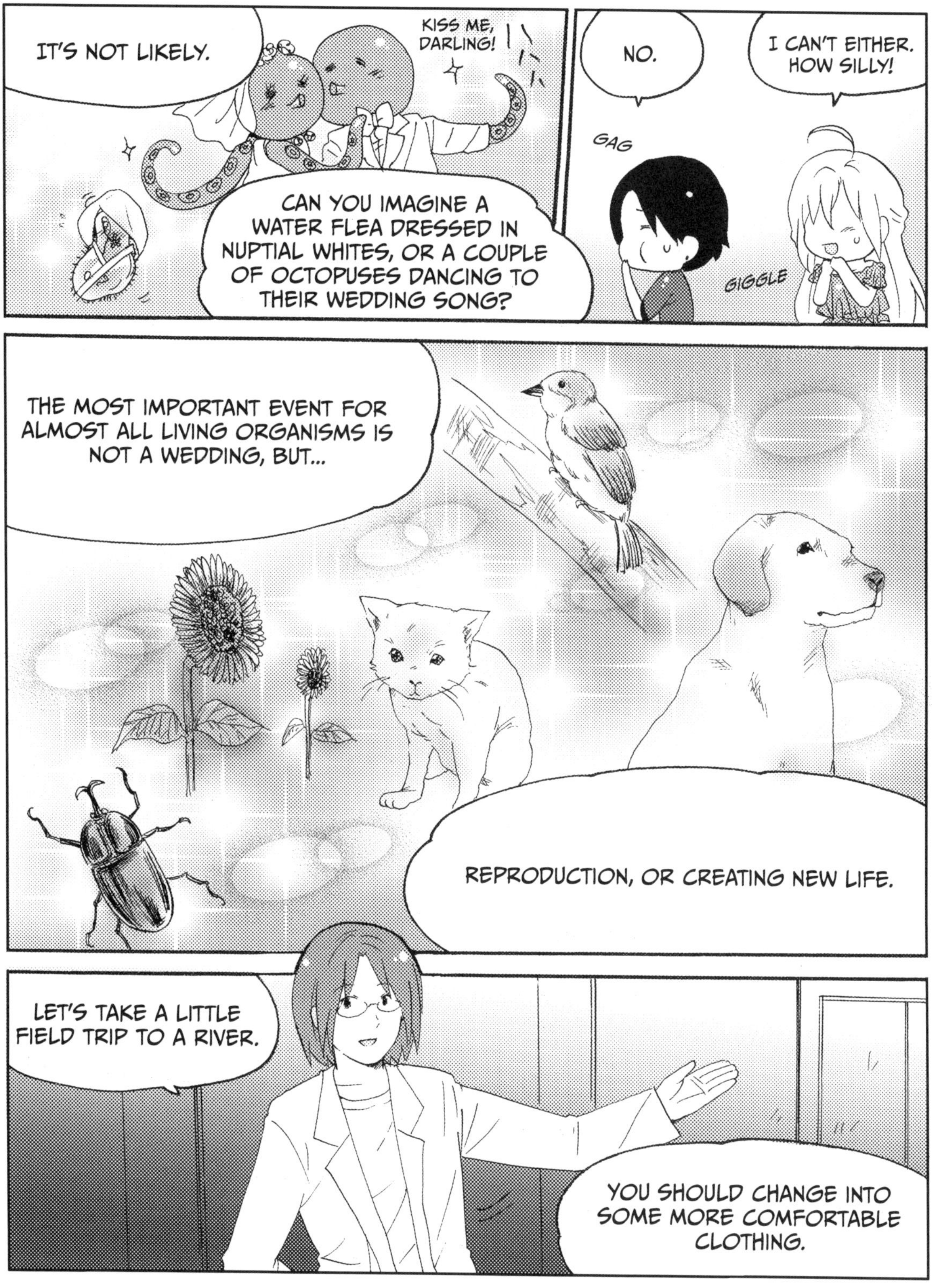
IT'S NOT LIKELY.
KISS ME, DARLING!
CAN YOU IMAGINE A WATER FLEA DRESSED IN NUPTIAL WHITES, OR A COUPLE OF OCTOPUSES DANCING TO THEIR WEDDING SONG?
NO.
I CAN'T EITHER. HOW SILLY!
GAG
GIGGLE
THE MOST IMPORTANT EVENT FOR ALMOST ALL LIVING ORGANISMS IS NOT A WEDDING, BUT...
REPRODUCTION, OR CREATING NEW LIFE.
LET'S TAKE A LITTLE FIELD TRIP TO A RIVER.
YOU SHOULD CHANGE INTO SOME MORE COMFORTABLE CLOTHING.

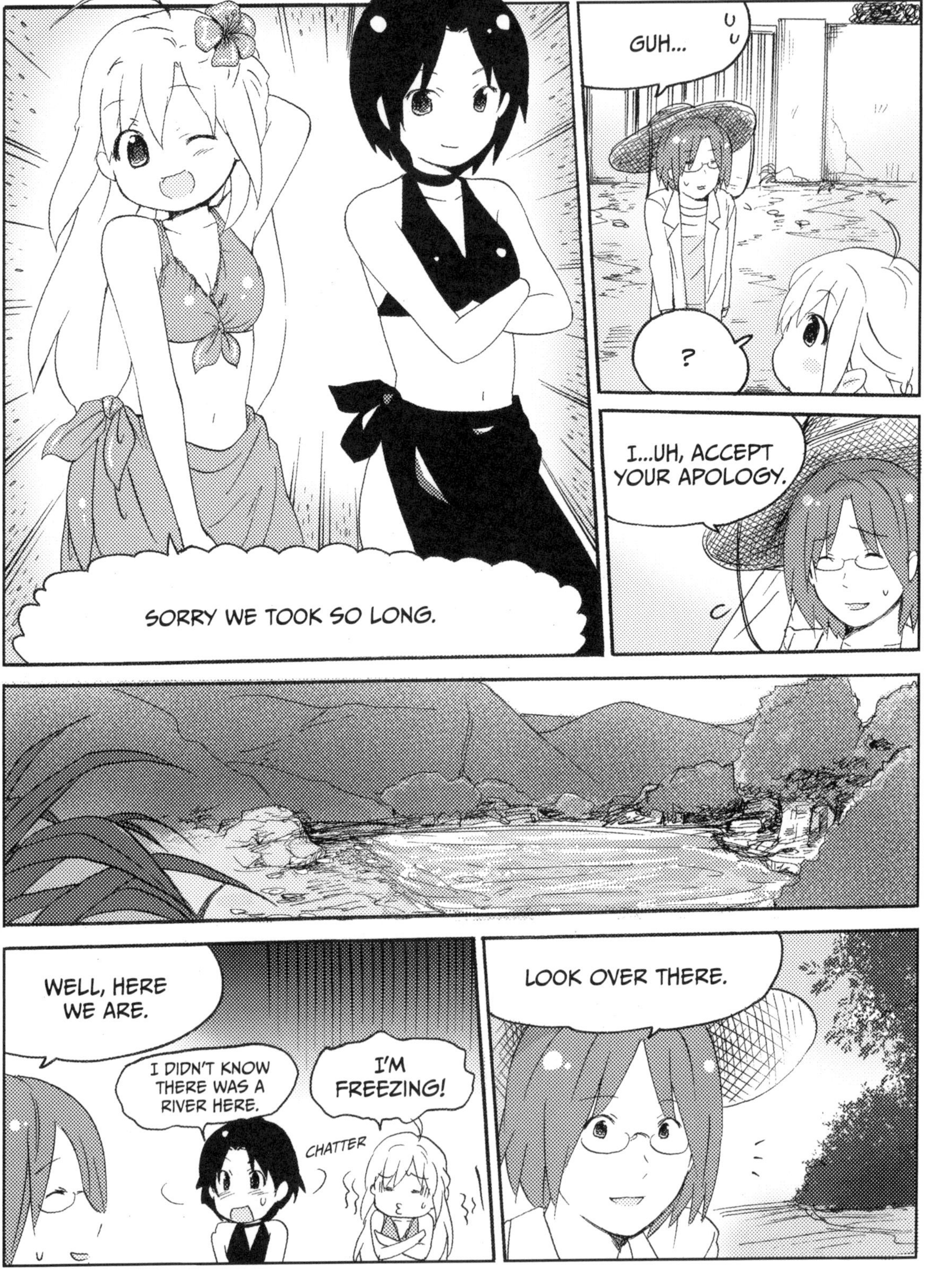
SORRY WE TOOK SO LONG.
GUH...
?
I...UH, ACCEPT YOUR APOLOGY.
WELL, HERE WE ARE.
I DIDN'T KNOW THERE WAS A RIVER HERE.
I'M FREEZING!
CHATTER
LOOK OVER THERE.

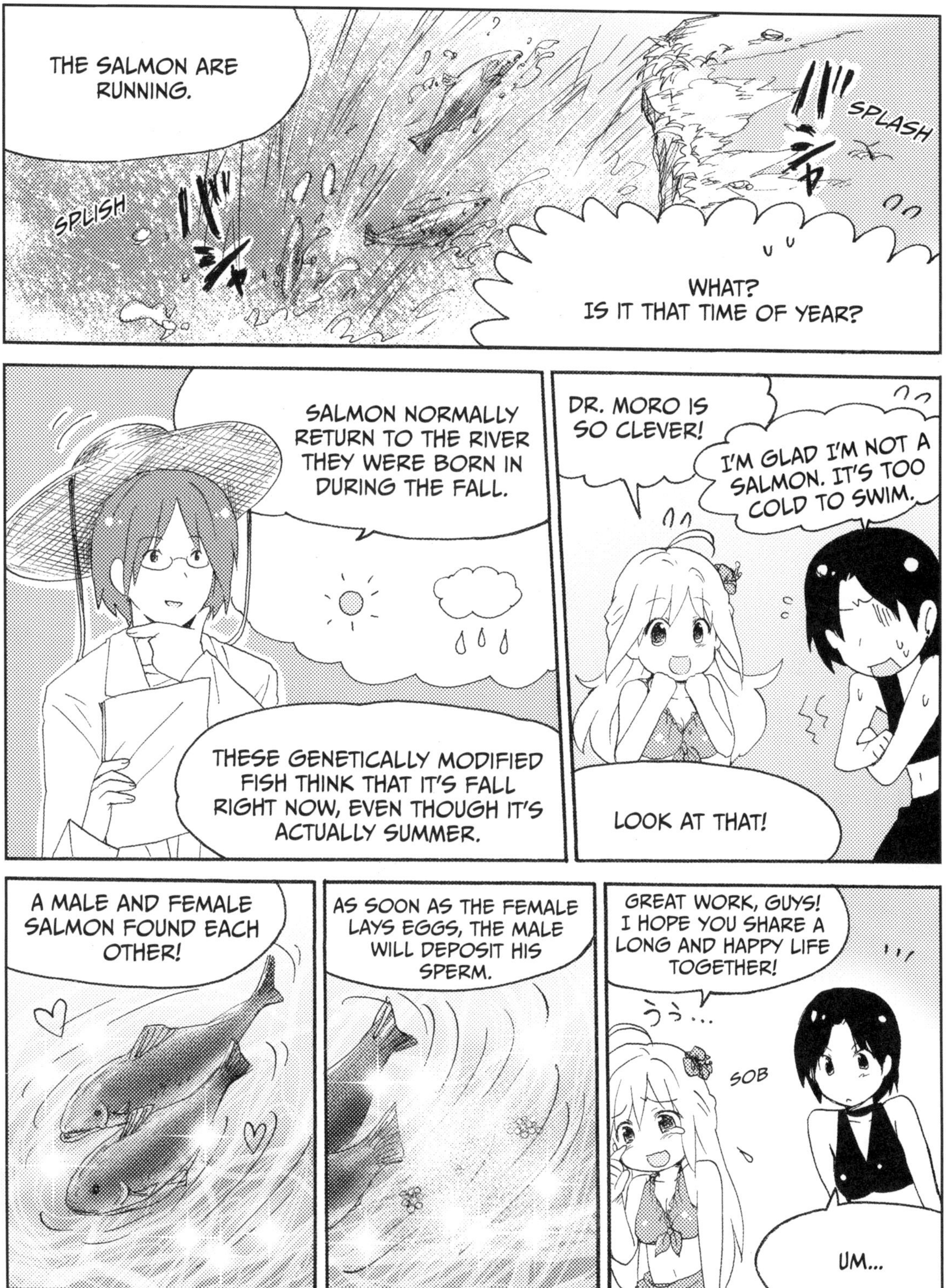
THE SALMON ARE RUNNING.
SPLISH
SPLASH
WHAT? IS IT THAT TIME OF YEAR?
SALMON NORMALLY RETURN TO THE RIVER THEY WERE BORN IN DURING THE FALL.
THESE GENETICALLY MODIFIED FISH THINK THAT IT'S FALL RIGHT NOW, EVEN THOUGH IT'S ACTUALLY SUMMER.
DR. MORO IS SO CLEVER!
I'M GLAD I'M NOT A SALMON. IT'S TOO COLD TO SWIM.
LOOK AT THAT!
A MALE AND FEMALE SALMON FOUND EACH OTHER!
AS SOON AS THE FEMALE LAYS EGGS, THE MALE WILL DEPOSIT HIS SPERM.
GREAT WORK, GUYS! I HOPE YOU SHARE A LONG AND HAPPY LIFE TOGETHER!
うう...
SOB
UM...

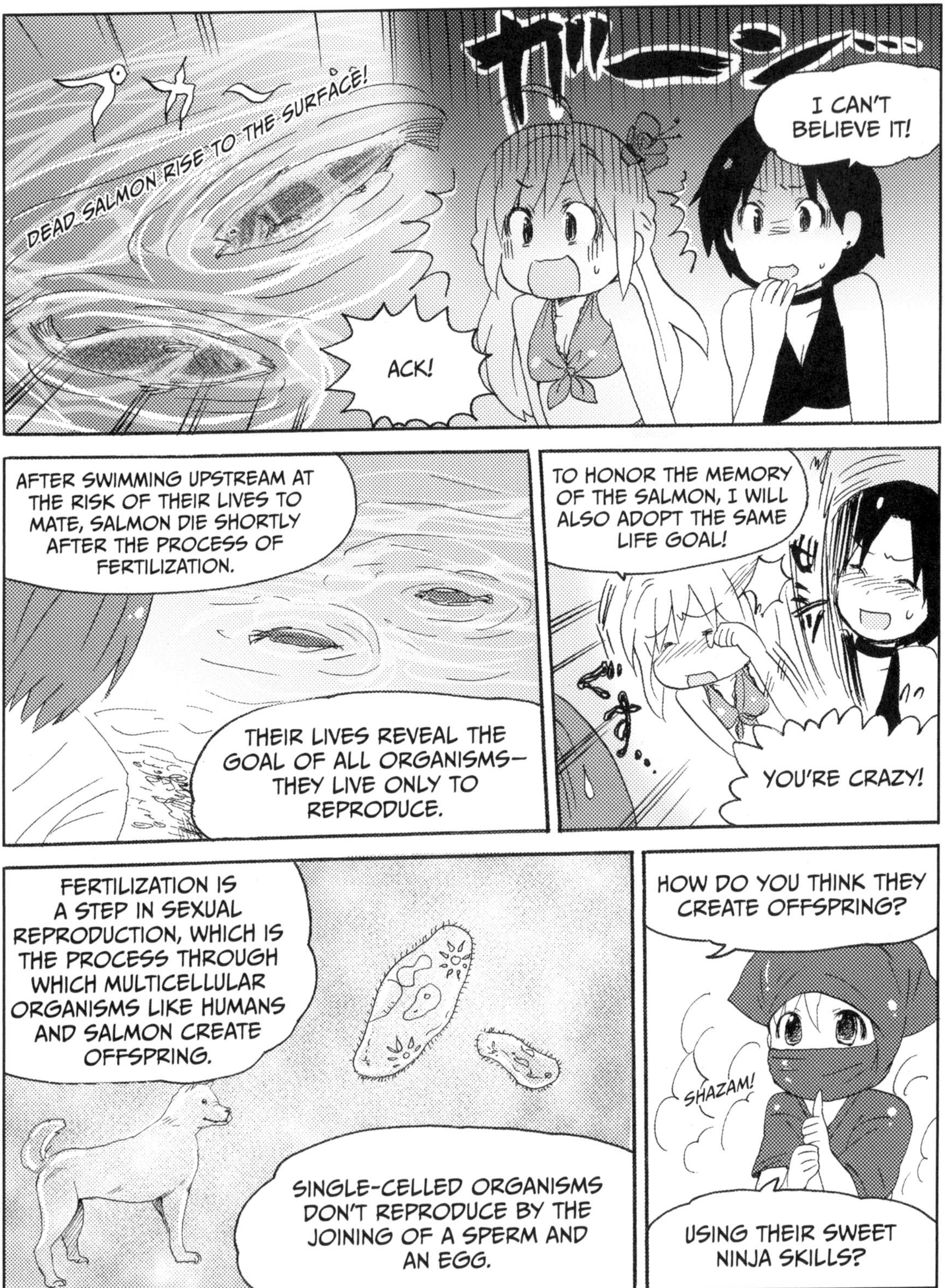
DEAD SALMON RISE TO THE SURFACE!
I CAN'T BELIEVE IT!
ACK!
AFTER SWIMMING UPSTREAM AT THE RISK OF THEIR LIVES TO MATE, SALMON DIE SHORTLY AFTER THE PROCESS OF FERTILIZATION.
THEIR LIVES REVEAL THE GOAL OF ALL ORGANISMS—THEY LIVE ONLY TO REPRODUCE.
TO HONOR THE MEMORY OF THE SALMON, I WILL ALSO ADOPT THE SAME LIFE GOAL!
YOU'RE CRAZY!
FERTILIZATION IS A STEP IN SEXUAL REPRODUCTION, WHICH IS THE PROCESS THROUGH WHICH MULTICELLULAR ORGANISMS LIKE HUMANS AND SALMON CREATE OFFSPRING.
SINGLE-CELLED ORGANISMS DON'T REPRODUCE BY THE JOINING OF A SPERM AND AN EGG.
HOW DO YOU THINK THEY CREATE OFFSPRING?
SHAZAM!
USING THEIR SWEET NINJA SKILLS?

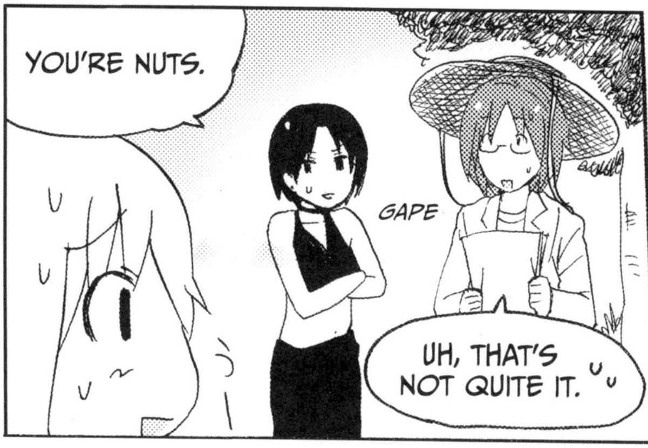
YOU'RE NUTS.
GAPE
UH, THAT'S NOT QUITE IT.

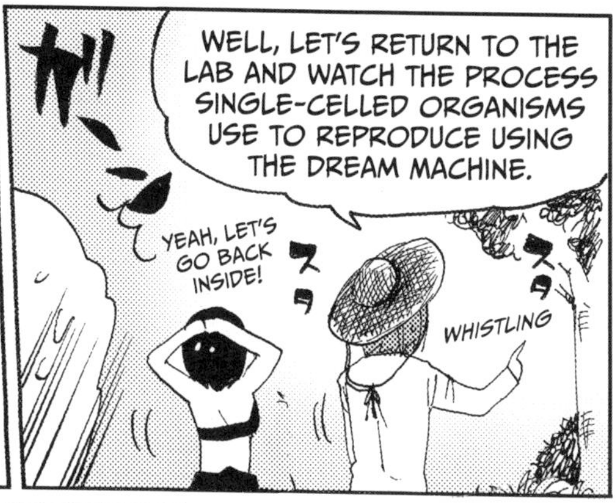
WELL, LET'S RETURN TO THE LAB AND WATCH THE PROCESS SINGLE-CELLED ORGANISMS USE TO REPRODUCE USING THE DREAM MACHINE.
YEAH, LET'S GO BACK INSIDE!
WHISTLING

CELL DIVISION: THE SIMPLEST WAY TO REPRODUCE

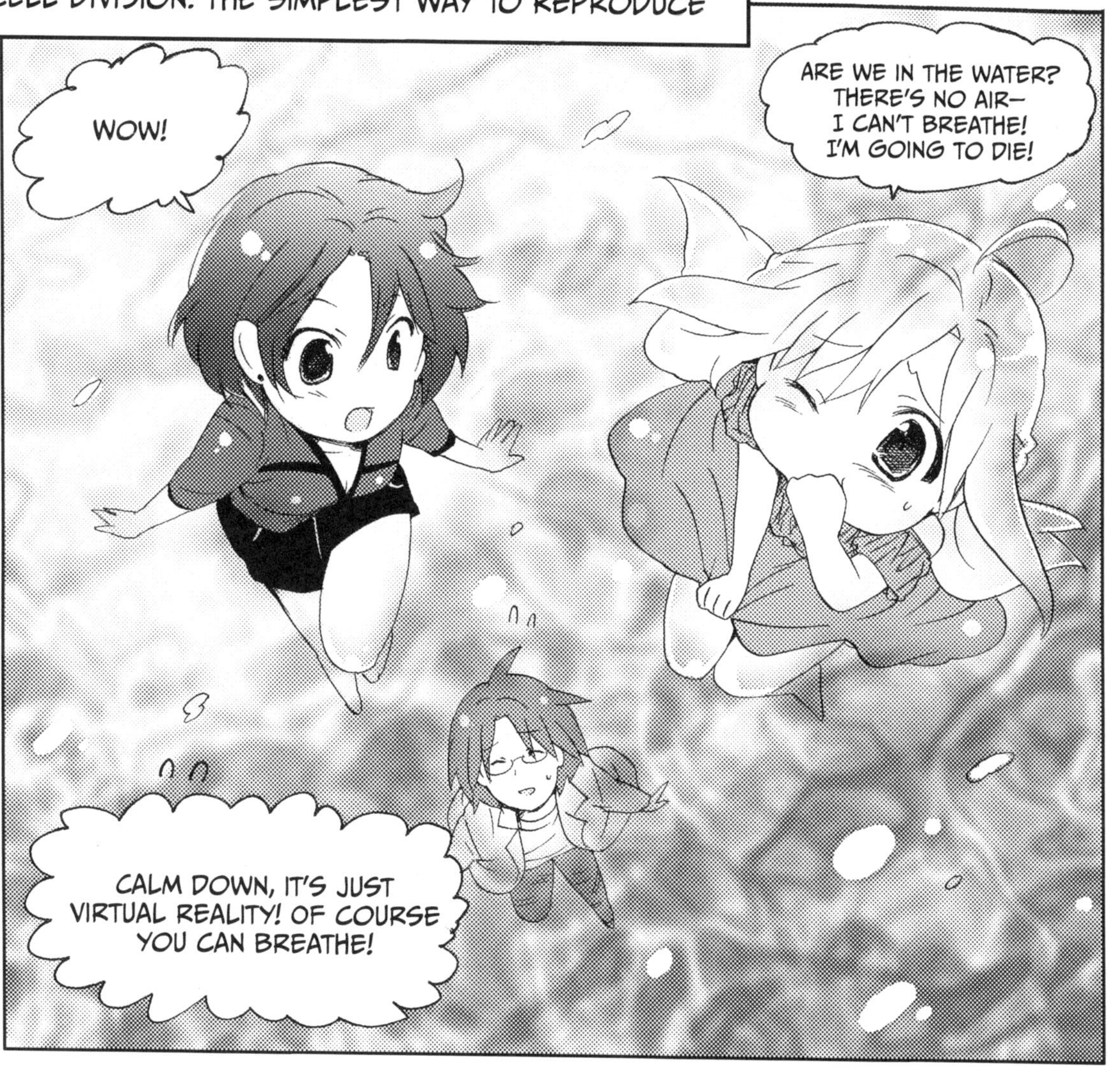
WOW!
ARE WE IN THE WATER? THERE'S NO AIR— I CAN'T BREATHE! I'M GOING TO DIE!
CALM DOWN, IT'S JUST VIRTUAL REALITY! OF COURSE YOU CAN BREATHE!

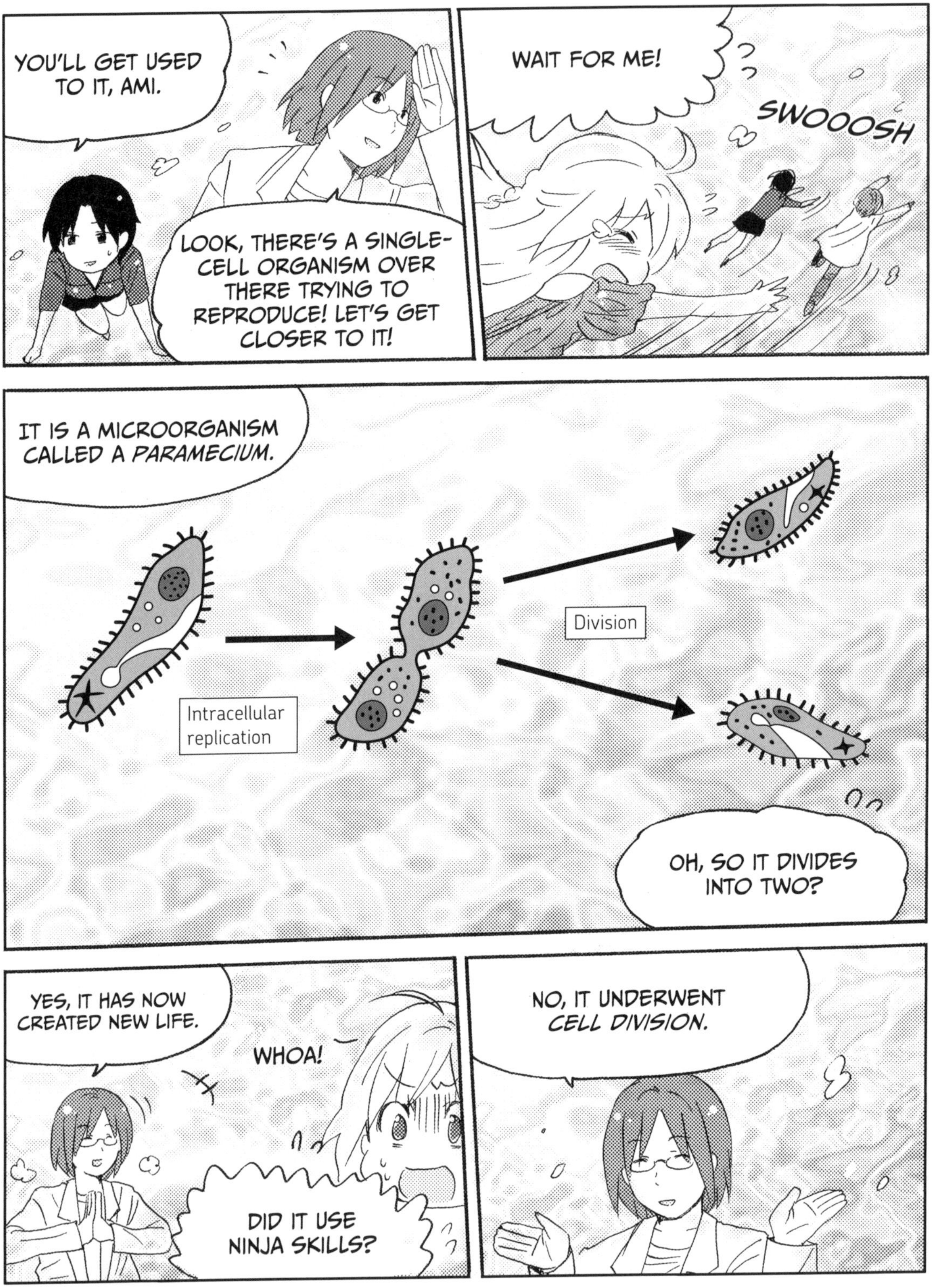
YOU'LL GET USED TO IT, AMI.
LOOK, THERE'S A SINGLE-CELL ORGANISM OVER THERE TRYING TO REPRODUCE! LET'S GET CLOSER TO IT!
WAIT FOR ME!
SWOOOSH
IT IS A MICROORGANISM CALLED A *PARAMECIUM.*
Intracellular replication
Division
OH, SO IT DIVIDES INTO TWO?
YES, IT HAS NOW CREATED NEW LIFE.
WHOA!
DID IT USE NINJA SKILLS?
NO, IT UNDERWENT *CELL DIVISION.*

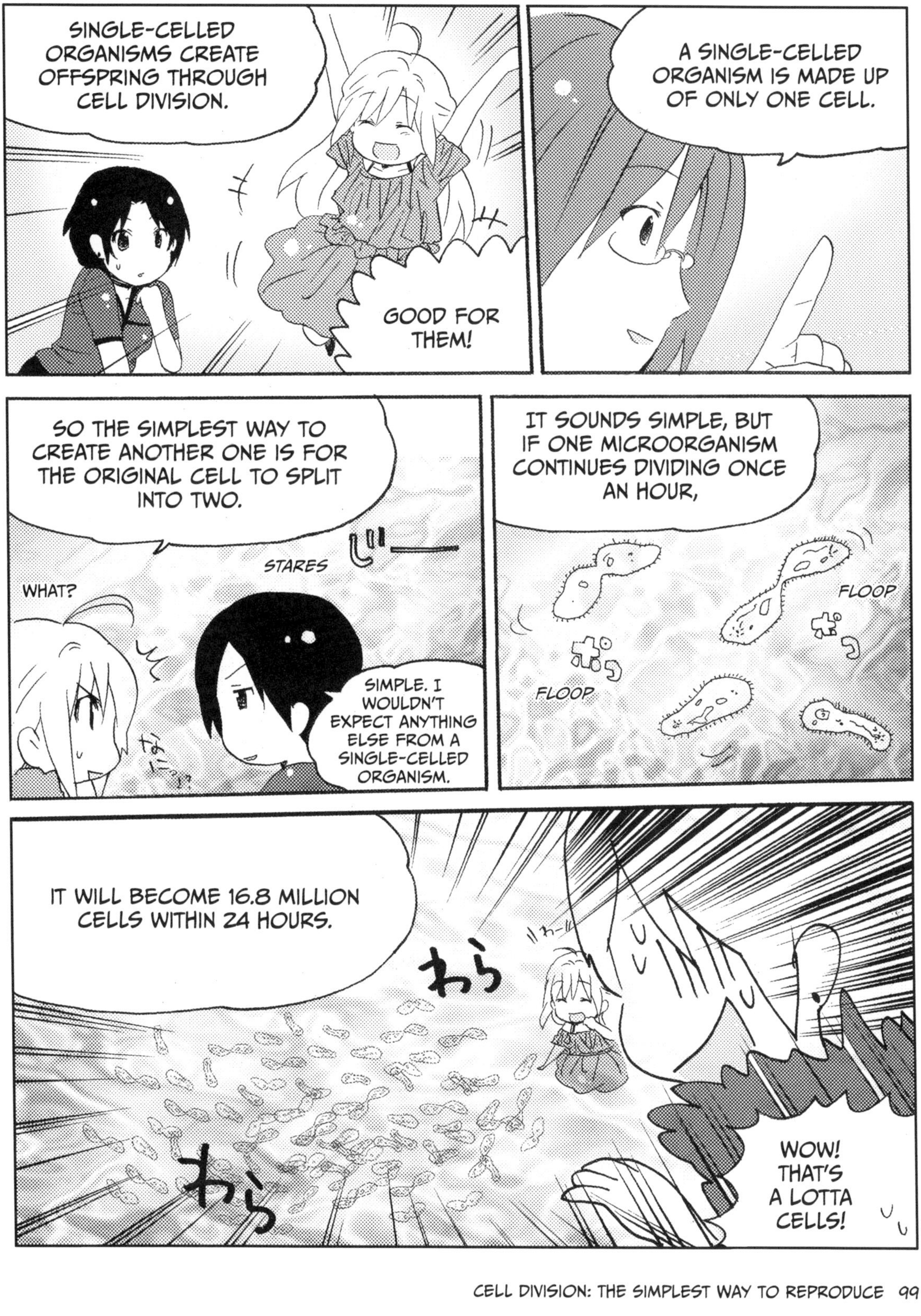
SINGLE-CELLED ORGANISMS CREATE OFFSPRING THROUGH CELL DIVISION.
GOOD FOR THEM!
A SINGLE-CELLED ORGANISM IS MADE UP OF ONLY ONE CELL.
SO THE SIMPLEST WAY TO CREATE ANOTHER ONE IS FOR THE ORIGINAL CELL TO SPLIT INTO TWO.
STARES
WHAT?
SIMPLE. I WOULDN'T EXPECT ANYTHING ELSE FROM A SINGLE-CELLED ORGANISM.
IT SOUNDS SIMPLE, BUT IF ONE MICROORGANISM CONTINUES DIVIDING ONCE AN HOUR,
FLOOP
FLOOP
IT WILL BECOME 16.8 MILLION CELLS WITHIN 24 HOURS.
わら
わら
WOW! THAT'S A LOTTA CELLS!

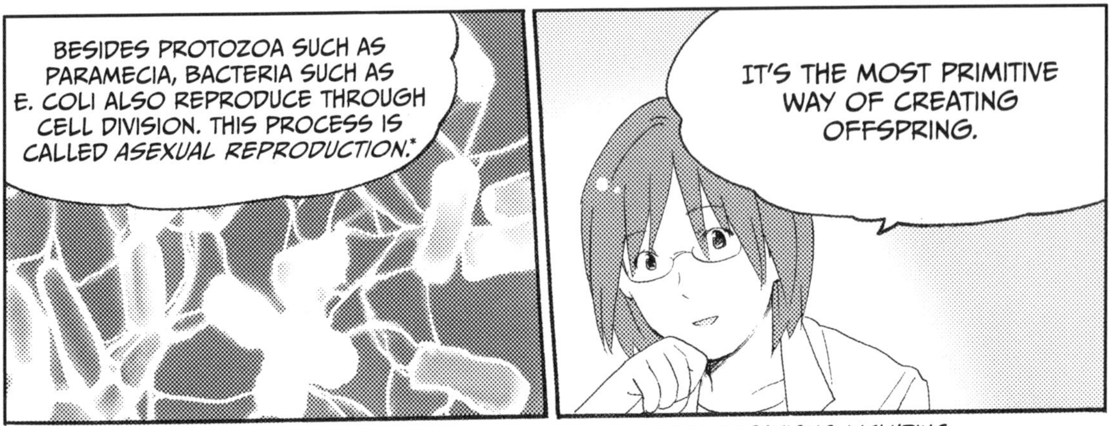
BESIDES PROTOZOA SUCH AS PARAMECIA, BACTERIA SUCH AS E. COLI ALSO REPRODUCE THROUGH CELL DIVISION. THIS PROCESS IS CALLED *ASEXUAL REPRODUCTION*.*
IT'S THE MOST PRIMITIVE WAY OF CREATING OFFSPRING.

* THERE ARE SOME EXCEPTIONS TO THIS RULE. SOME SINGLE-CELLED ORGANISMS, INCLUDING PARAMECIA, EXCHANGE SOME OF THEIR GENES WITHOUT REPRODUCING.

CELL DIVISION OCCURS IN THE BODIES OF MULTICELLULAR ORGANISMS

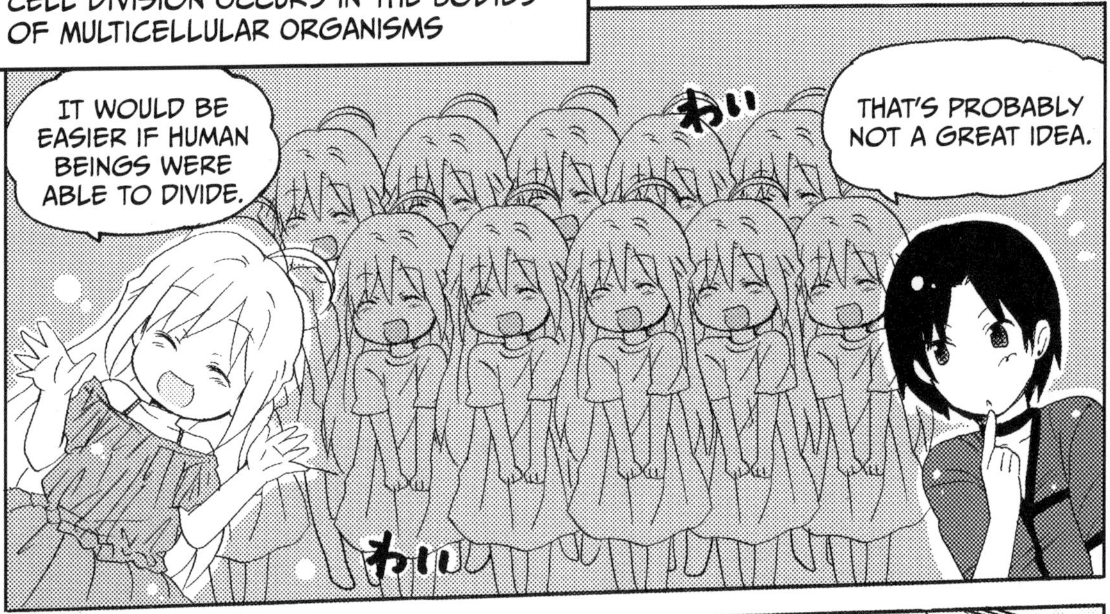
IT WOULD BE EASIER IF HUMAN BEINGS WERE ABLE TO DIVIDE.
わい
THAT'S PROBABLY NOT A GREAT IDEA.
わい

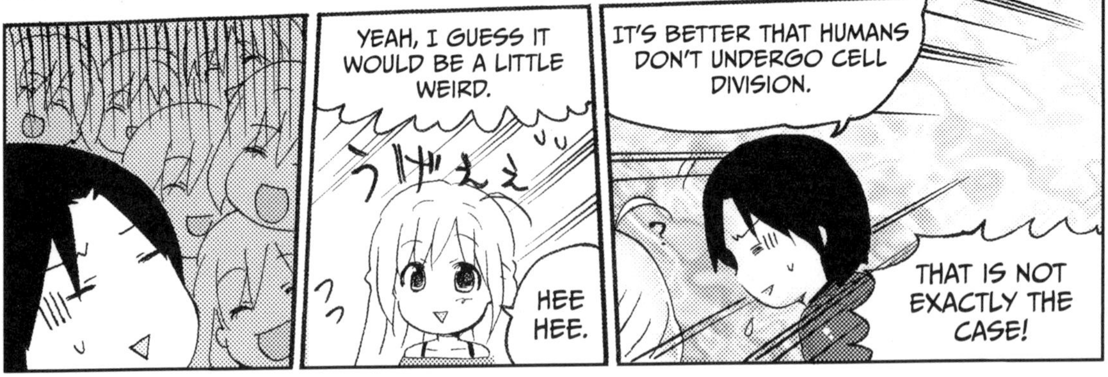
YEAH, I GUESS IT WOULD BE A LITTLE WEIRD.
HEE HEE.
IT'S BETTER THAT HUMANS DON'T UNDERGO CELL DIVISION.
THAT IS NOT EXACTLY THE CASE!

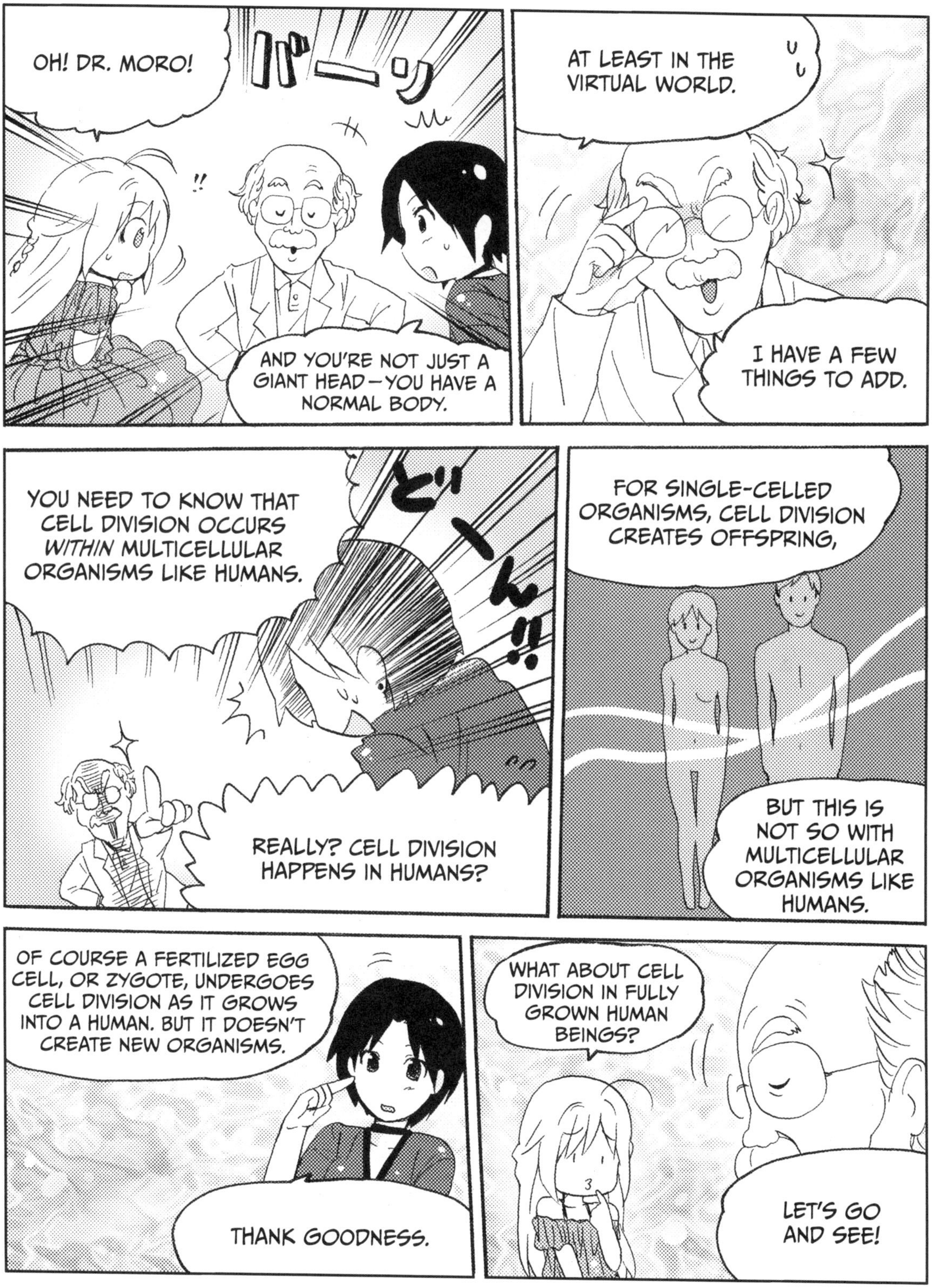
OH! DR. MORO!
バーッ
AND YOU'RE NOT JUST A GIANT HEAD—YOU HAVE A NORMAL BODY.
AT LEAST IN THE VIRTUAL WORLD.
I HAVE A FEW THINGS TO ADD.
YOU NEED TO KNOW THAT CELL DIVISION OCCURS *WITHIN* MULTICELLULAR ORGANISMS LIKE HUMANS.
どーん!!
REALLY? CELL DIVISION HAPPENS IN HUMANS?
FOR SINGLE-CELLED ORGANISMS, CELL DIVISION CREATES OFFSPRING,
BUT THIS IS NOT SO WITH MULTICELLULAR ORGANISMS LIKE HUMANS.
OF COURSE A FERTILIZED EGG CELL, OR ZYGOTE, UNDERGOES CELL DIVISION AS IT GROWS INTO A HUMAN. BUT IT DOESN'T CREATE NEW ORGANISMS.
THANK GOODNESS.
WHAT ABOUT CELL DIVISION IN FULLY GROWN HUMAN BEINGS?
LET'S GO AND SEE!

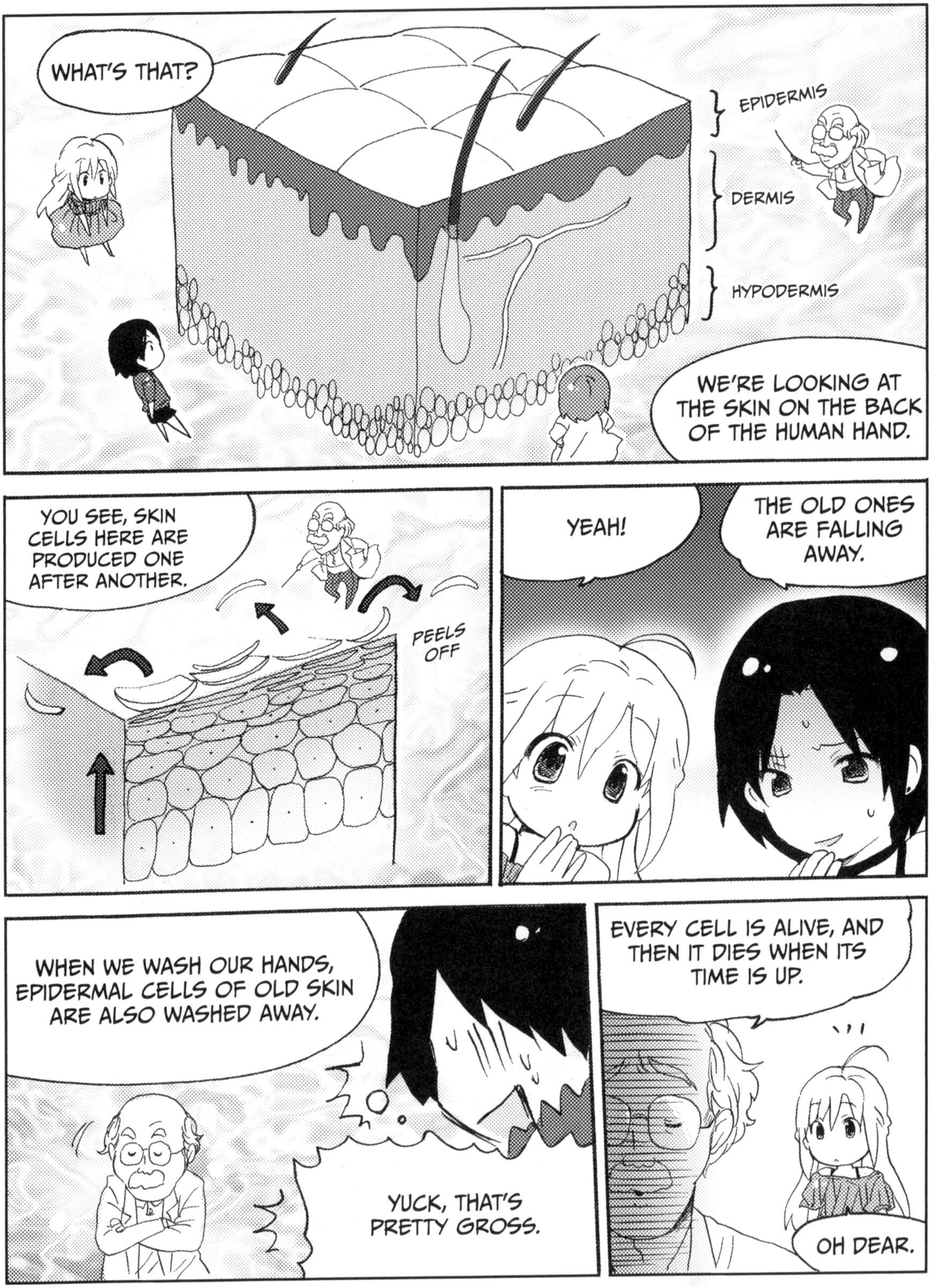
WHAT'S THAT?
EPIDERMIS
DERMIS
HYPODERMIS
WE'RE LOOKING AT THE SKIN ON THE BACK OF THE HUMAN HAND.
YOU SEE, SKIN CELLS HERE ARE PRODUCED ONE AFTER ANOTHER.
PEELS OFF
YEAH!
THE OLD ONES ARE FALLING AWAY.
WHEN WE WASH OUR HANDS, EPIDERMAL CELLS OF OLD SKIN ARE ALSO WASHED AWAY.
YUCK, THAT'S PRETTY GROSS.
EVERY CELL IS ALIVE, AND THEN IT DIES WHEN ITS TIME IS UP.
OH DEAR.

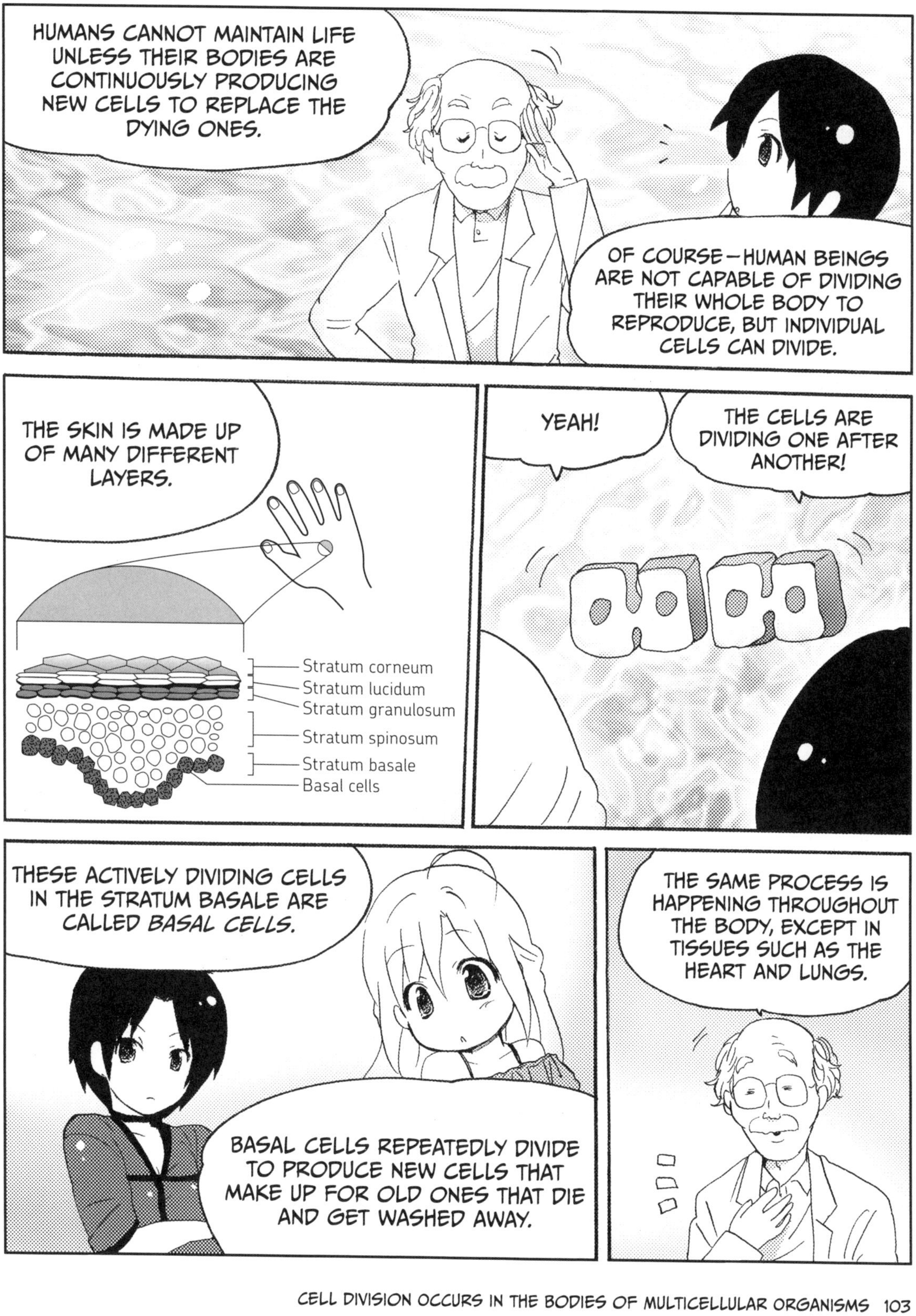
HUMANS CANNOT MAINTAIN LIFE UNLESS THEIR BODIES ARE CONTINUOUSLY PRODUCING NEW CELLS TO REPLACE THE DYING ONES.
OF COURSE—HUMAN BEINGS ARE NOT CAPABLE OF DIVIDING THEIR WHOLE BODY TO REPRODUCE, BUT INDIVIDUAL CELLS CAN DIVIDE.
THE SKIN IS MADE UP OF MANY DIFFERENT LAYERS.
Stratum corneum
Stratum lucidum
Stratum granulosum
Stratum spinosum
Stratum basale
Basal cells
YEAH!
THE CELLS ARE DIVIDING ONE AFTER ANOTHER!
THESE ACTIVELY DIVIDING CELLS IN THE STRATUM BASALE ARE CALLED *BASAL CELLS*.
BASAL CELLS REPEATEDLY DIVIDE TO PRODUCE NEW CELLS THAT MAKE UP FOR OLD ONES THAT DIE AND GET WASHED AWAY.
THE SAME PROCESS IS HAPPENING THROUGHOUT THE BODY, EXCEPT IN TISSUES SUCH AS THE HEART AND LUNGS.

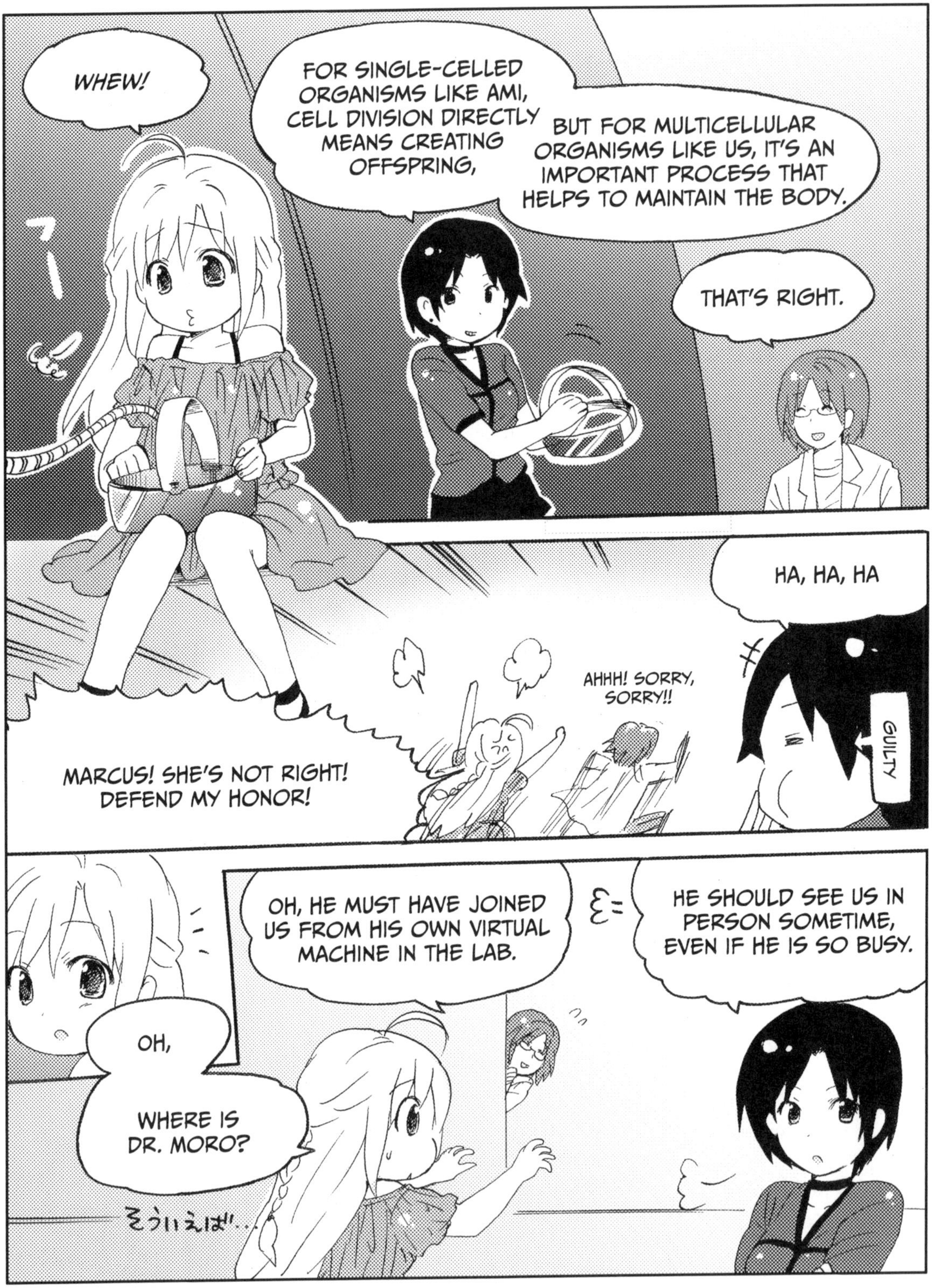
WHEW!
FOR SINGLE-CELLED ORGANISMS LIKE AMI, CELL DIVISION DIRECTLY MEANS CREATING OFFSPRING,
BUT FOR MULTICELLULAR ORGANISMS LIKE US, IT'S AN IMPORTANT PROCESS THAT HELPS TO MAINTAIN THE BODY.
THAT'S RIGHT.
HA, HA, HA
AHHH! SORRY, SORRY!!
MARCUS! SHE'S NOT RIGHT! DEFEND MY HONOR!
GUILTY
OH,
WHERE IS DR. MORO?
そういえば…
OH, HE MUST HAVE JOINED US FROM HIS OWN VIRTUAL MACHINE IN THE LAB.
HE SHOULD SEE US IN PERSON SOMETIME, EVEN IF HE IS SO BUSY.

DNA IS REPLICATED BEFORE CELL DIVISION

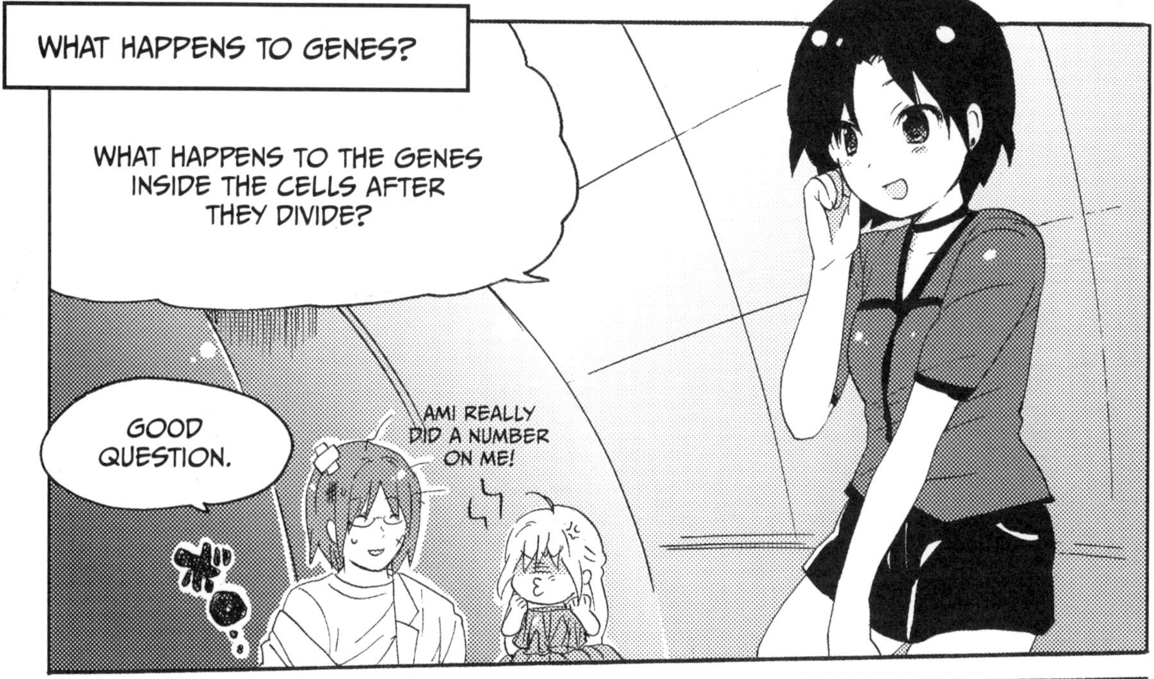
WHAT HAPPENS TO GENES?
WHAT HAPPENS TO THE GENES INSIDE THE CELLS AFTER THEY DIVIDE?
GOOD QUESTION.
AMI REALLY DID A NUMBER ON ME!

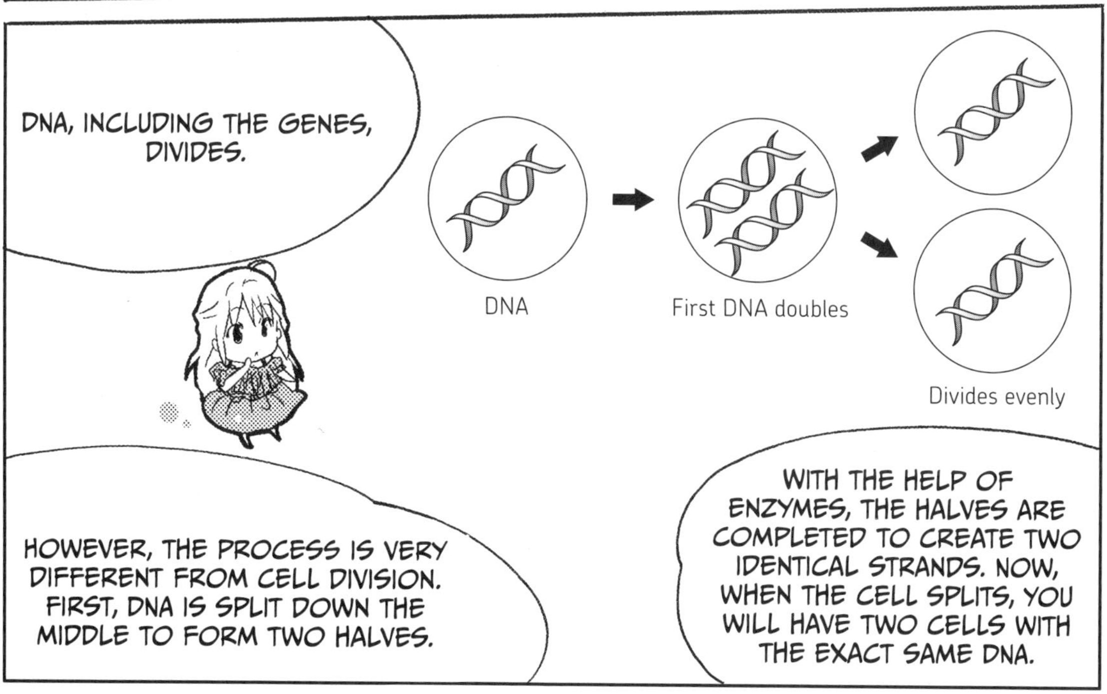
DNA, INCLUDING THE GENES, DIVIDES.
DNA
First DNA doubles
Divides evenly
HOWEVER, THE PROCESS IS VERY DIFFERENT FROM CELL DIVISION. FIRST, DNA IS SPLIT DOWN THE MIDDLE TO FORM TWO HALVES.
WITH THE HELP OF ENZYMES, THE HALVES ARE COMPLETED TO CREATE TWO IDENTICAL STRANDS. NOW, WHEN THE CELL SPLITS, YOU WILL HAVE TWO CELLS WITH THE EXACT SAME DNA.

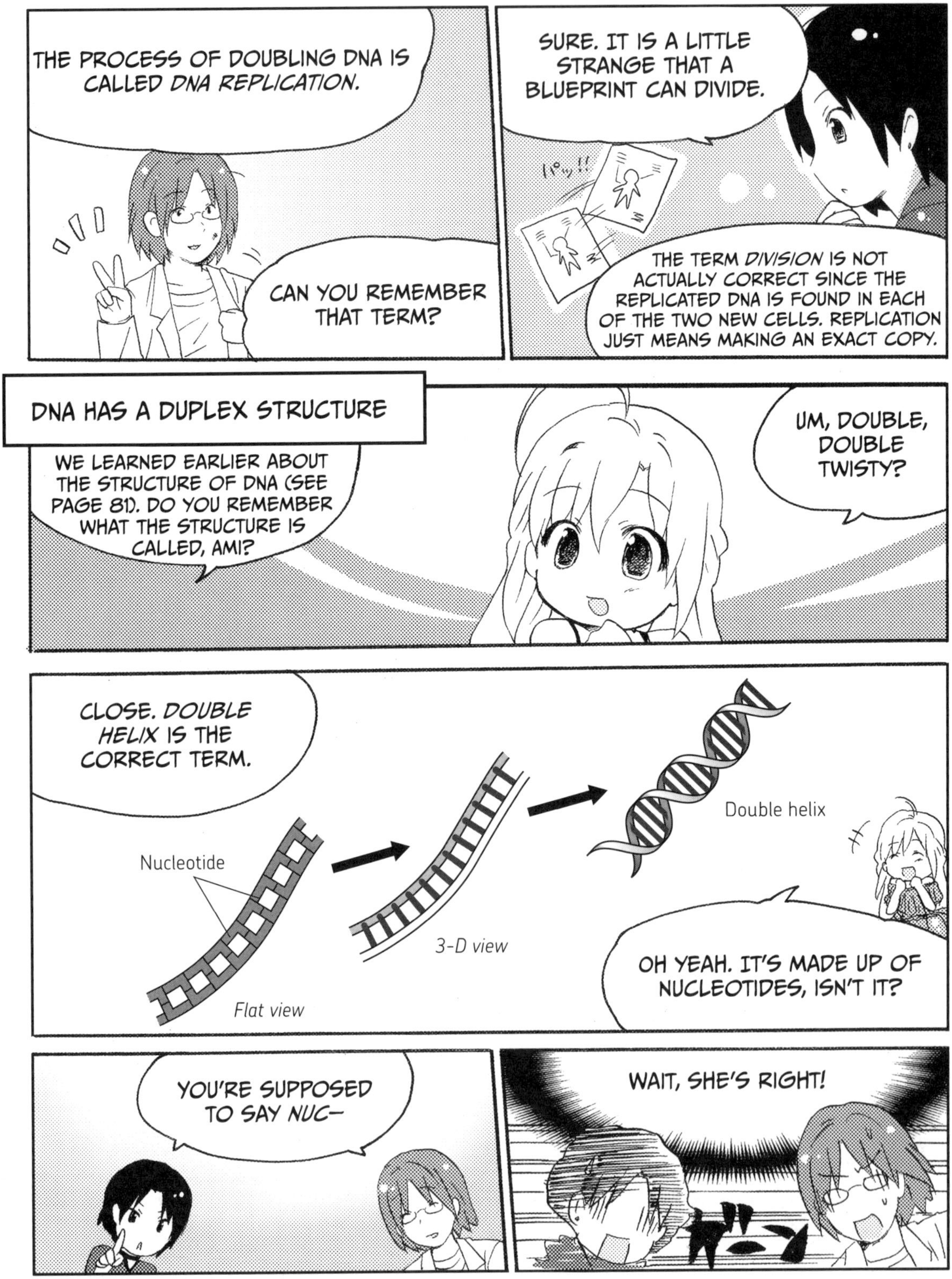
THE PROCESS OF DOUBLING DNA IS CALLED *DNA REPLICATION.*
CAN YOU REMEMBER THAT TERM?
SURE. IT IS A LITTLE STRANGE THAT A BLUEPRINT CAN DIVIDE.
パッ!!
THE TERM *DIVISION* IS NOT ACTUALLY CORRECT SINCE THE REPLICATED DNA IS FOUND IN EACH OF THE TWO NEW CELLS. REPLICATION JUST MEANS MAKING AN EXACT COPY.
DNA HAS A DUPLEX STRUCTURE
WE LEARNED EARLIER ABOUT THE STRUCTURE OF DNA (SEE PAGE 81). DO YOU REMEMBER WHAT THE STRUCTURE IS CALLED, AMI?
UM, DOUBLE, DOUBLE TWISTY?
CLOSE. *DOUBLE HELIX* IS THE CORRECT TERM.
Nucleotide
Flat view
3-D view
Double helix
OH YEAH. IT'S MADE UP OF NUCLEOTIDES, ISN'T IT?
YOU'RE SUPPOSED TO SAY *NUC*—
WAIT, SHE'S RIGHT!

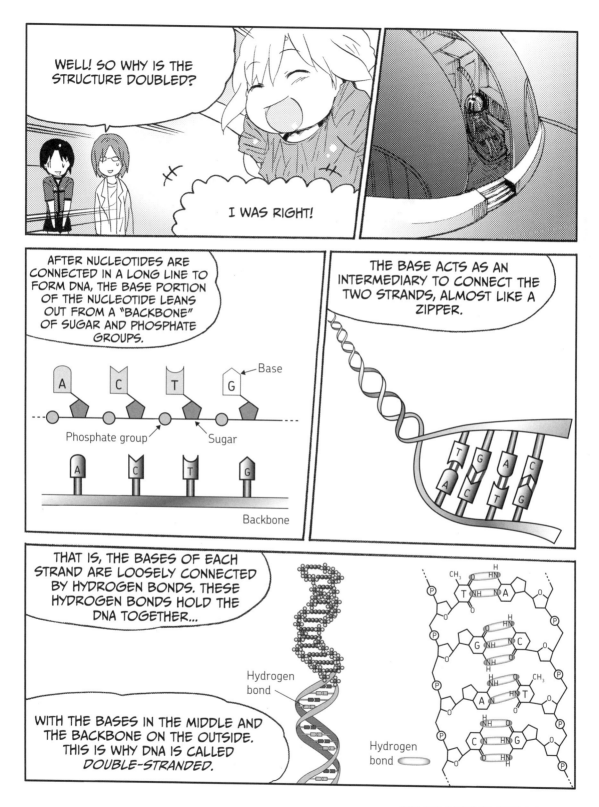

WELL! SO WHY IS THE STRUCTURE DOUBLED?

I WAS RIGHT!

AFTER NUCLEOTIDES ARE CONNECTED IN A LONG LINE TO FORM DNA, THE BASE PORTION OF THE NUCLEOTIDE LEANS OUT FROM A "BACKBONE" OF SUGAR AND PHOSPHATE GROUPS.

Base

A C T G

Phosphate group

Sugar

A C T G

Backbone

THE BASE ACTS AS AN INTERMEDIARY TO CONNECT THE TWO STRANDS, ALMOST LIKE A ZIPPER.

THAT IS, THE BASES OF EACH STRAND ARE LOOSELY CONNECTED BY HYDROGEN BONDS. THESE HYDROGEN BONDS HOLD THE DNA TOGETHER...

Hydrogen bond

WITH THE BASES IN THE MIDDLE AND THE BACKBONE ON THE OUTSIDE. THIS IS WHY DNA IS CALLED DOUBLE-STRANDED.

Hydrogen bond

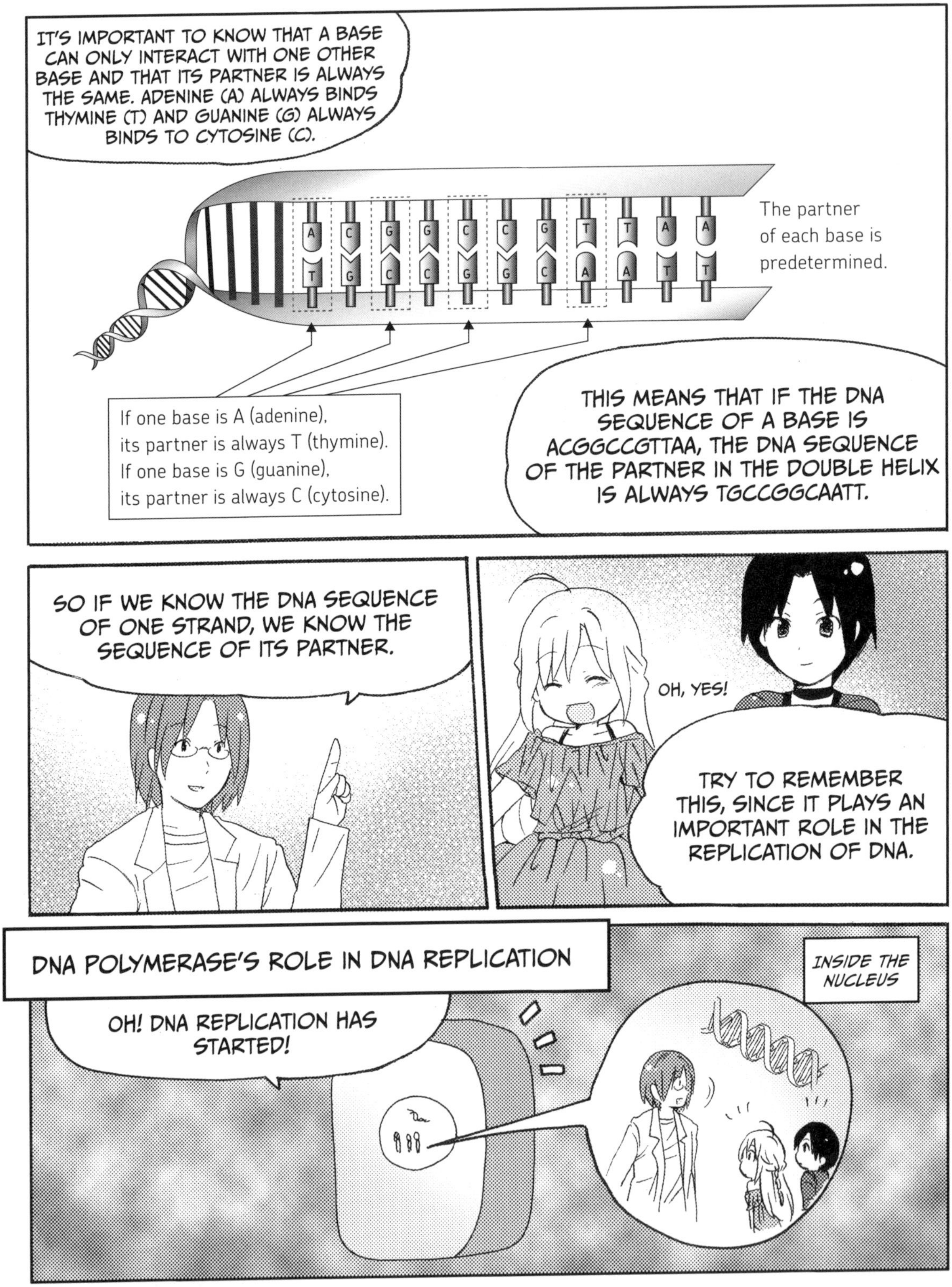
IT'S IMPORTANT TO KNOW THAT A BASE CAN ONLY INTERACT WITH ONE OTHER BASE AND THAT ITS PARTNER IS ALWAYS THE SAME. ADENINE (A) ALWAYS BINDS THYMINE (T) AND GUANINE (G) ALWAYS BINDS TO CYTOSINE (C).
A C G G C C G T T A A
T G C C G G C A A T T
The partner of each base is predetermined.
If one base is A (adenine), its partner is always T (thymine). If one base is G (guanine), its partner is always C (cytosine).
THIS MEANS THAT IF THE DNA SEQUENCE OF A BASE IS ACGGCCGTTAA, THE DNA SEQUENCE OF THE PARTNER IN THE DOUBLE HELIX IS ALWAYS TGCCGGCAATT.
SO IF WE KNOW THE DNA SEQUENCE OF ONE STRAND, WE KNOW THE SEQUENCE OF ITS PARTNER.
OH, YES!
TRY TO REMEMBER THIS, SINCE IT PLAYS AN IMPORTANT ROLE IN THE REPLICATION OF DNA.
DNA POLYMERASE'S ROLE IN DNA REPLICATION
INSIDE THE NUCLEUS
OH! DNA REPLICATION HAS STARTED!

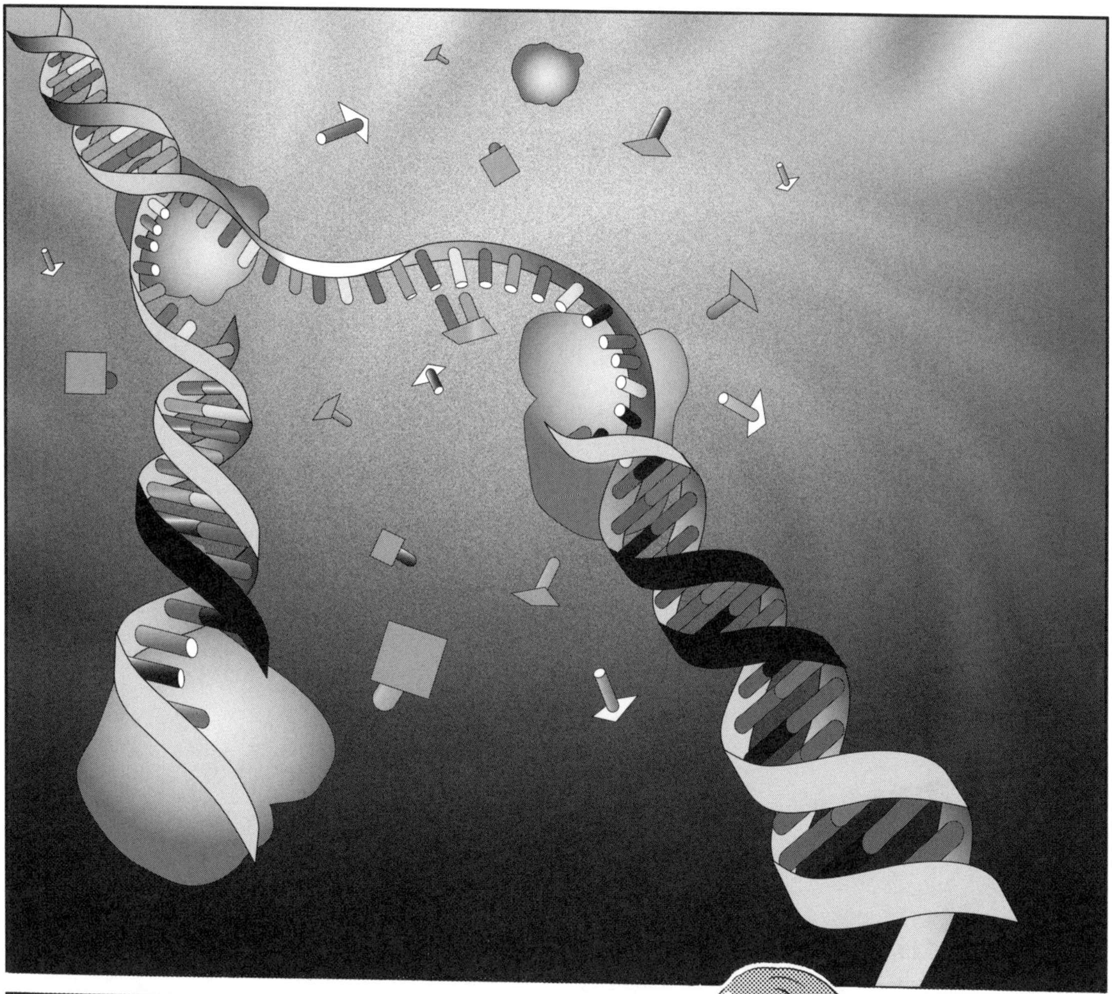

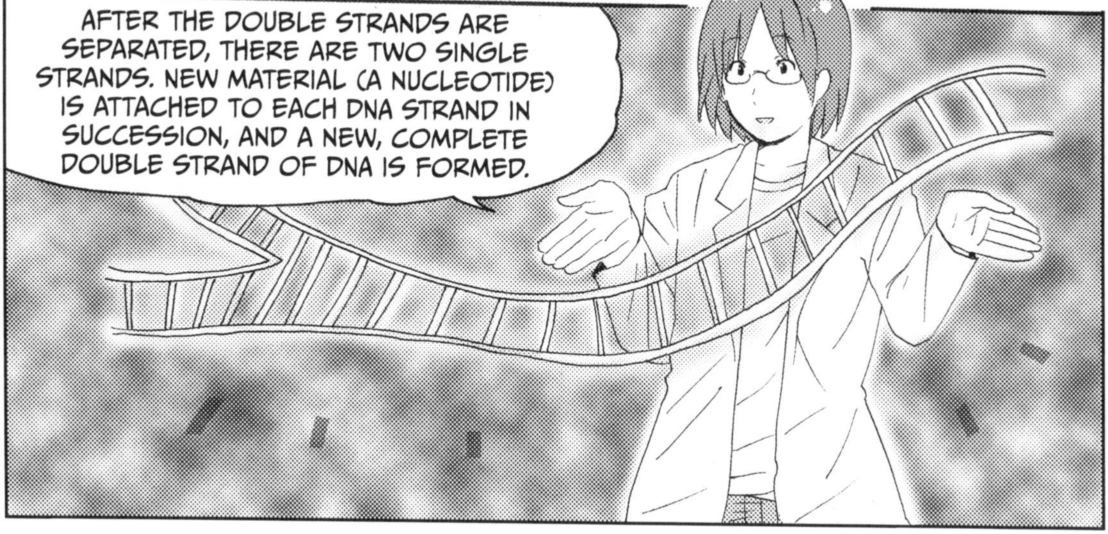
AFTER THE DOUBLE STRANDS ARE SEPARATED, THERE ARE TWO SINGLE STRANDS. NEW MATERIAL (A NUCLEOTIDE) IS ATTACHED TO EACH DNA STRAND IN SUCCESSION, AND A NEW, COMPLETE DOUBLE STRAND OF DNA IS FORMED.

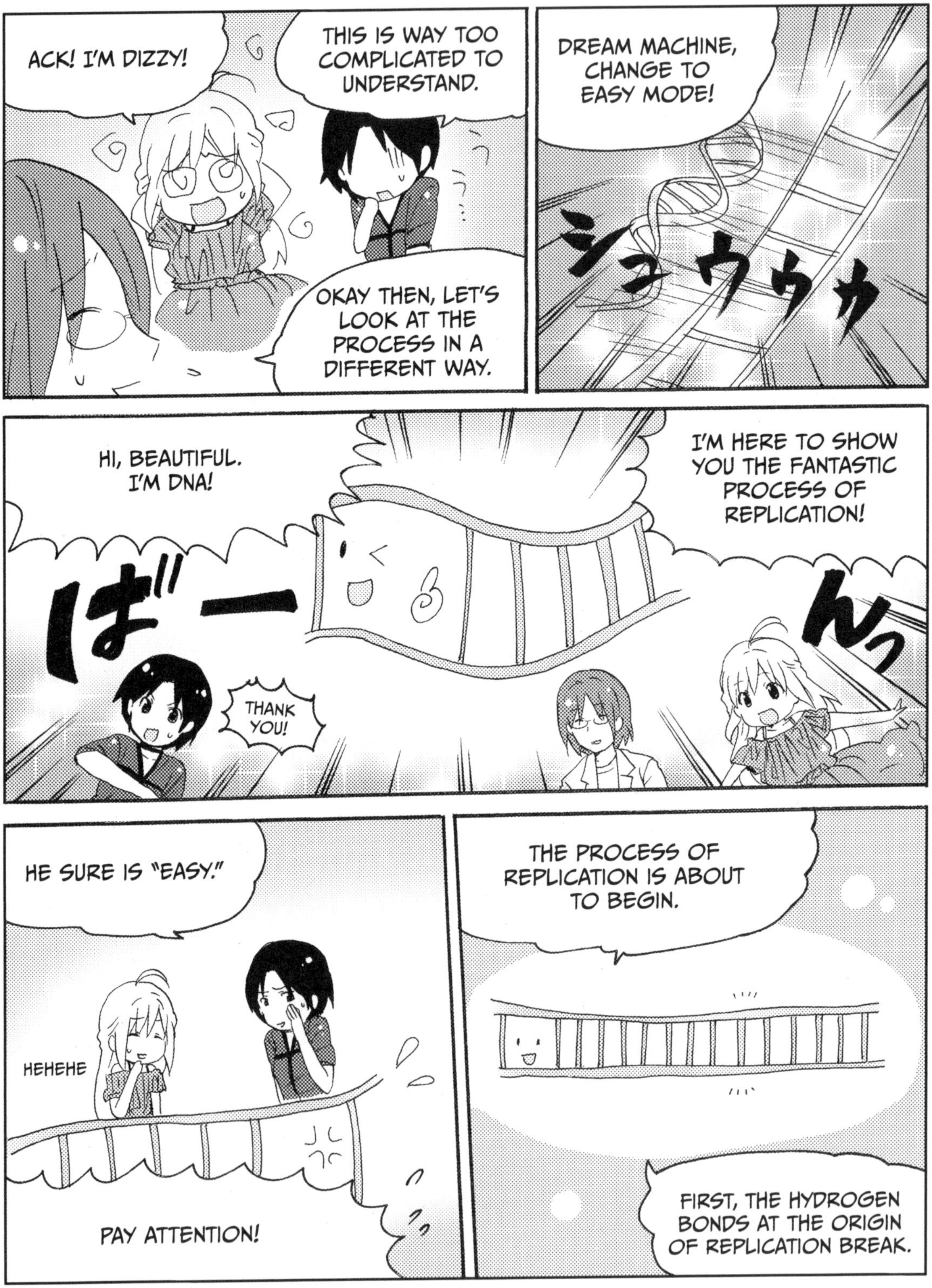
ACK! I'M DIZZY!
THIS IS WAY TOO COMPLICATED TO UNDERSTAND.
OKAY THEN, LET'S LOOK AT THE PROCESS IN A DIFFERENT WAY.
DREAM MACHINE, CHANGE TO EASY MODE!
HI, BEAUTIFUL. I'M DNA!
ばーんっ
THANK YOU!
I'M HERE TO SHOW YOU THE FANTASTIC PROCESS OF REPLICATION!
HE SURE IS "EASY."
HEHEHE
PAY ATTENTION!
THE PROCESS OF REPLICATION IS ABOUT TO BEGIN.
FIRST, THE HYDROGEN BONDS AT THE ORIGIN OF REPLICATION BREAK.

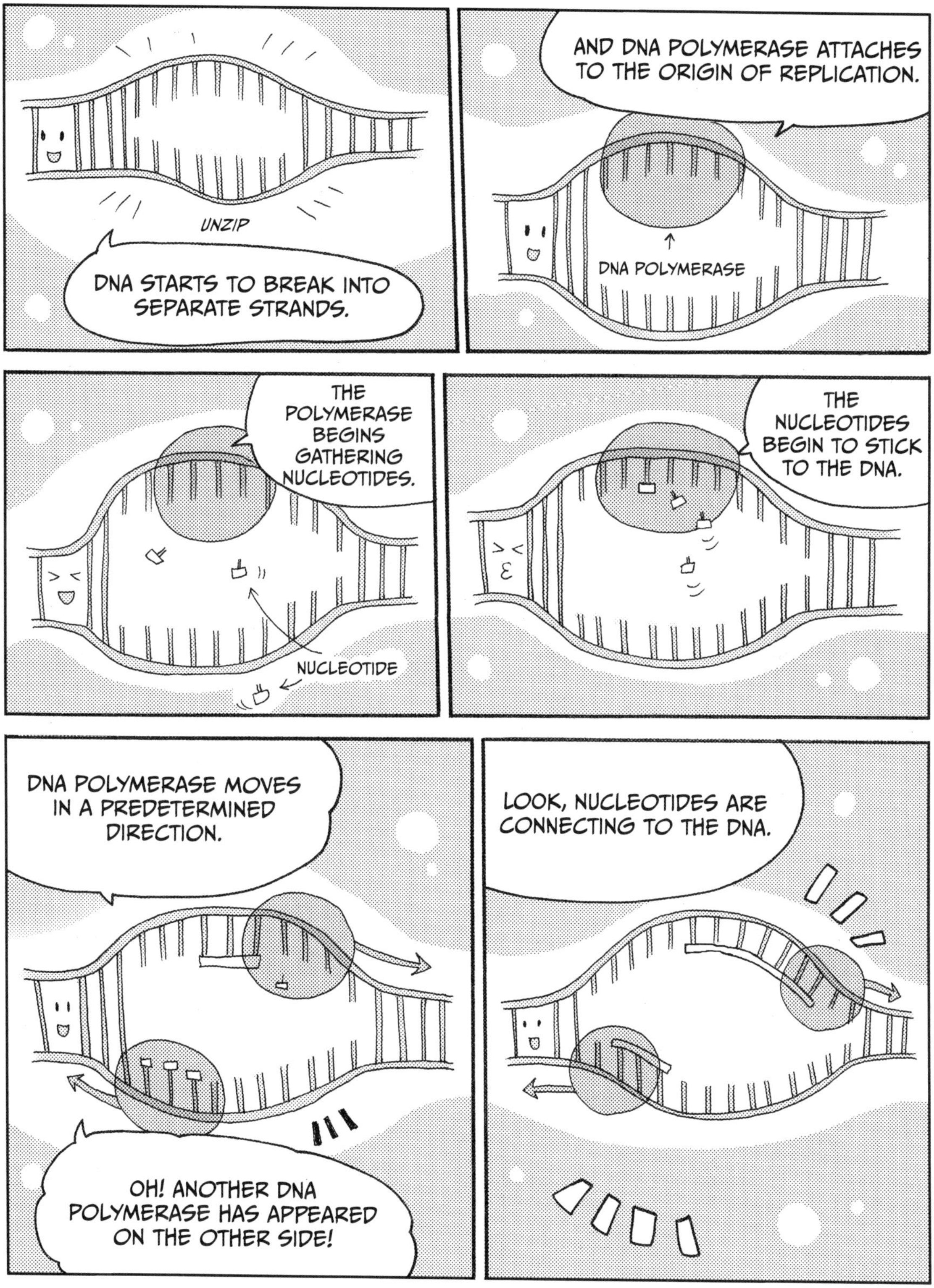
UNZIP
DNA STARTS TO BREAK INTO SEPARATE STRANDS.
AND DNA POLYMERASE ATTACHES TO THE ORIGIN OF REPLICATION.
DNA POLYMERASE
THE POLYMERASE BEGINS GATHERING NUCLEOTIDES.
NUCLEOTIDE
THE NUCLEOTIDES BEGIN TO STICK TO THE DNA.
DNA POLYMERASE MOVES IN A PREDETERMINED DIRECTION.
OH! ANOTHER DNA POLYMERASE HAS APPEARED ON THE OTHER SIDE!
LOOK, NUCLEOTIDES ARE CONNECTING TO THE DNA.

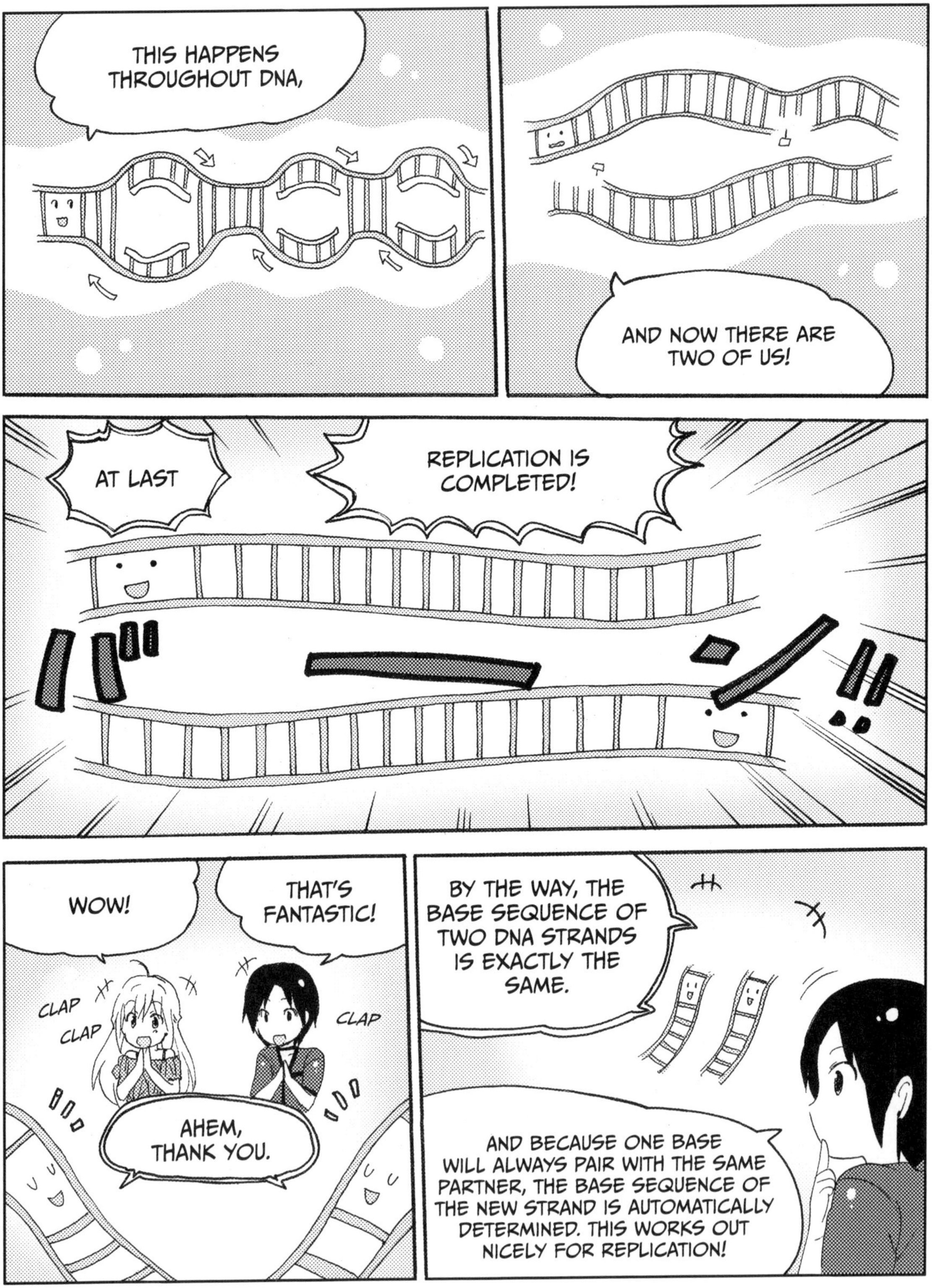
THIS HAPPENS THROUGHOUT DNA,
AND NOW THERE ARE TWO OF US!
AT LAST
REPLICATION IS COMPLETED!
WOW!
THAT'S FANTASTIC!
CLAP
CLAP
CLAP
AHEM, THANK YOU.
BY THE WAY, THE BASE SEQUENCE OF TWO DNA STRANDS IS EXACTLY THE SAME.
AND BECAUSE ONE BASE WILL ALWAYS PAIR WITH THE SAME PARTNER, THE BASE SEQUENCE OF THE NEW STRAND IS AUTOMATICALLY DETERMINED. THIS WORKS OUT NICELY FOR REPLICATION!

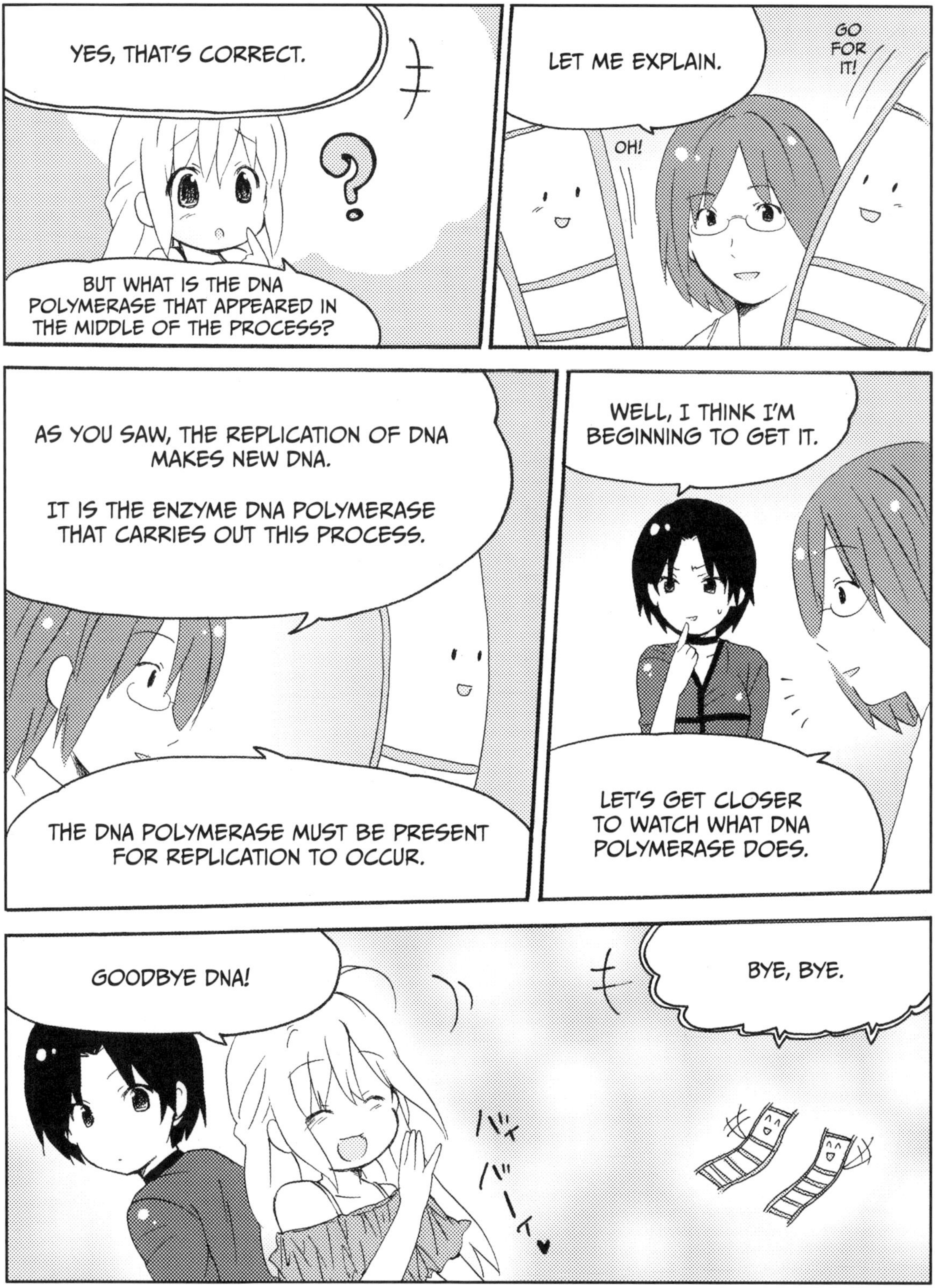
YES, THAT'S CORRECT.
BUT WHAT IS THE DNA POLYMERASE THAT APPEARED IN THE MIDDLE OF THE PROCESS?
LET ME EXPLAIN.
GO FOR IT!
OH!
AS YOU SAW, THE REPLICATION OF DNA MAKES NEW DNA.
IT IS THE ENZYME DNA POLYMERASE THAT CARRIES OUT THIS PROCESS.
THE DNA POLYMERASE MUST BE PRESENT FOR REPLICATION TO OCCUR.
WELL, I THINK I'M BEGINNING TO GET IT.
LET'S GET CLOSER TO WATCH WHAT DNA POLYMERASE DOES.
GOODBYE DNA!
BYE, BYE.
バィバーイ♥

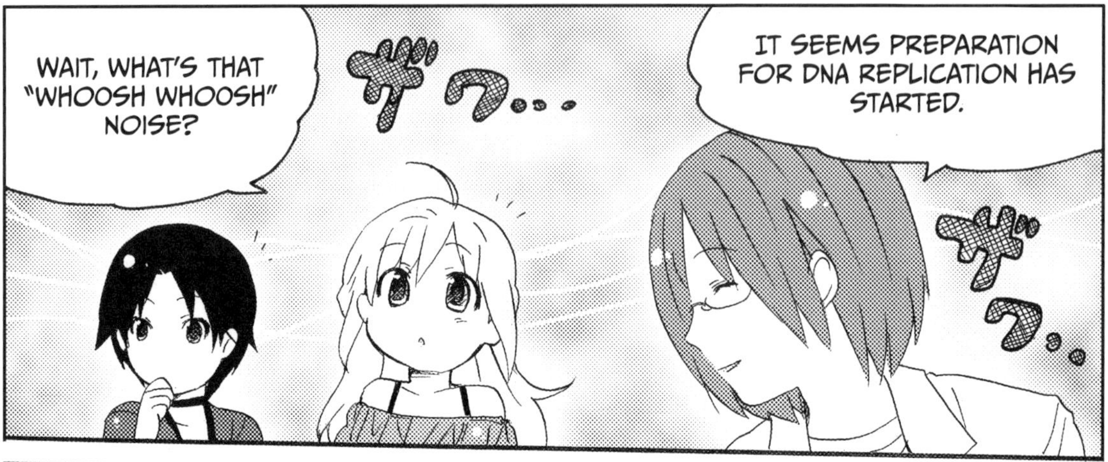
WAIT, WHAT'S THAT "WHOOSH WHOOSH" NOISE?
ザワ...
IT SEEMS PREPARATION FOR DNA REPLICATION HAS STARTED.
ザワ...

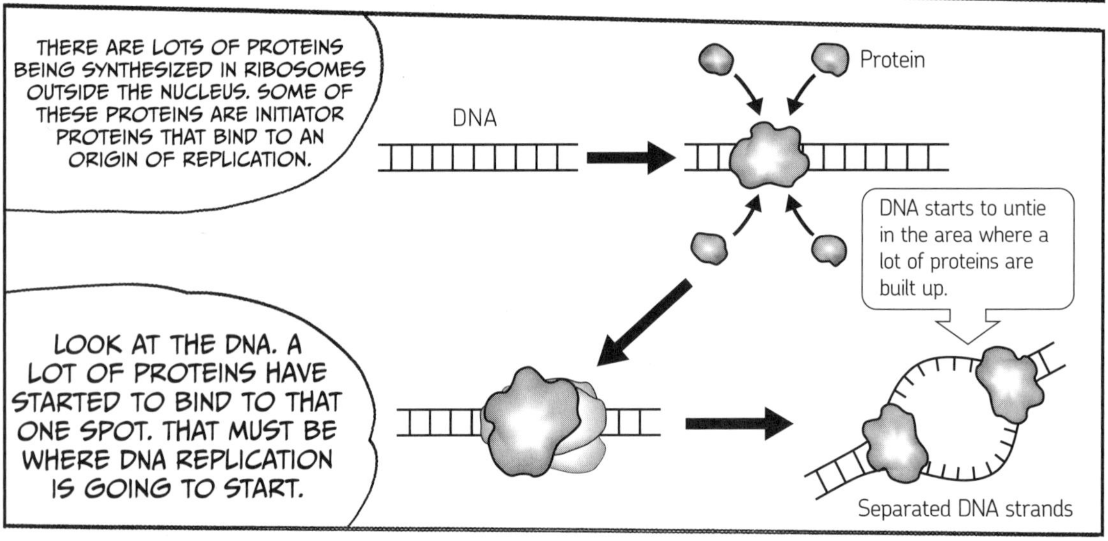
THERE ARE LOTS OF PROTEINS BEING SYNTHESIZED IN RIBOSOMES OUTSIDE THE NUCLEUS. SOME OF THESE PROTEINS ARE INITIATOR PROTEINS THAT BIND TO AN ORIGIN OF REPLICATION.
Protein
DNA
DNA starts to untie in the area where a lot of proteins are built up.
LOOK AT THE DNA. A LOT OF PROTEINS HAVE STARTED TO BIND TO THAT ONE SPOT. THAT MUST BE WHERE DNA REPLICATION IS GOING TO START.
Separated DNA strands

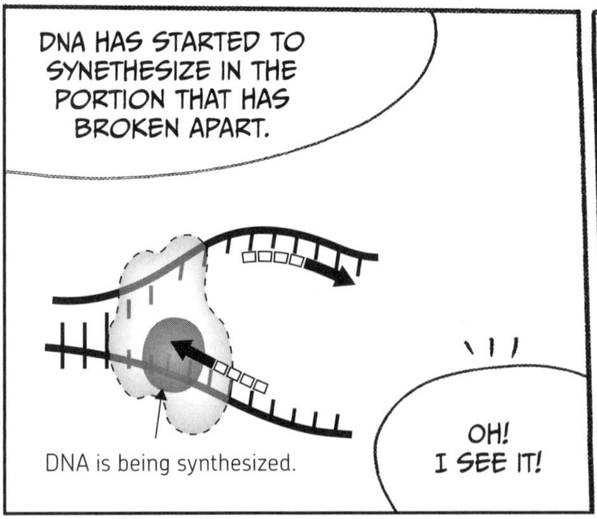
DNA HAS STARTED TO SYNETHESIZE IN THE PORTION THAT HAS BROKEN APART.
DNA is being synthesized.
OH! I SEE IT!

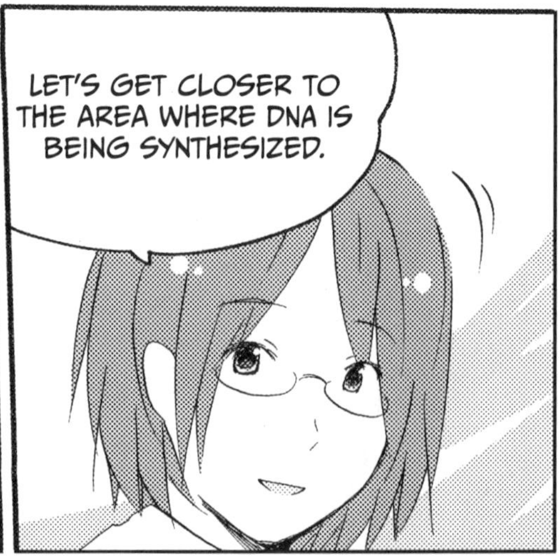
LET'S GET CLOSER TO THE AREA WHERE DNA IS BEING SYNTHESIZED.

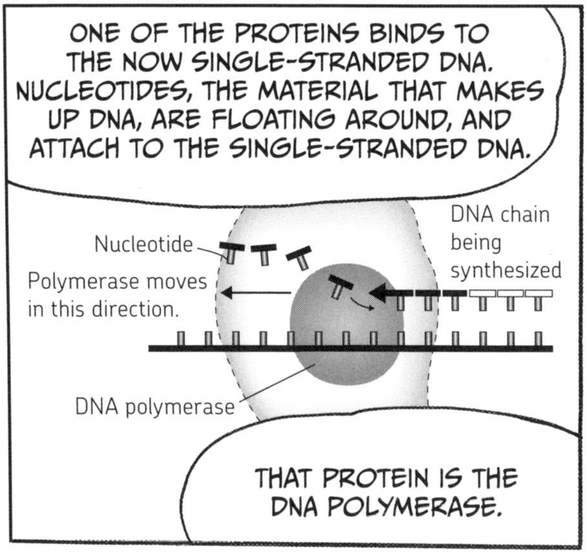
ONE OF THE PROTEINS BINDS TO THE NOW SINGLE-STRANDED DNA. NUCLEOTIDES, THE MATERIAL THAT MAKES UP DNA, ARE FLOATING AROUND, AND ATTACH TO THE SINGLE-STRANDED DNA.
Nucleotide
Polymerase moves in this direction.
DNA chain being synthesized
DNA polymerase
THAT PROTEIN IS THE DNA POLYMERASE.

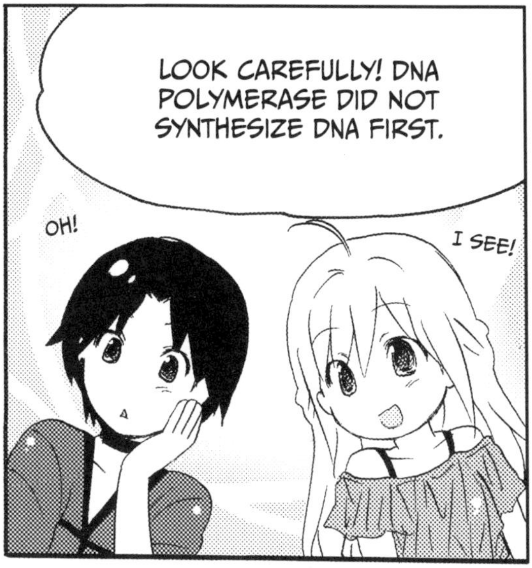
LOOK CAREFULLY! DNA POLYMERASE DID NOT SYNTHESIZE DNA FIRST.
OH!
I SEE!

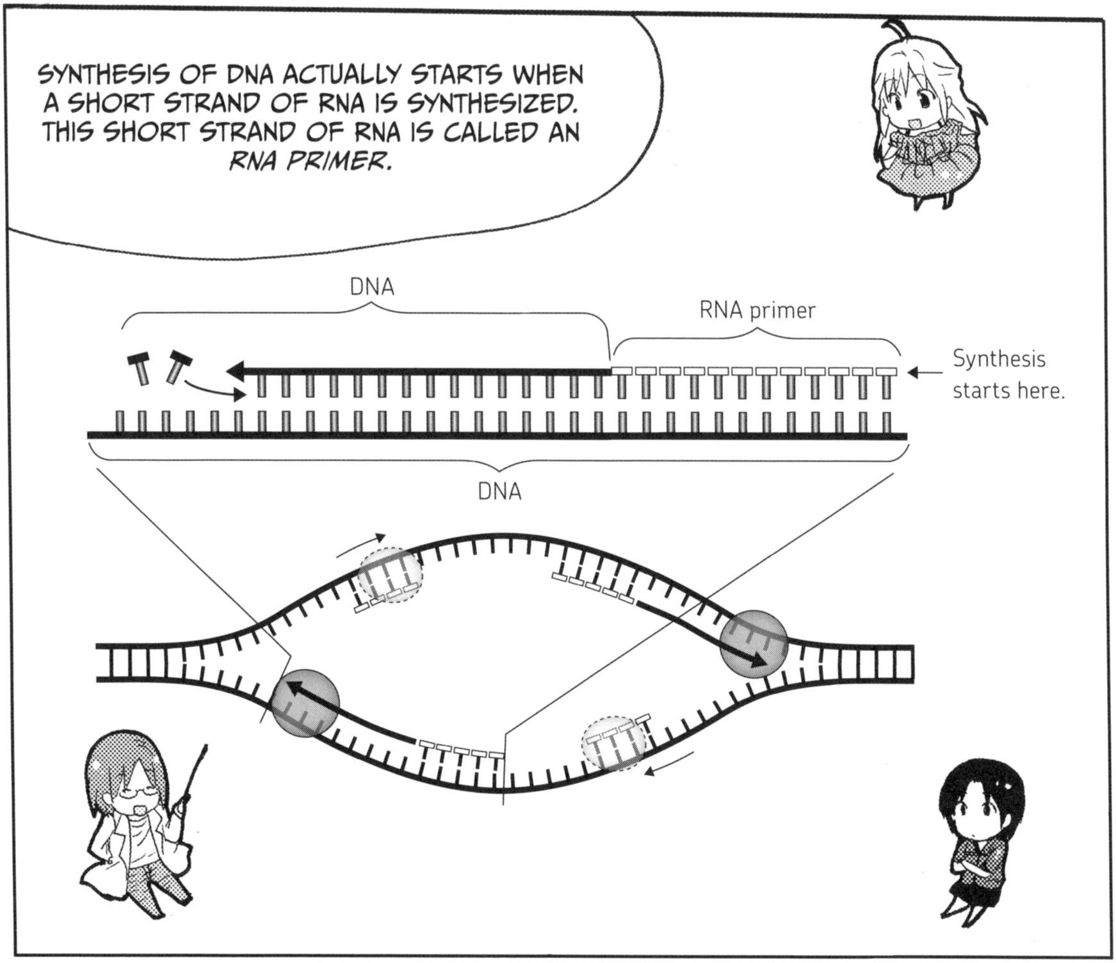
SYNTHESIS OF DNA ACTUALLY STARTS WHEN A SHORT STRAND OF RNA IS SYNTHESIZED. THIS SHORT STRAND OF RNA IS CALLED AN *RNA PRIMER*.
DNA
RNA primer
Synthesis starts here.
DNA

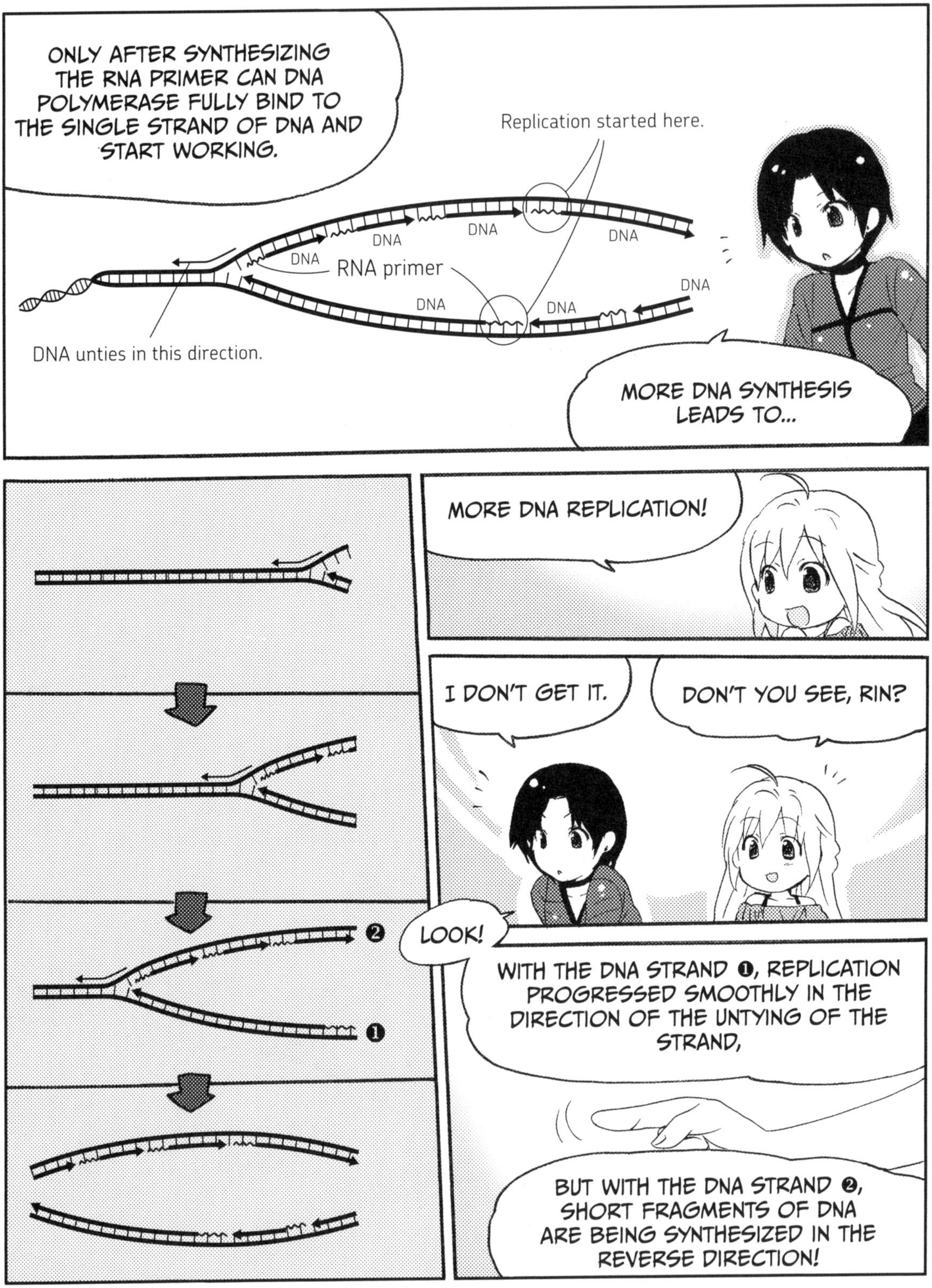
ONLY AFTER SYNTHESIZING THE RNA PRIMER CAN DNA POLYMERASE FULLY BIND TO THE SINGLE STRAND OF DNA AND START WORKING.
Replication started here.
DNA
DNA
DNA
DNA
RNA primer
DNA
DNA
DNA
DNA unties in this direction.
MORE DNA SYNTHESIS LEADS TO...
❷
❶
MORE DNA REPLICATION!
I DON'T GET IT.
DON'T YOU SEE, RIN?
LOOK!
WITH THE DNA STRAND ❶, REPLICATION PROGRESSED SMOOTHLY IN THE DIRECTION OF THE UNTYING OF THE STRAND,
BUT WITH THE DNA STRAND ❷, SHORT FRAGMENTS OF DNA ARE BEING SYNTHESIZED IN THE REVERSE DIRECTION!

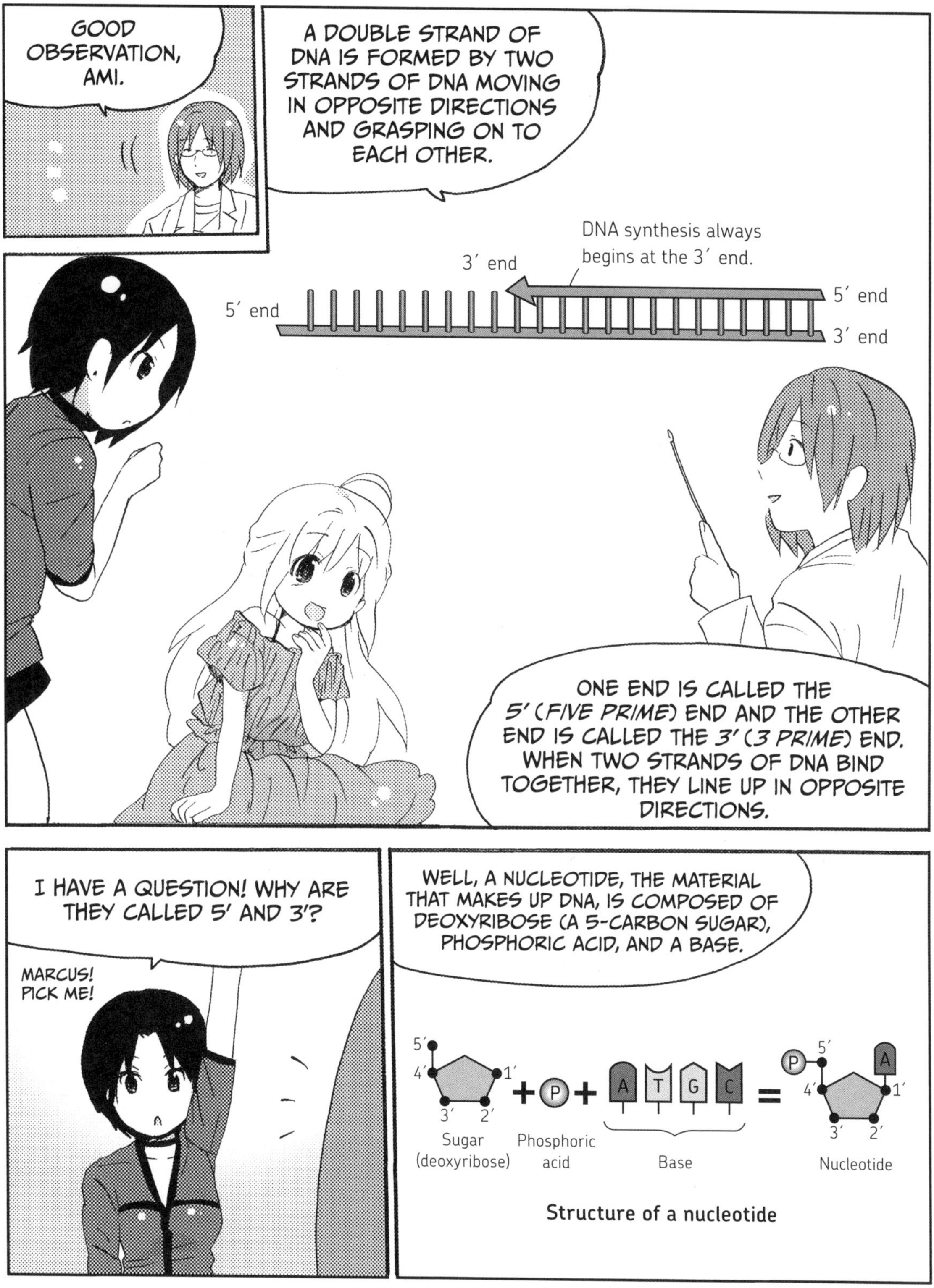
GOOD OBSERVATION, AMI.
A DOUBLE STRAND OF DNA IS FORMED BY TWO STRANDS OF DNA MOVING IN OPPOSITE DIRECTIONS AND GRASPING ON TO EACH OTHER.
DNA synthesis always begins at the 3′ end.
3′ end
5′ end
5′ end
3′ end
ONE END IS CALLED THE *5′ (FIVE PRIME)* END AND THE OTHER END IS CALLED THE *3′ (3 PRIME)* END. WHEN TWO STRANDS OF DNA BIND TOGETHER, THEY LINE UP IN OPPOSITE DIRECTIONS.
I HAVE A QUESTION! WHY ARE THEY CALLED 5′ AND 3′?
MARCUS! PICK ME!
WELL, A NUCLEOTIDE, THE MATERIAL THAT MAKES UP DNA, IS COMPOSED OF DEOXYRIBOSE (A 5-CARBON SUGAR), PHOSPHORIC ACID, AND A BASE.
5′
4′
1′
3′
2′
P
A
T
G
C
P
5′
A
4′
1′
3′
2′
Sugar (deoxyribose)
Phosphoric acid
Base
Nucleotide
Structure of a nucleotide

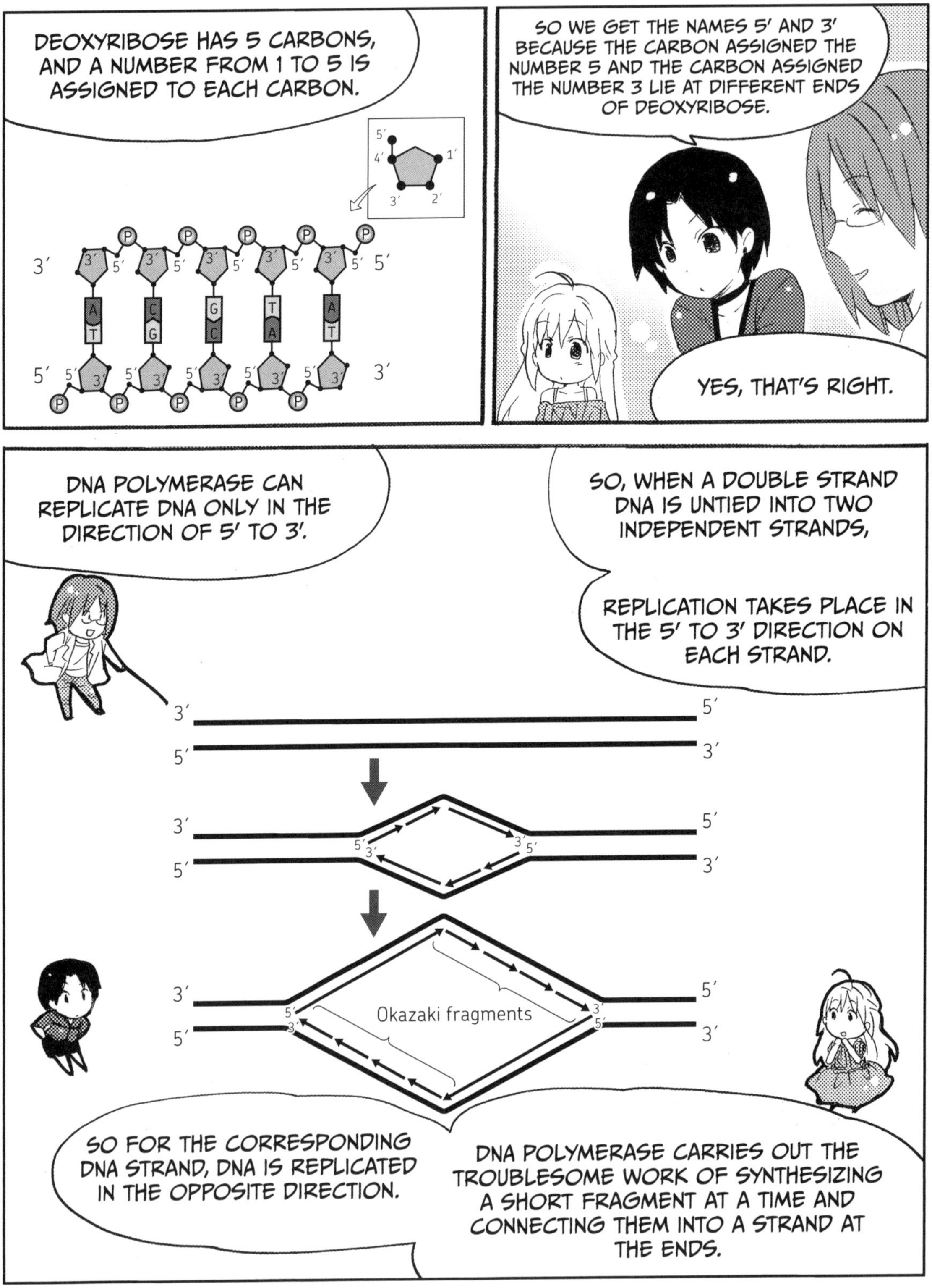
DEOXYRIBOSE HAS 5 CARBONS, AND A NUMBER FROM 1 TO 5 IS ASSIGNED TO EACH CARBON.
5′ 4′ 3′ 2′ 1′
3′ 5′ 5′ 5′ 5′ 5′ 5′
A T C G G C T A A T
5′ 5′ 5′ 5′ 5′ 5′ 3′
P
SO WE GET THE NAMES 5′ AND 3′ BECAUSE THE CARBON ASSIGNED THE NUMBER 5 AND THE CARBON ASSIGNED THE NUMBER 3 LIE AT DIFFERENT ENDS OF DEOXYRIBOSE.
YES, THAT'S RIGHT.
DNA POLYMERASE CAN REPLICATE DNA ONLY IN THE DIRECTION OF 5′ TO 3′.
SO, WHEN A DOUBLE STRAND DNA IS UNTIED INTO TWO INDEPENDENT STRANDS,
REPLICATION TAKES PLACE IN THE 5′ TO 3′ DIRECTION ON EACH STRAND.
3′ 5′
5′ 3′
Okazaki fragments
SO FOR THE CORRESPONDING DNA STRAND, DNA IS REPLICATED IN THE OPPOSITE DIRECTION.
DNA POLYMERASE CARRIES OUT THE TROUBLESOME WORK OF SYNTHESIZING A SHORT FRAGMENT AT A TIME AND CONNECTING THEM INTO A STRAND AT THE ENDS.

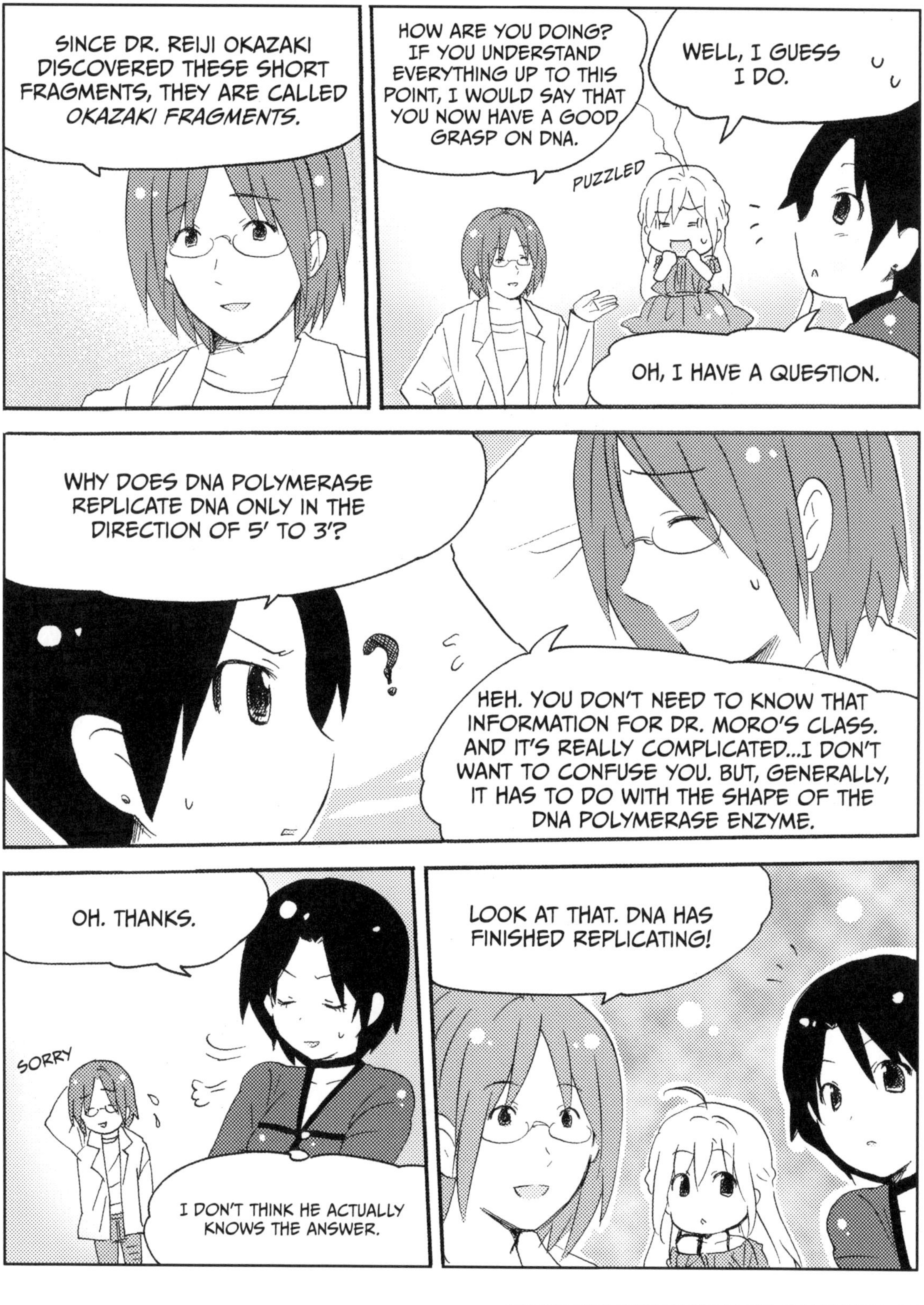
SINCE DR. REIJI OKAZAKI DISCOVERED THESE SHORT FRAGMENTS, THEY ARE CALLED *OKAZAKI FRAGMENTS.*
HOW ARE YOU DOING? IF YOU UNDERSTAND EVERYTHING UP TO THIS POINT, I WOULD SAY THAT YOU NOW HAVE A GOOD GRASP ON DNA.
WELL, I GUESS I DO.
PUZZLED
OH, I HAVE A QUESTION.
WHY DOES DNA POLYMERASE REPLICATE DNA ONLY IN THE DIRECTION OF 5′ TO 3′?
HEH. YOU DON'T NEED TO KNOW THAT INFORMATION FOR DR. MORO'S CLASS. AND IT'S REALLY COMPLICATED...I DON'T WANT TO CONFUSE YOU. BUT, GENERALLY, IT HAS TO DO WITH THE SHAPE OF THE DNA POLYMERASE ENZYME.
OH. THANKS.
SORRY
I DON'T THINK HE ACTUALLY KNOWS THE ANSWER.
LOOK AT THAT. DNA HAS FINISHED REPLICATING!

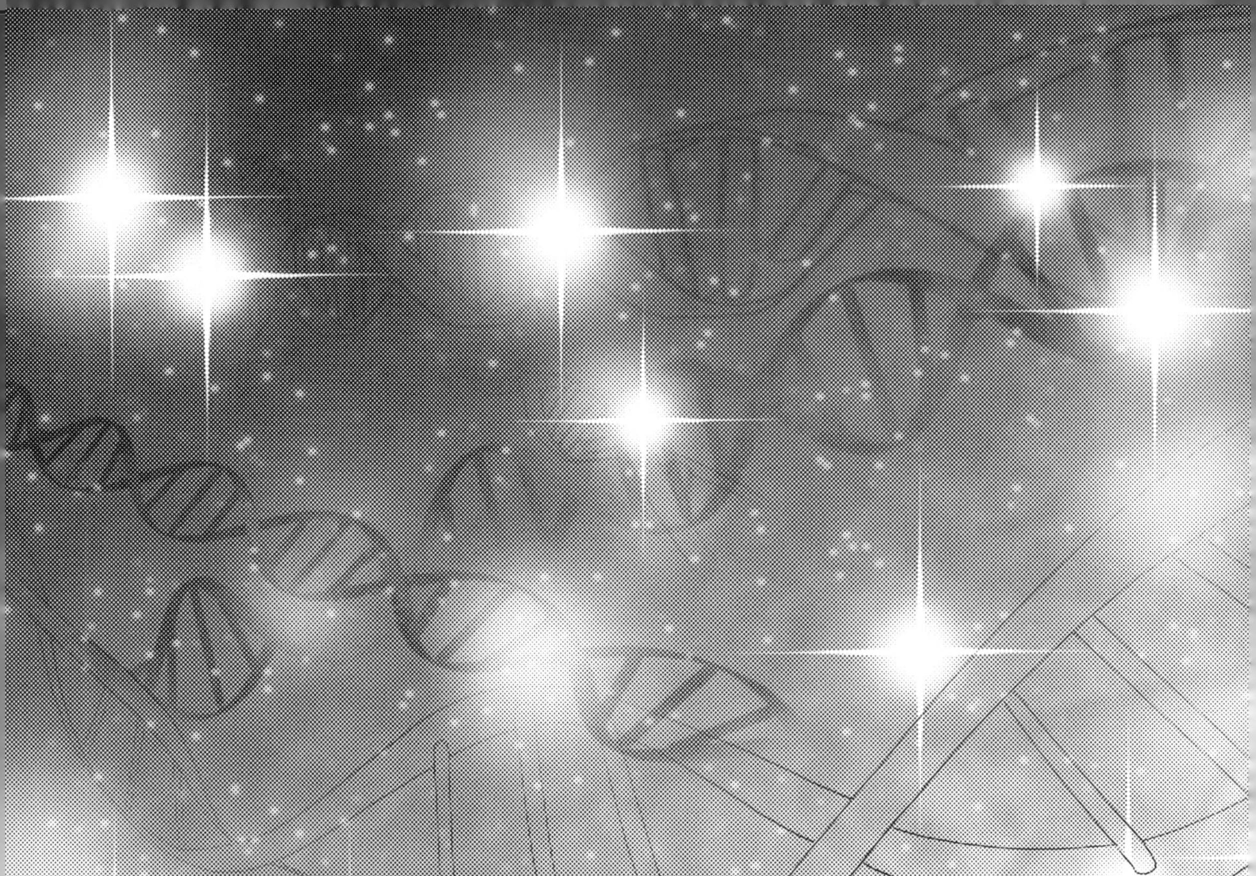

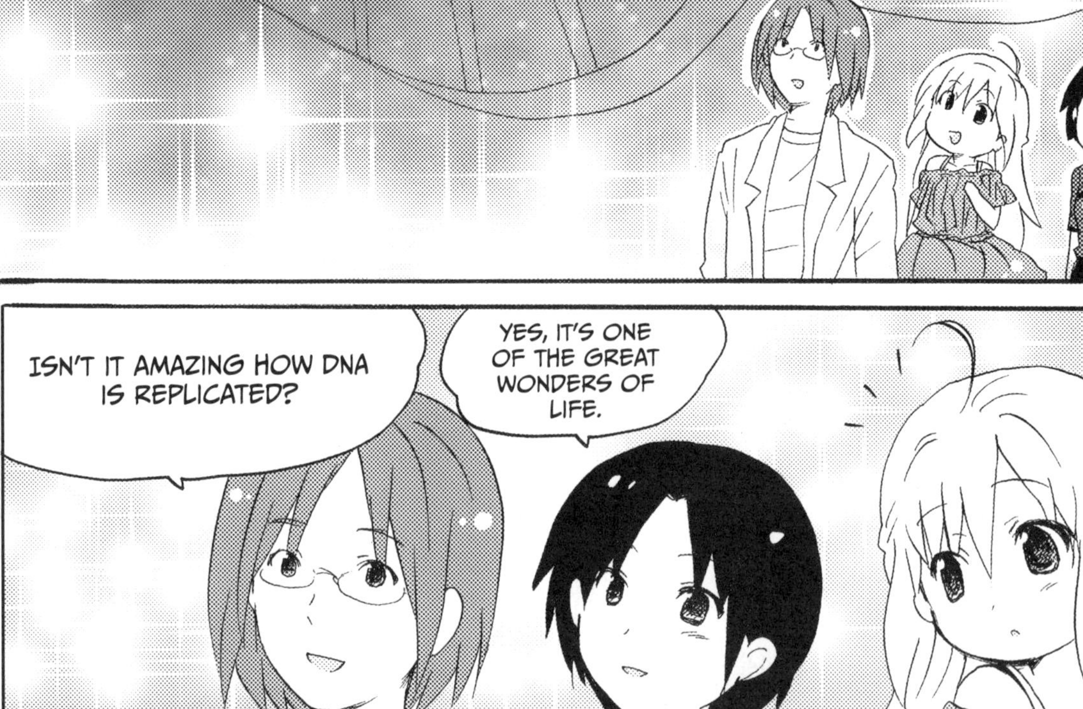
ISN'T IT AMAZING HOW DNA IS REPLICATED?
YES, IT'S ONE OF THE GREAT WONDERS OF LIFE.

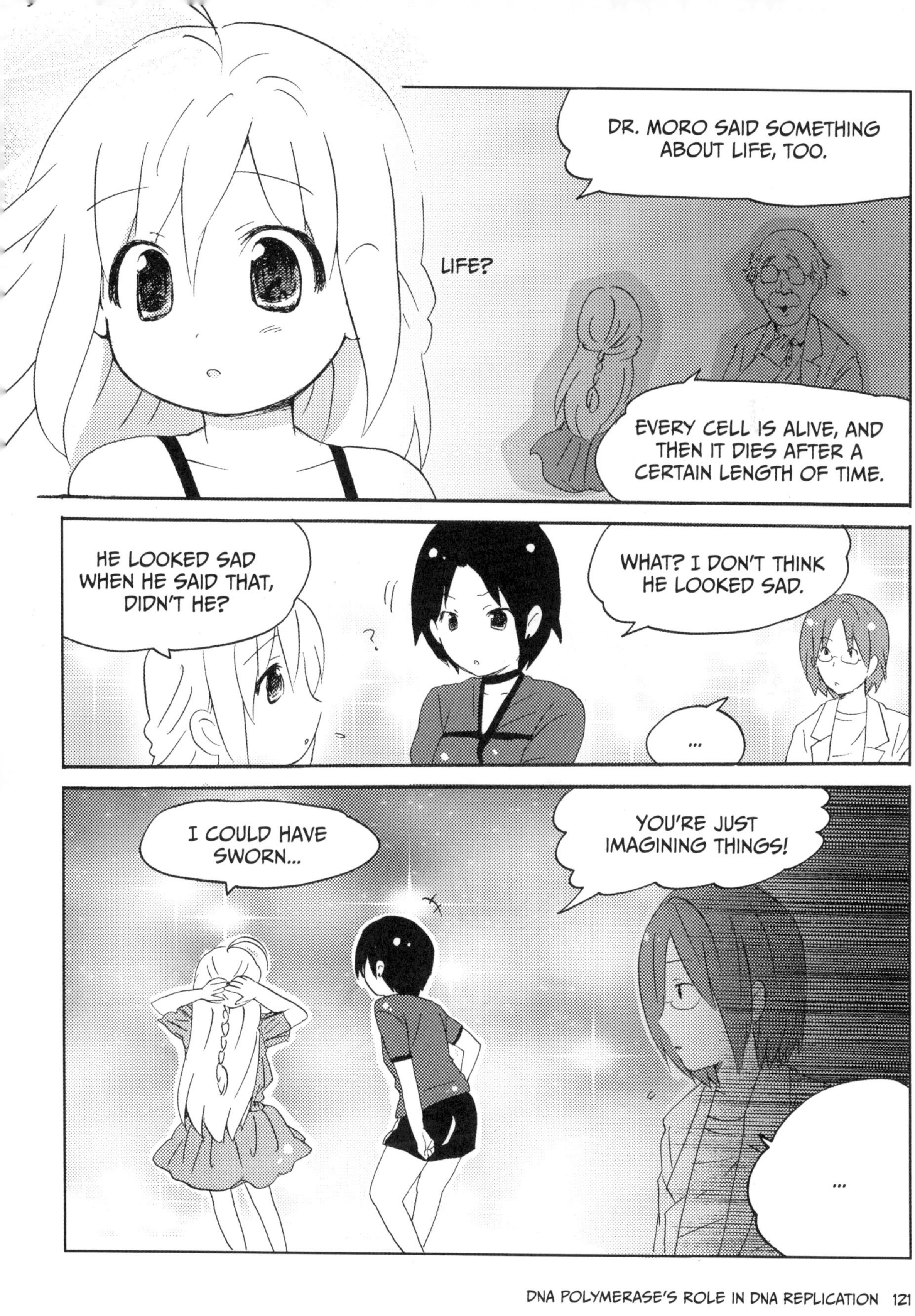
DR. MORO SAID SOMETHING ABOUT LIFE, TOO.
LIFE?
EVERY CELL IS ALIVE, AND THEN IT DIES AFTER A CERTAIN LENGTH OF TIME.
HE LOOKED SAD WHEN HE SAID THAT, DIDN'T HE?
WHAT? I DON'T THINK HE LOOKED SAD.
...
I COULD HAVE SWORN...
YOU'RE JUST IMAGINING THINGS!
...

WHAT IS A CHROMOSOME?

When we talk about cell division, we cannot avoid talking about chromosomes.

What's a chromosome?

A *chromosome* is DNA in a specific form—it contains genetic information. Before cell division starts, chromosomes gather together at the center of a cell. When cell division begins, chromosomes are torn apart into two pieces.

A chromosome is made up of one long string of a substance called *chromatin*. Do you remember the beaded chain that you learned about in Chapter 1? That is chromatin. It is made up of proteins called *histones*, DNA, and a few other molecules. Each bead in the chain is made up of 1.7 turns of DNA wound around a histone. A single strand of DNA winds around bead after bead, forming the thick thread of *chromatin*.

A histone is actually a set of proteins, eight molecules bound together, with two molecules of each of the following four types: H2A, H2B, H3, and H4.

Chromosomes, each a long strand of chromatin, are normally scattered throughout the nucleus and are invisible even with a microscope. Only when they condense into a compact shape for cell division do they become visible under the microscope.

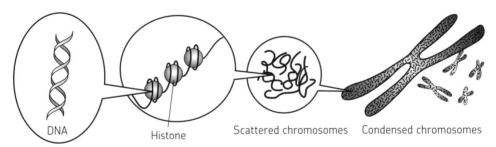

DNA   Histone   Scattered chromosomes   Condensed chromosomes

DNA and chromosomes

Chromosomes, which were discovered in the 19th century, are named after their tendency to stain easily with certain dyes (*chroma* means *color* in Greek—you've probably seen it in other words like chromatic and monochrome).

THE HUMAN BODY CONTAINS 24 CHROMOSOMES

All human cells (except sex cells) contain 24 chromosomes.

But the number of chromosomes varies among living organisms. Higher order animals do not necessarily have more chromosomes than lower order animals. For example, a goldfish has about 100 chromosomes!

22 out of the 24 types are *autosomes* and are unrelated to gender. Each cell has two copies of each autosome. Why two? Because one is inherited from the father and the other from the mother.

A number from 1 to 22 is assigned to each autosome, with the largest chromosome being number 1. They are referred to as chromosome 3, chromosome 16, and so on. The two chromosomes which are not autosomes are *sex chromosomes*. They consist of an X chromosome and a Y chromosome.

These chromosomes determine the sex of a human being. Male cells have an X chromosome and a Y chromosome (XY), while female cells have two X chromosomes (XX).

Male
Female
Autosome
Y
X
Sex chromosome
X
X
Sex chromosome

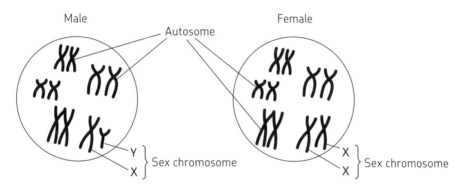

CHROMOSOMES ARE ONLY VISIBLE AT THE TIME OF CELL DIVISION

Since chromosomes only thicken when replicated DNA and histones have completely condensed, they appear only when cells divide.

When cutting a large cloth into two pieces, for example, you can save time by folding it a few times before cutting it. In the same way, it is easier for cells to divide if DNA is condensed rather than dispersed inside the nucleus.

Now let's look at cell division.

DYNAMIC CELL DIVISION

DNA is now replicated. The next phase starts.

As replication of DNA is completed, a cell begins to prepare to divide itself entirely into two. Cell division takes place in two general steps: mitosis and cytokinesis.

MITOSIS

Cell division starts in the nucleus, where DNA is contained. The process of dividing the contents of the cell's nucleus into two is called *mitosis*.

Is the division of a nucleus similar to atomic fission, which generates radioactivity and is used for atomic power generation?

All that happens inside the body?!

NOOOoooooo!!! The nucleus of a cell is not the same thing as the nucleus of an atom. They have the same name, but they are very, very different things. Mitosis has nothing to do with nuclear fission!

Well, that's a relief!

In mitosis, replicated DNA is condensed tightly within the chromosomes, and the the chromosomes start to form a thick, condensed *X* shape. Substances called *centrosomes*, which used to be on the sides of the nucleus, start moving toward both poles of the cell.

Prophase

Nucleus
Nuclear envelope
Replicated DNA
Centrosome
Cell

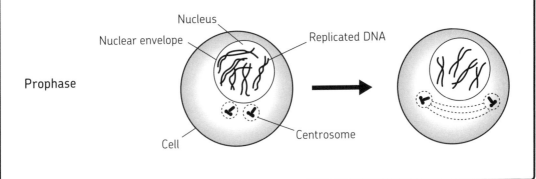

Although the process eventually creates two nuclei, the nucleus isn't just split in two. The original nuclear envelope is actually broken apart at the beginning of mitosis and then reformed after the cell divides.

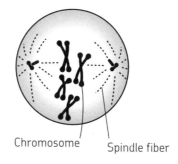

Chromosome   Spindle fiber

 Look! Something is changing around the centrosome!

 You're right. String-like substances called *spindle fibers* start extending from the centrosomes, each of which is now located on opposite poles. These spindle fibers are made up of long thin substances called *microtubules*.

After the membrane has disappeared, replicated DNA that has started to condense is flung out to the sea of cytoplasm. This process is important.

After the condensation of chromosomes has finished and the formation of their shape is almost completed, the string-like spindle fiber being extended from the centrosomes on each pole reaches near the center of each chromosome and sticks tight there. The existence of the nuclear envelope would be a hindrance to the spindle fiber if it hadn't disappeared.

Spindle fiber

Metaphase

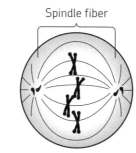

The spindle fibers move the chromosomes to the middle of the cell. This is called the *mitotic spindle* or *spindle apparatus*.

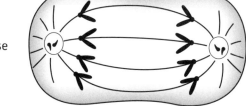

Anaphase

During anaphase, the spindle fibers stuck to each chromosome pull toward both poles of the cell. As a result, the chromosomes being arranged at the center are drawn apart to each pole.

Some time after being drawn apart, the chromosomes start to scatter and unfold to restore their original state. They become invisible even with a microscope. Then formation of the nuclear envelope, currently scattered in the cytoplasm, starts, and the shape of a nucleus begins appearing in each pole.

Formation of a nucleus

Telophase

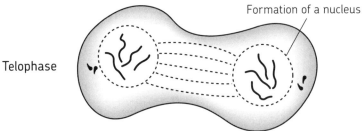

In order for cells to divide and survive, the DNA has to replicate and be divided evenly into two new nuclei rather than just haphazardly splitting the original nucleus in two.

Because of the appearance of the mitotic spindle, this step is referred to as *mitotic division*.

CYTOKINESIS

Is the process of division the same for animals and plants?

It is almost the same as far as mitosis is concerned. In cytokinesis, which occurs later, the entire cell starts to divide into two. There is a big difference between animal cells and plant cells in this process.

In animal cells, the portion near the center of a cell constricts, and this constriction progresses until the cell is divided equally into two. The important part about cytokinesis is that all of the organelles, cytoplasm, and nutrients in the cell are also divided evenly between the two new cells. It's just like making cookies; you're trying to split cookie dough into smaller pieces with an equal number of chocolate chips and nuts in each piece.

Animal cell

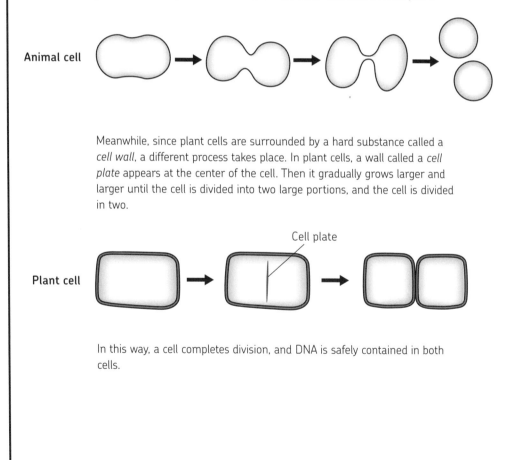

Meanwhile, since plant cells are surrounded by a hard substance called a *cell wall*, a different process takes place. In plant cells, a wall called a *cell plate* appears at the center of the cell. Then it gradually grows larger and larger until the cell is divided into two large portions, and the cell is divided in two.

Plant cell

Cell plate

In this way, a cell completes division, and DNA is safely contained in both cells.

WHAT IS A CELL CYCLE?

Some cells continue dividing indefinitely and some do not. A cell's function will determine how often it replicates. Basal cells (see page 103), which you saw deep in the skin, are a type of cell that continues dividing endlessly, as skin cells are constantly replaced. Basal cells have limited lives. Older basal cells at the end of their lives are believed to stop dividing.

As you learned in this chapter, cell division follows predetermined steps: DNA is replicated, chromatin condenses into chromosomes, the nuclear envelope disappears, the cell's contents are split in two, and an entire cell divides.

A cell has many checkpoints during replication to make sure no errors occur during this important process.

This ring indicates a
check point in the cell cycle.

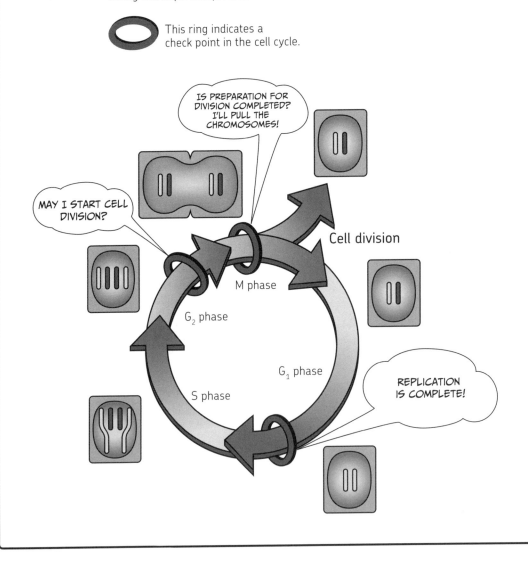

IS PREPARATION FOR DIVISION COMPLETED? I'LL PULL THE CHROMOSOMES!

MAY I START CELL DIVISION?

Cell division

M phase

G$_2$ phase

G$_1$ phase

S phase

REPLICATION IS COMPLETE!

Cells that divide many times repeat these steps again and again. The series of steps in which a cell completes one division is called a *cell cycle*. The cell cycle is divided into four phases:

- **G_1 phase** The cell prepares for DNA replication in this step. Enzymes that are needed for DNA replication (the S phase) are produced at this time, along with other proteins the cell needs. You can think of G_1 as standing for the first growth stage of the cell.

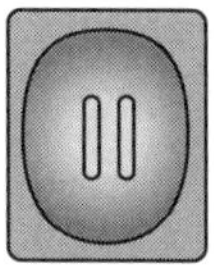

- **S phase** DNA is replicated in this step. DNA is synthesized, so this step is called the S phase.

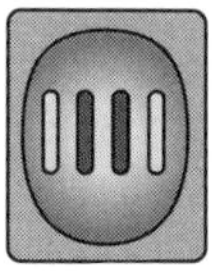

- **G_2 phase** The cell prepares for division in this step. Microtubules that are needed during the M phase are synthesized at this time. This is called G_2, as it's the second period of growth and protein synthesis. Collectively, the G_1, S, and G_2 phases are referred to as *interphase*. As you may expect, a cell's preparation for division (interphase) is a much longer process than cell division itself (the M phase).

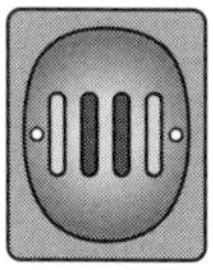

- **M phase** Mitosis and cytokinesis are carried out in this step. It is called the M phase because this is when mitosis occurs.

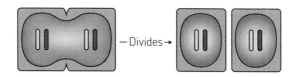

WHAT CAUSES CANCER?

A *cancer cell* is one that was once functioning normally but then suddenly gets a mutation in its genes that make the cell go "mad." These changes in its genes make cancer cells multiply without regard for other cells, growing out of control and stealing energy and nutrients from other cells in the tissue. If cancer cells continue multiplying, they can become visible under a microscope, or even to human eyes in the form of a tumor.

There are many ways a healthy cell can become a cancer cell, but no matter the reason, all cancer cells carry on multiplying because the functions of their genes that regulate cell division have gone out of control.

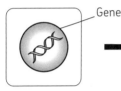

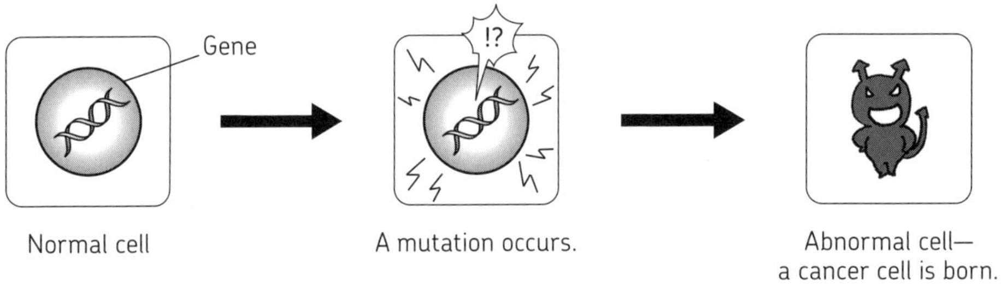

Gene
!?
Normal cell
A mutation occurs.
Abnormal cell—
a cancer cell is born.

With normal cells, tumor suppressor genes act like a brake; these genes apply the break at the checkpoints in the cell cycle to prevent arbitrary replication of the cell. However, in some cells, these tumor suppressor genes become mutated and lose their ability to function. As a result, the brake is disabled, allowing the cell to divide over and over again.

I'M GOING TO DIVIDE.
NO, I WON'T ALLOW IT!
Nucleus
Tumor suppressor gene
Division!
Disabled
Tumor
Cancer cells continue multiplying.

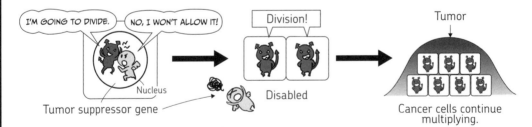

In some cancer cells, genes that work to accelerate and promote cell division go awry, and the growth and division of the cell become *super accelerated*. The tumor suppressor genes cannot do their job if this happens. The cells will repeatedly divide to become cancerous and interfere with the normal functions of tissues and organs.

A multicellular organism is like a society made up of cells. The cells don't normally do things that disturb the order of the society. The cells divide as needed and carry out their special assignments. However, cancer cells are like gangs who steal and recruit new members, spreading through society and disrupting order.

4

HOW IS A
PROTEIN MADE?

A GENE BECOMES USEFUL AFTER TRANSCRIPTION

HOW A PROTEIN IS MADE

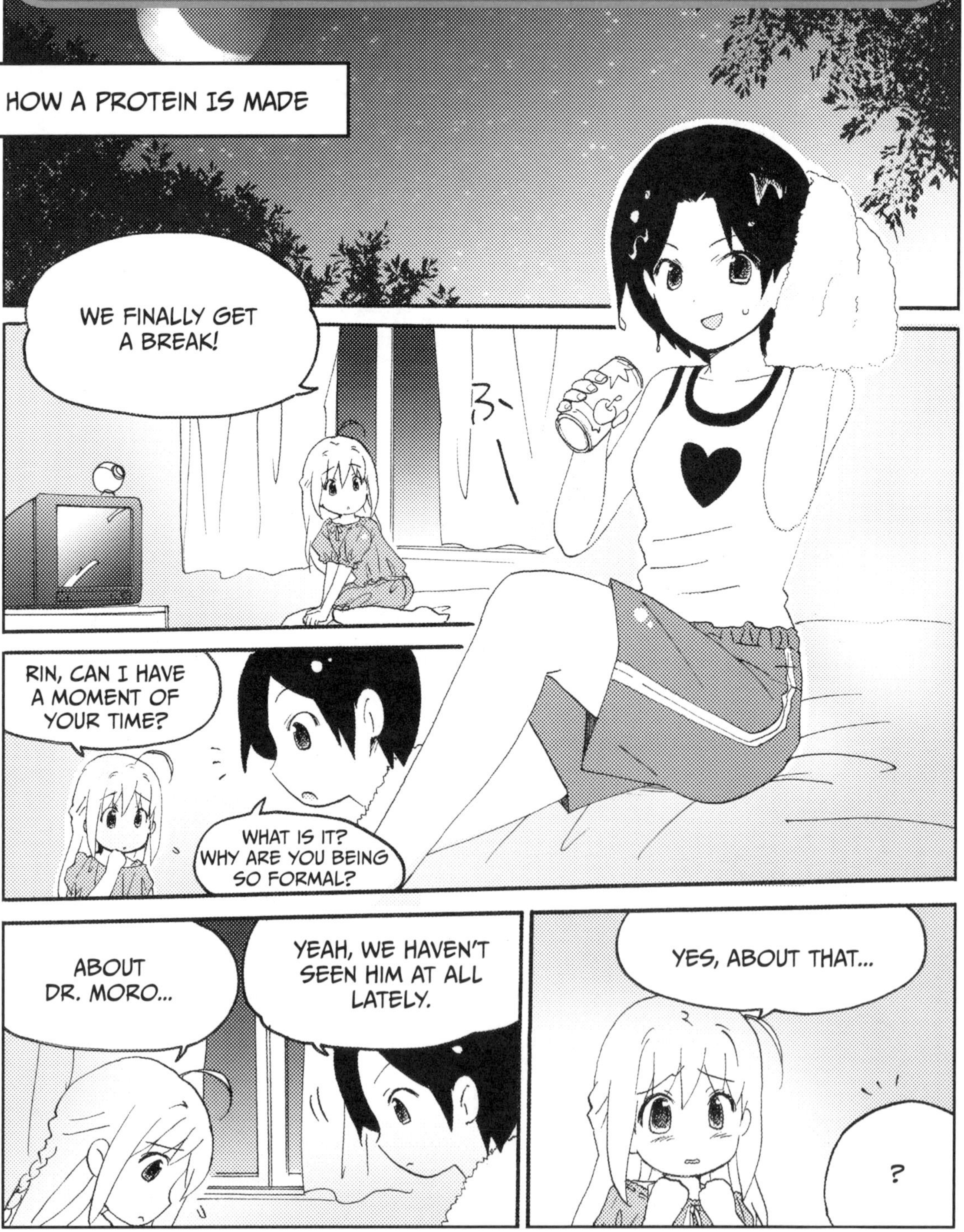

WE FINALLY GET A BREAK!
RIN, CAN I HAVE A MOMENT OF YOUR TIME?
WHAT IS IT? WHY ARE YOU BEING SO FORMAL?
ABOUT DR. MORO...
YEAH, WE HAVEN'T SEEN HIM AT ALL LATELY.
YES, ABOUT THAT...
?

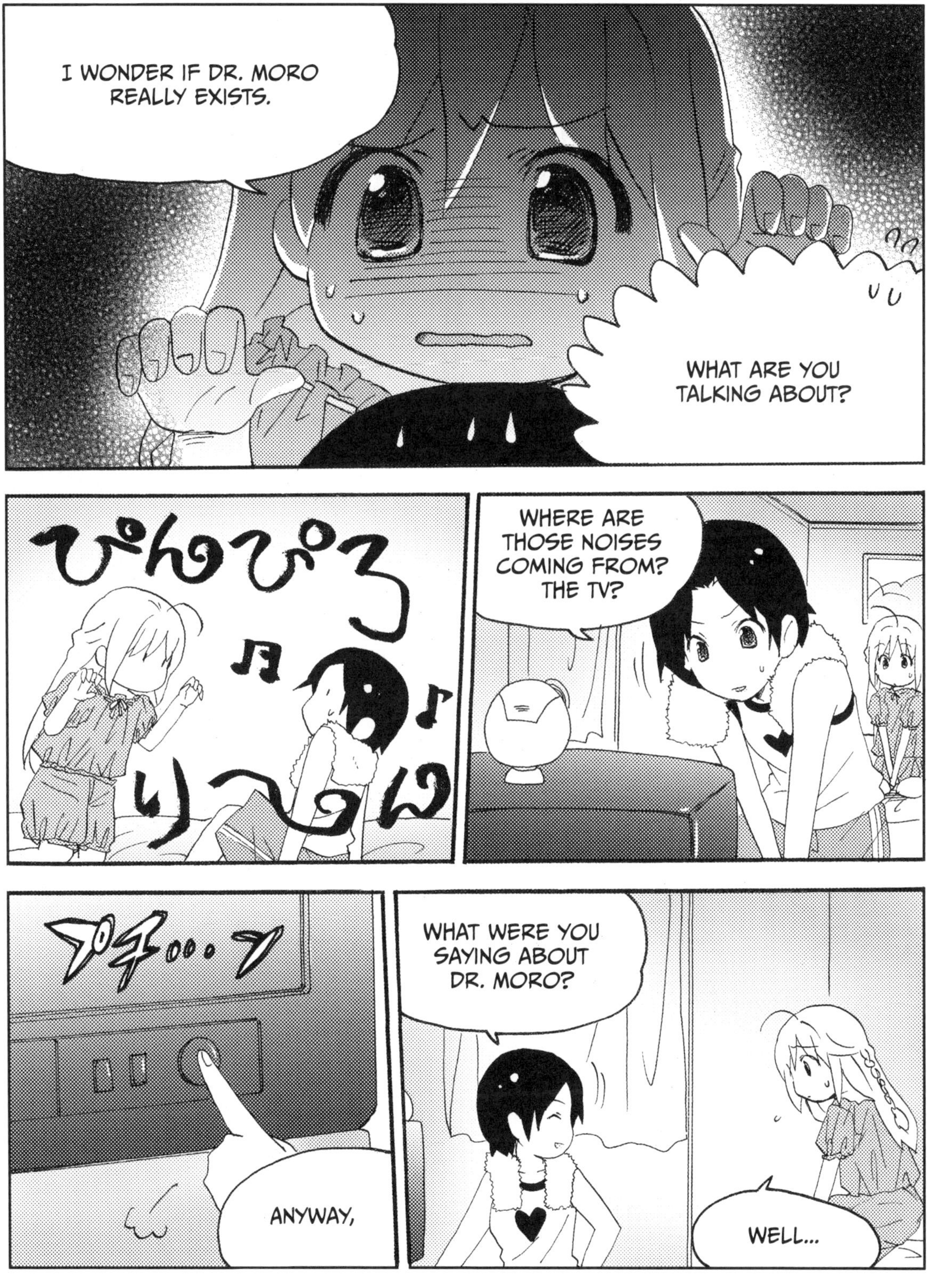
I WONDER IF DR. MORO REALLY EXISTS.
WHAT ARE YOU TALKING ABOUT?
ぴんぴろりーん
WHERE ARE THOSE NOISES COMING FROM? THE TV?
プチ…ッ
ANYWAY,
WHAT WERE YOU SAYING ABOUT DR. MORO?
WELL...

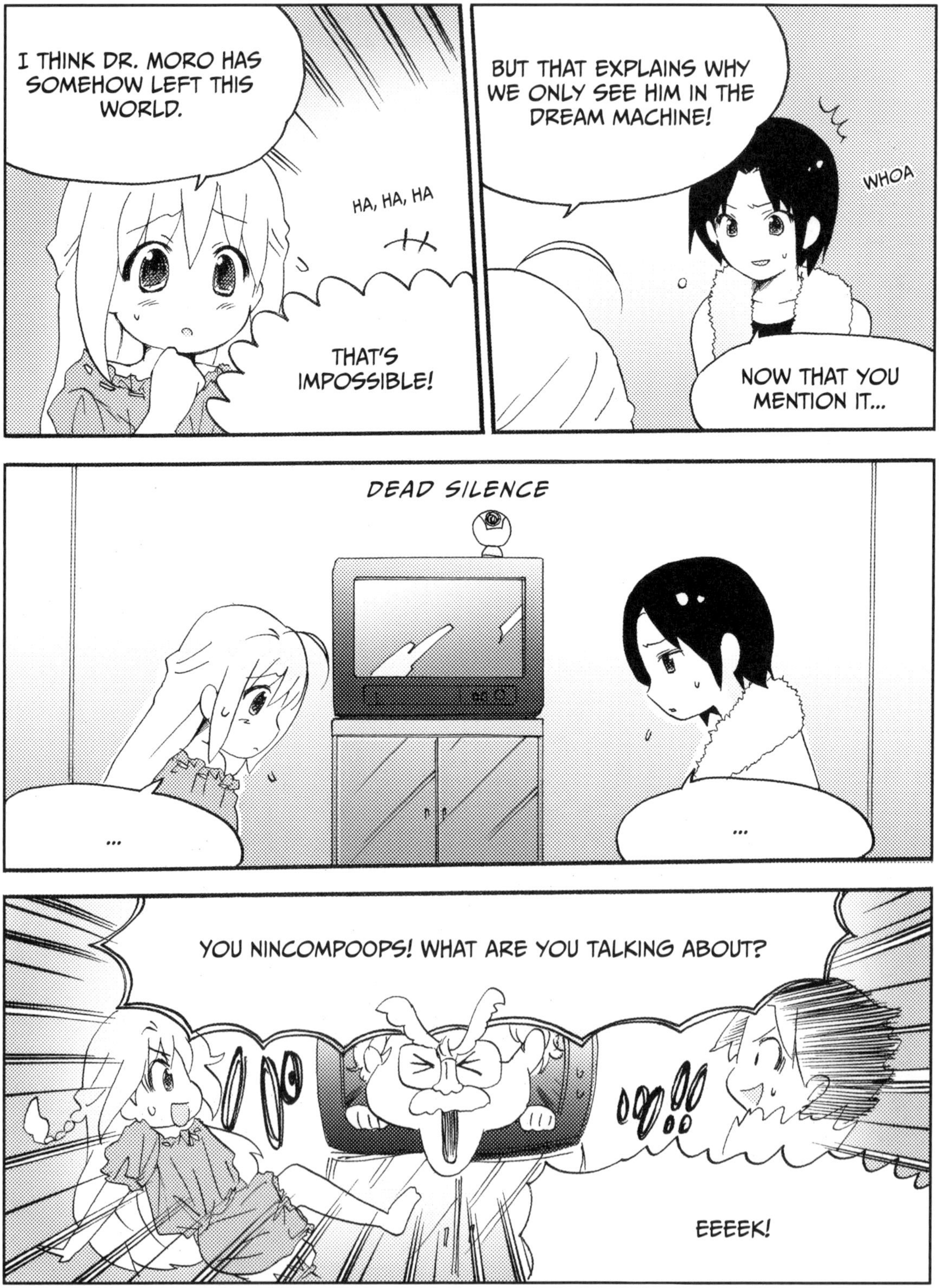
I THINK DR. MORO HAS SOMEHOW LEFT THIS WORLD.
HA, HA, HA
THAT'S IMPOSSIBLE!
BUT THAT EXPLAINS WHY WE ONLY SEE HIM IN THE DREAM MACHINE!
WHOA
NOW THAT YOU MENTION IT...
DEAD SILENCE
...
...
YOU NINCOMPOOPS! WHAT ARE YOU TALKING ABOUT?
EEEEK!

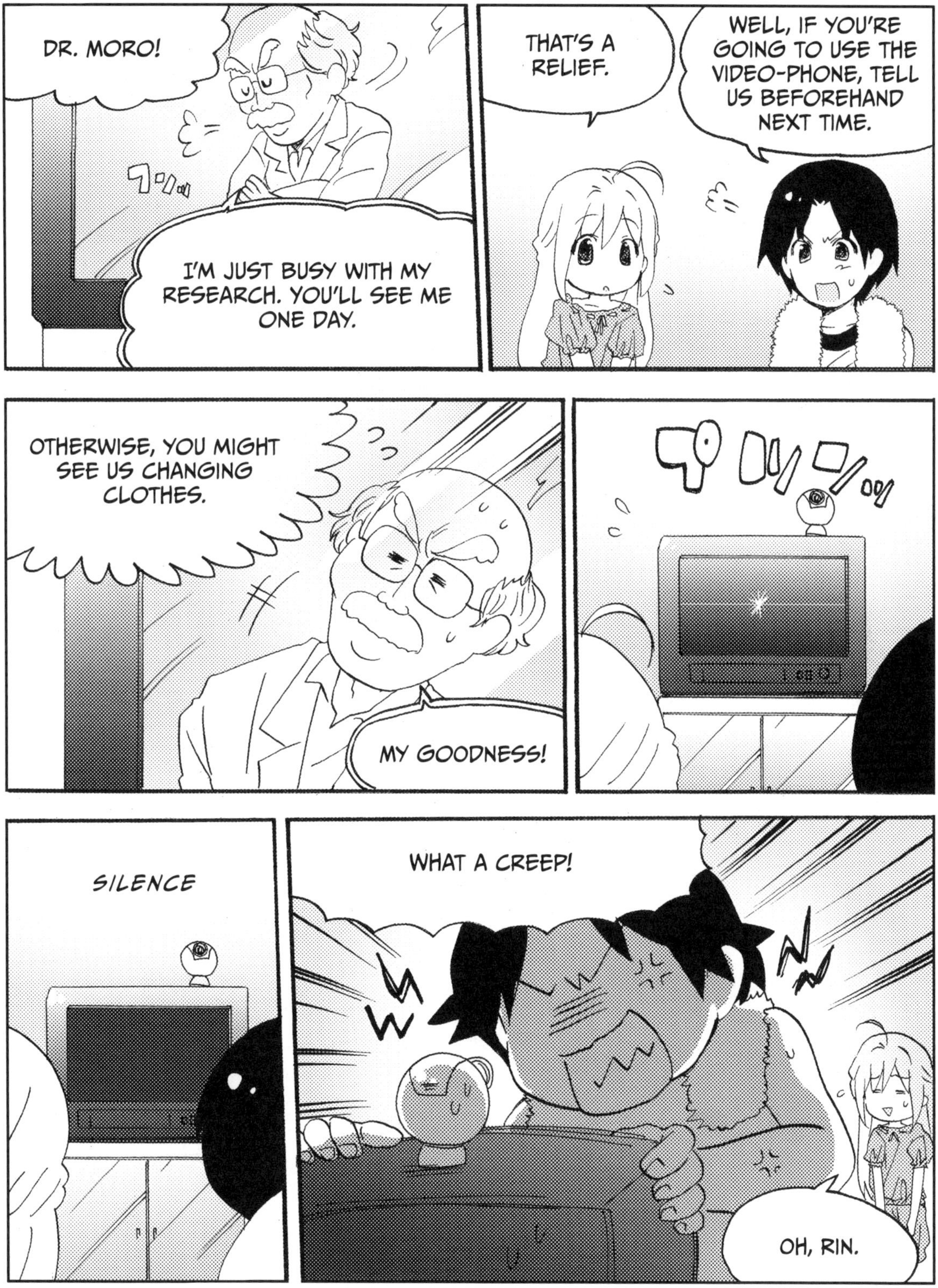
DR. MORO!
I'M JUST BUSY WITH MY RESEARCH. YOU'LL SEE ME ONE DAY.
THAT'S A RELIEF.
WELL, IF YOU'RE GOING TO USE THE VIDEO-PHONE, TELL US BEFOREHAND NEXT TIME.
OTHERWISE, YOU MIGHT SEE US CHANGING CLOTHES.
MY GOODNESS!
SILENCE
WHAT A CREEP!
OH, RIN.

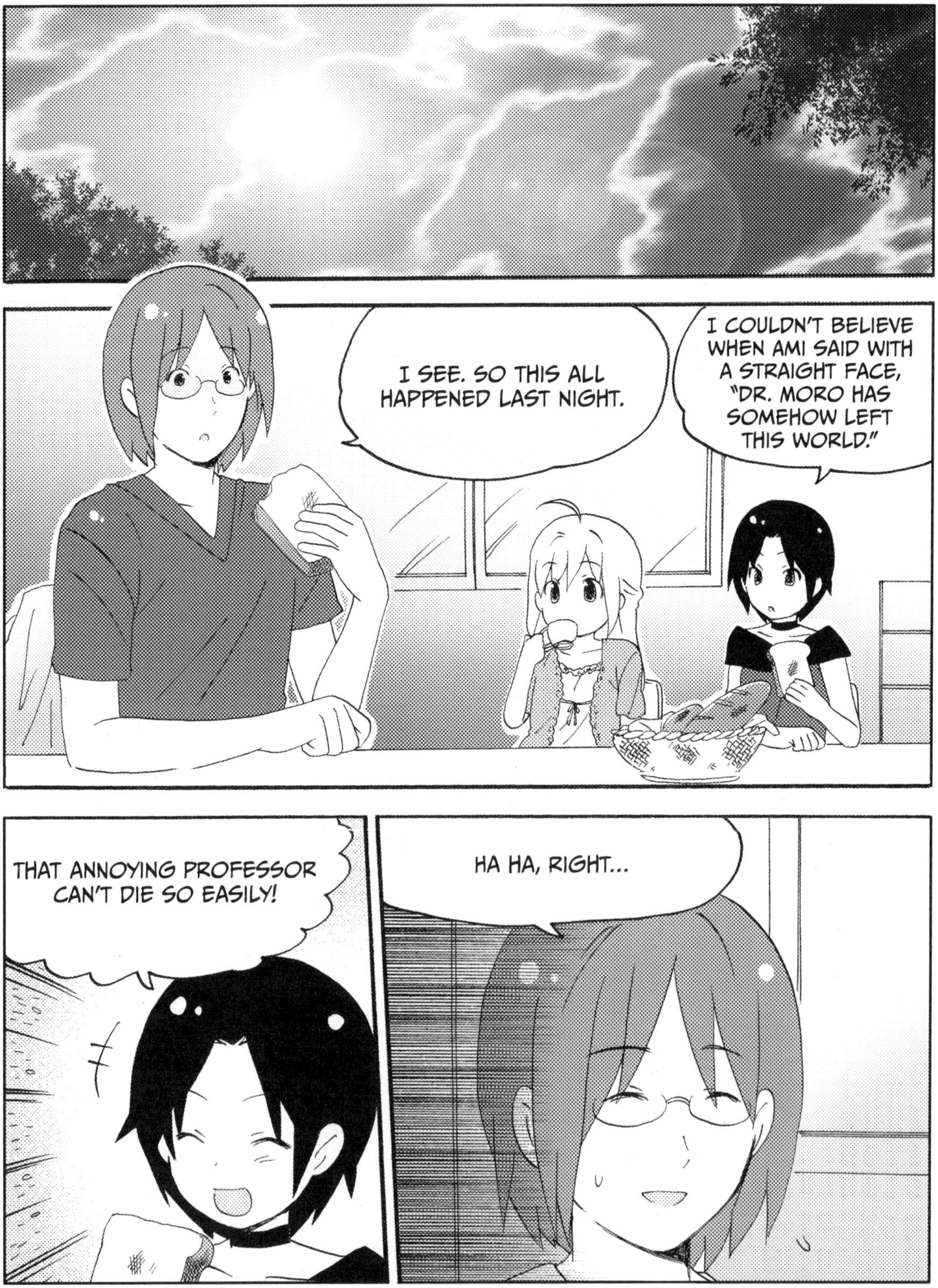
I SEE. SO THIS ALL HAPPENED LAST NIGHT.
I COULDN'T BELIEVE WHEN AMI SAID WITH A STRAIGHT FACE, "DR. MORO HAS SOMEHOW LEFT THIS WORLD."
THAT ANNOYING PROFESSOR CAN'T DIE SO EASILY!
HA HA, RIGHT...

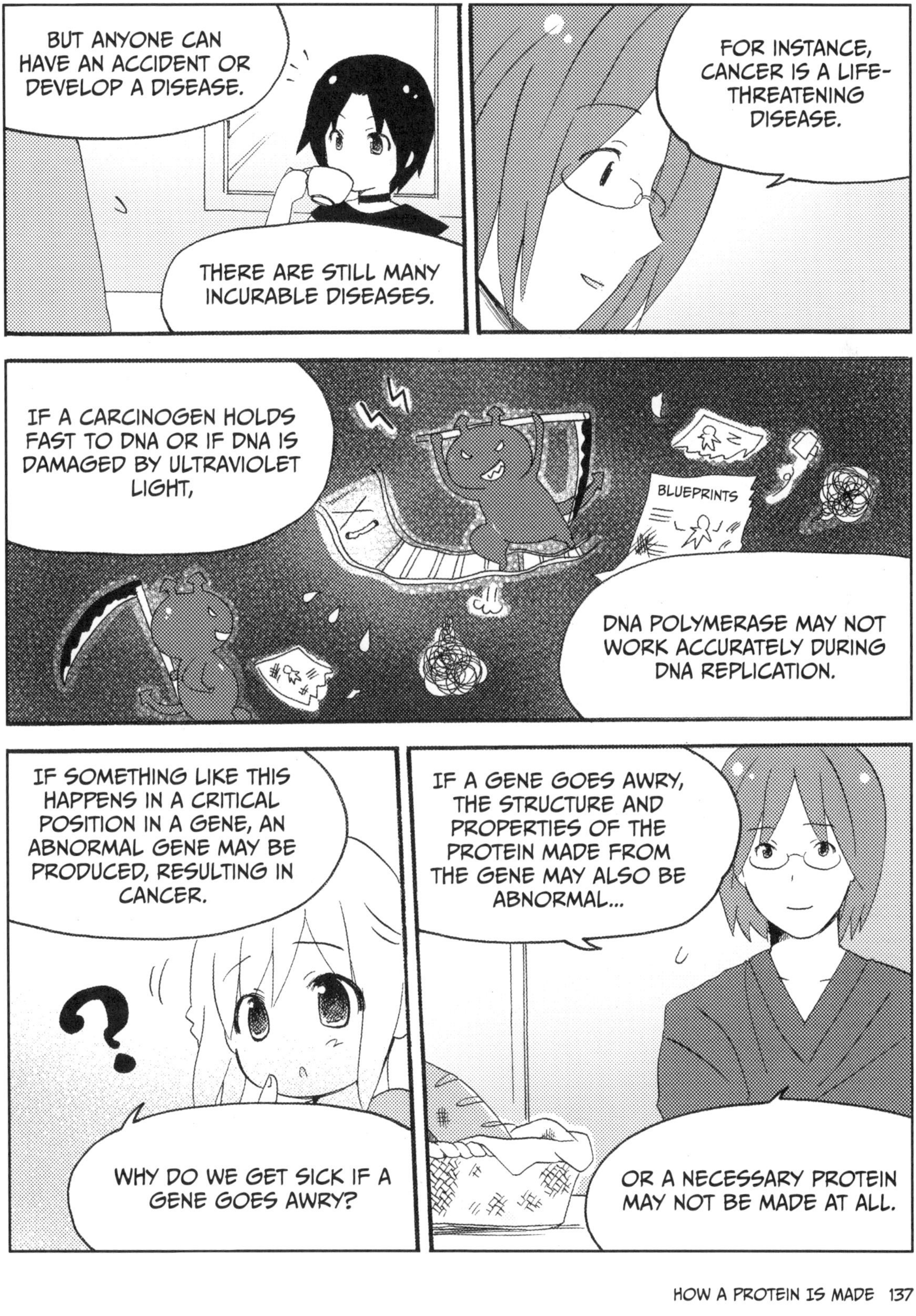
BUT ANYONE CAN HAVE AN ACCIDENT OR DEVELOP A DISEASE.
THERE ARE STILL MANY INCURABLE DISEASES.
FOR INSTANCE, CANCER IS A LIFE-THREATENING DISEASE.
IF A CARCINOGEN HOLDS FAST TO DNA OR IF DNA IS DAMAGED BY ULTRAVIOLET LIGHT,
BLUEPRINTS
DNA POLYMERASE MAY NOT WORK ACCURATELY DURING DNA REPLICATION.
IF SOMETHING LIKE THIS HAPPENS IN A CRITICAL POSITION IN A GENE, AN ABNORMAL GENE MAY BE PRODUCED, RESULTING IN CANCER.
WHY DO WE GET SICK IF A GENE GOES AWRY?
IF A GENE GOES AWRY, THE STRUCTURE AND PROPERTIES OF THE PROTEIN MADE FROM THE GENE MAY ALSO BE ABNORMAL...
OR A NECESSARY PROTEIN MAY NOT BE MADE AT ALL.

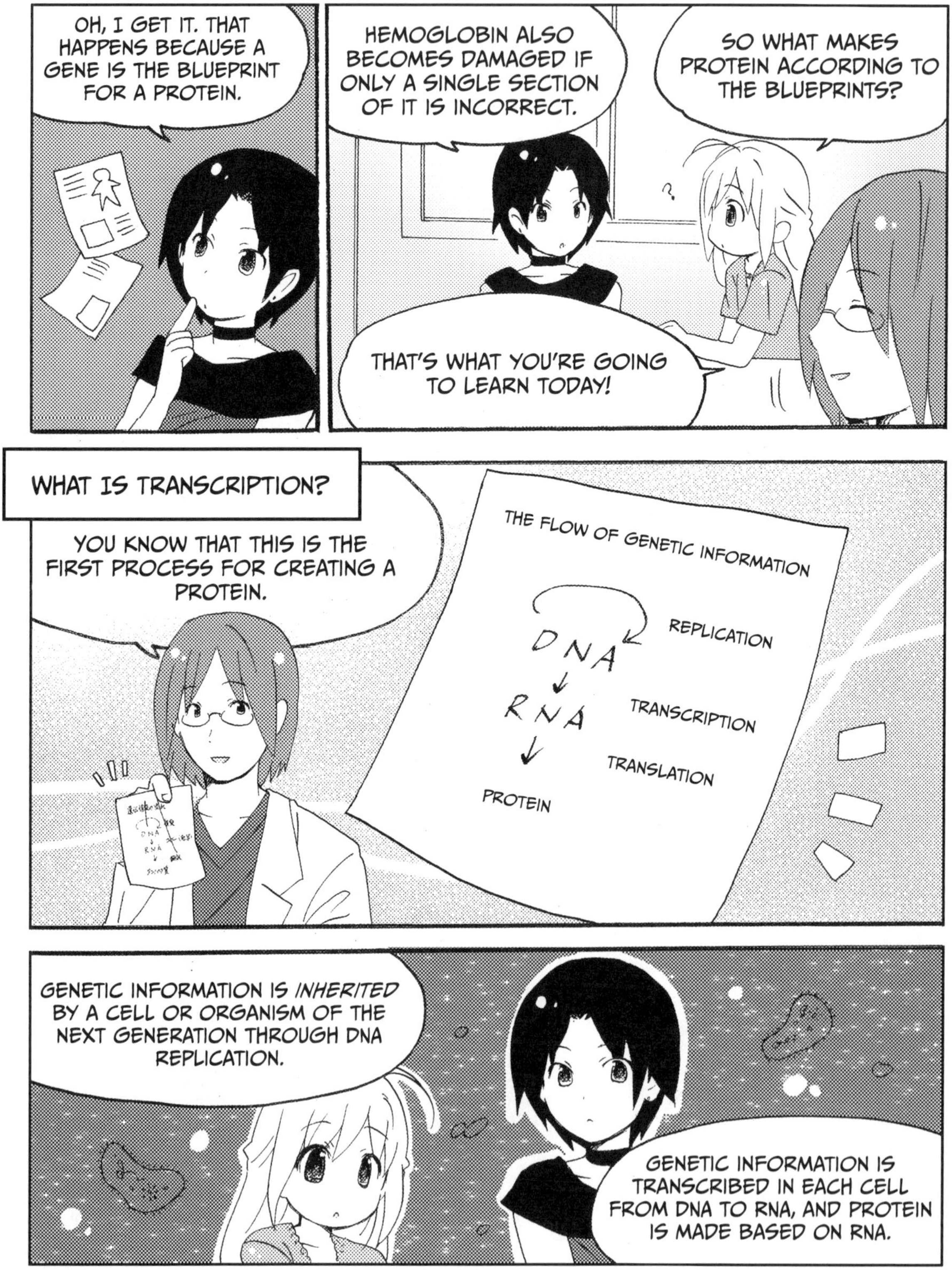
OH, I GET IT. THAT HAPPENS BECAUSE A GENE IS THE BLUEPRINT FOR A PROTEIN.
HEMOGLOBIN ALSO BECOMES DAMAGED IF ONLY A SINGLE SECTION OF IT IS INCORRECT.
SO WHAT MAKES PROTEIN ACCORDING TO THE BLUEPRINTS?
THAT'S WHAT YOU'RE GOING TO LEARN TODAY!
WHAT IS TRANSCRIPTION?
YOU KNOW THAT THIS IS THE FIRST PROCESS FOR CREATING A PROTEIN.
THE FLOW OF GENETIC INFORMATION
DNA
REPLICATION
RNA
TRANSCRIPTION
TRANSLATION
PROTEIN
GENETIC INFORMATION IS *INHERITED* BY A CELL OR ORGANISM OF THE NEXT GENERATION THROUGH DNA REPLICATION.
GENETIC INFORMATION IS TRANSCRIBED IN EACH CELL FROM DNA TO RNA, AND PROTEIN IS MADE BASED ON RNA.

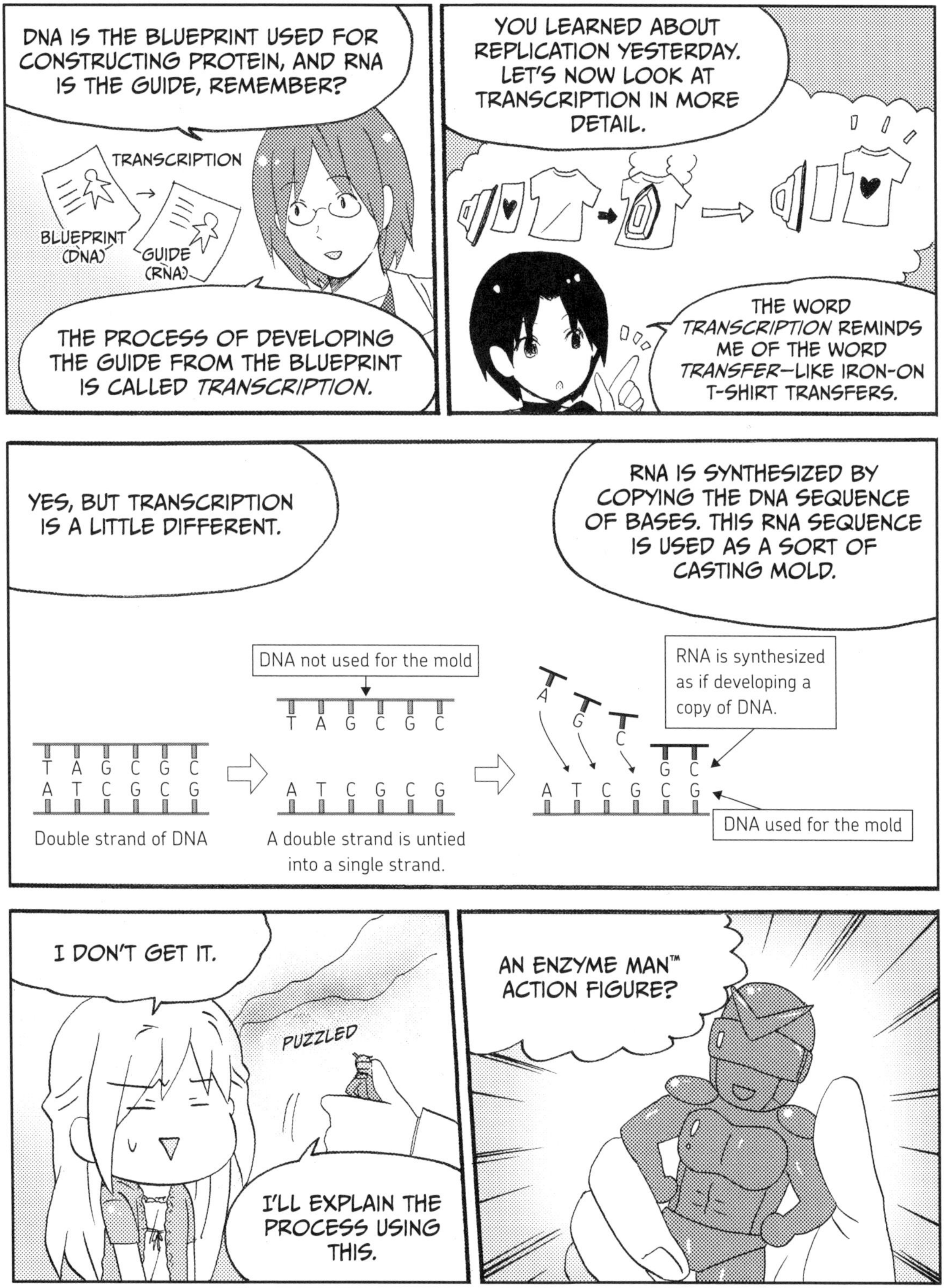
DNA IS THE BLUEPRINT USED FOR CONSTRUCTING PROTEIN, AND RNA IS THE GUIDE, REMEMBER?
TRANSCRIPTION
BLUEPRINT (DNA)
GUIDE (RNA)
THE PROCESS OF DEVELOPING THE GUIDE FROM THE BLUEPRINT IS CALLED *TRANSCRIPTION*.
YOU LEARNED ABOUT REPLICATION YESTERDAY. LET'S NOW LOOK AT TRANSCRIPTION IN MORE DETAIL.
THE WORD *TRANSCRIPTION* REMINDS ME OF THE WORD *TRANSFER*—LIKE IRON-ON T-SHIRT TRANSFERS.
YES, BUT TRANSCRIPTION IS A LITTLE DIFFERENT.
RNA IS SYNTHESIZED BY COPYING THE DNA SEQUENCE OF BASES. THIS RNA SEQUENCE IS USED AS A SORT OF CASTING MOLD.
T A G C G C
A T C G C G
Double strand of DNA
DNA not used for the mold
T A G C G C
A T C G C G
A double strand is untied into a single strand.
A G C
G C
A T C G C G
RNA is synthesized as if developing a copy of DNA.
DNA used for the mold
I DON'T GET IT.
PUZZLED
I'LL EXPLAIN THE PROCESS USING THIS.
AN ENZYME MAN™ ACTION FIGURE?

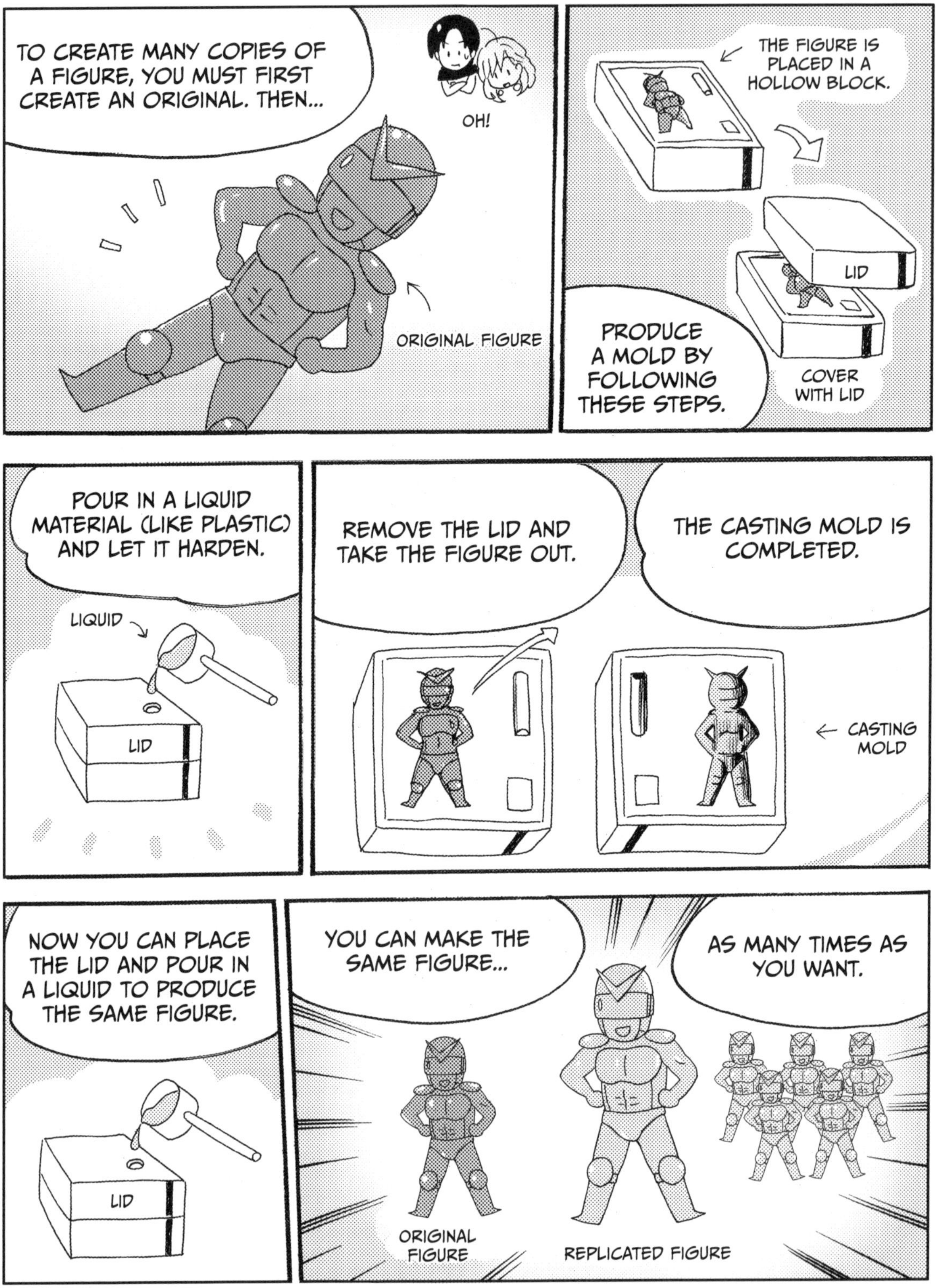
TO CREATE MANY COPIES OF A FIGURE, YOU MUST FIRST CREATE AN ORIGINAL. THEN...
OH!
ORIGINAL FIGURE
THE FIGURE IS PLACED IN A HOLLOW BLOCK.
LID
PRODUCE A MOLD BY FOLLOWING THESE STEPS.
COVER WITH LID
POUR IN A LIQUID MATERIAL (LIKE PLASTIC) AND LET IT HARDEN.
LIQUID
LID
REMOVE THE LID AND TAKE THE FIGURE OUT.
THE CASTING MOLD IS COMPLETED.
CASTING MOLD
NOW YOU CAN PLACE THE LID AND POUR IN A LIQUID TO PRODUCE THE SAME FIGURE.
LID
YOU CAN MAKE THE SAME FIGURE...
AS MANY TIMES AS YOU WANT.
ORIGINAL FIGURE
REPLICATED FIGURE

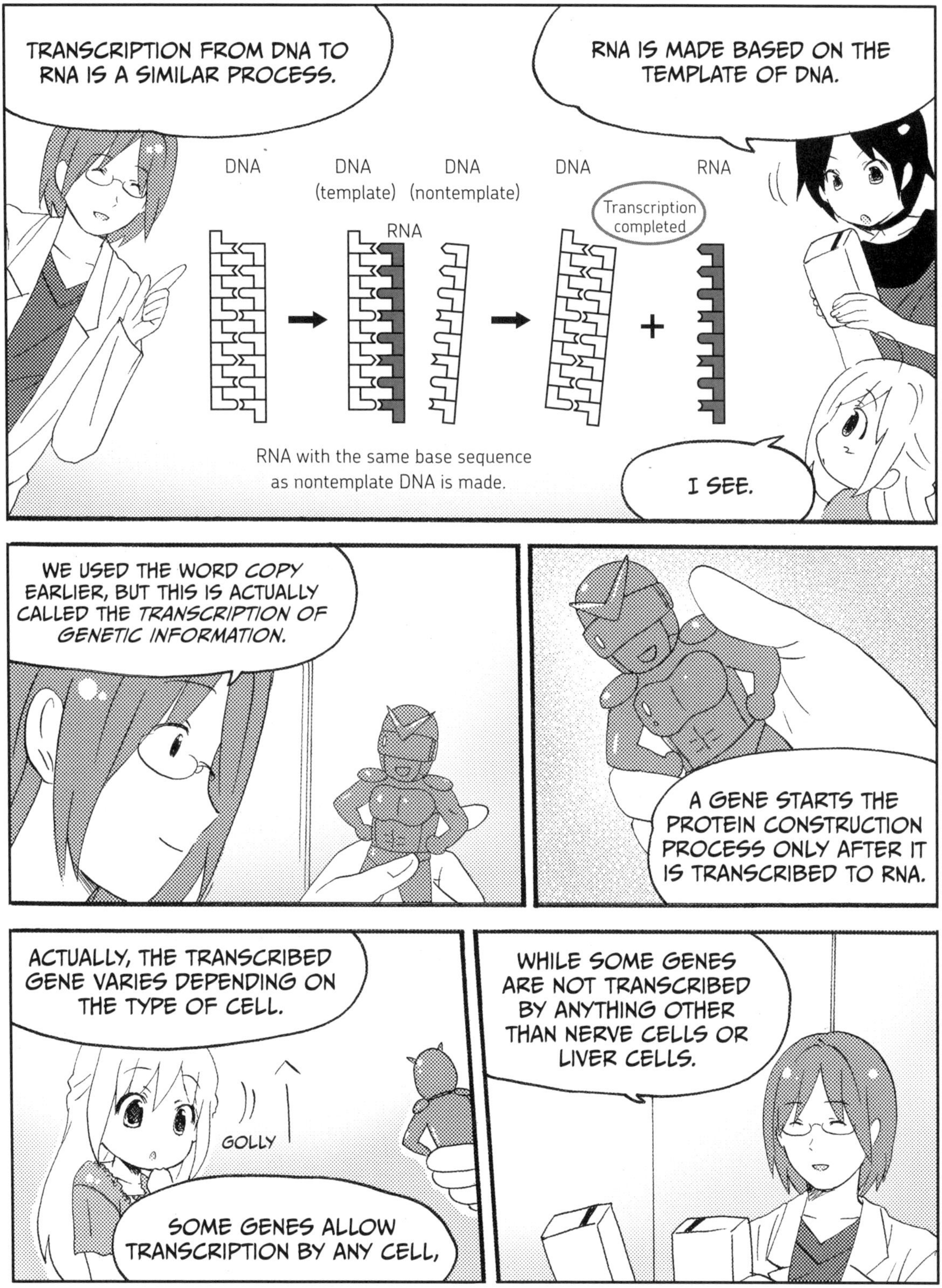
TRANSCRIPTION FROM DNA TO RNA IS A SIMILAR PROCESS.
RNA IS MADE BASED ON THE TEMPLATE OF DNA.
DNA
DNA (template)
RNA
DNA (nontemplate)
DNA
Transcription completed
RNA
+
RNA with the same base sequence as nontemplate DNA is made.
I SEE.
WE USED THE WORD *COPY* EARLIER, BUT THIS IS ACTUALLY CALLED THE *TRANSCRIPTION OF GENETIC INFORMATION.*
A GENE STARTS THE PROTEIN CONSTRUCTION PROCESS ONLY AFTER IT IS TRANSCRIBED TO RNA.
ACTUALLY, THE TRANSCRIBED GENE VARIES DEPENDING ON THE TYPE OF CELL.
GOLLY
SOME GENES ALLOW TRANSCRIPTION BY ANY CELL,
WHILE SOME GENES ARE NOT TRANSCRIBED BY ANYTHING OTHER THAN NERVE CELLS OR LIVER CELLS.

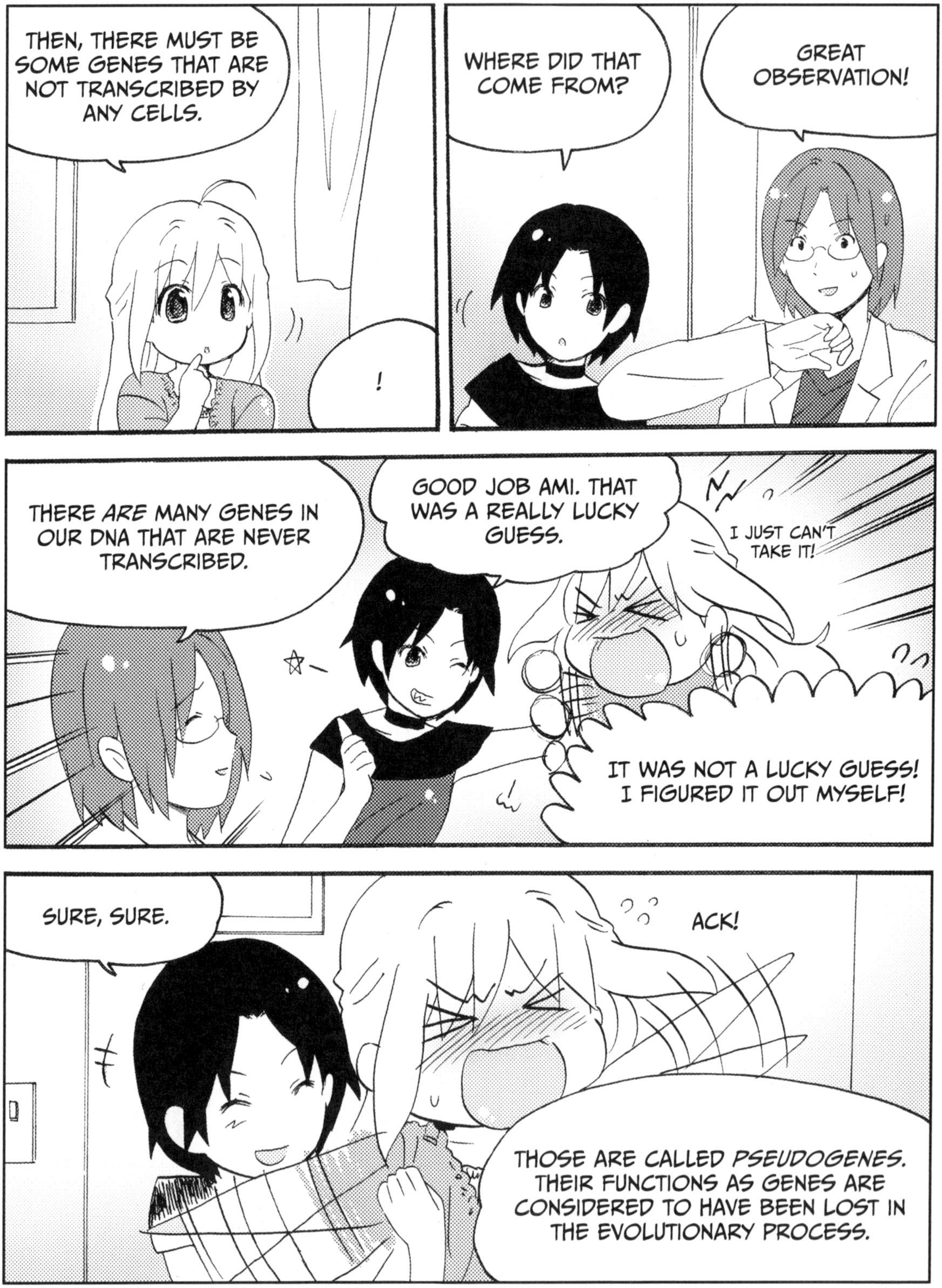
THEN, THERE MUST BE SOME GENES THAT ARE NOT TRANSCRIBED BY ANY CELLS.
!
WHERE DID THAT COME FROM?
GREAT OBSERVATION!
THERE *ARE* MANY GENES IN OUR DNA THAT ARE NEVER TRANSCRIBED.
GOOD JOB AMI. THAT WAS A REALLY LUCKY GUESS.
I JUST CAN'T TAKE IT!
IT WAS NOT A LUCKY GUESS! I FIGURED IT OUT MYSELF!
SURE, SURE.
ACK!
THOSE ARE CALLED *PSEUDOGENES*. THEIR FUNCTIONS AS GENES ARE CONSIDERED TO HAVE BEEN LOST IN THE EVOLUTIONARY PROCESS.

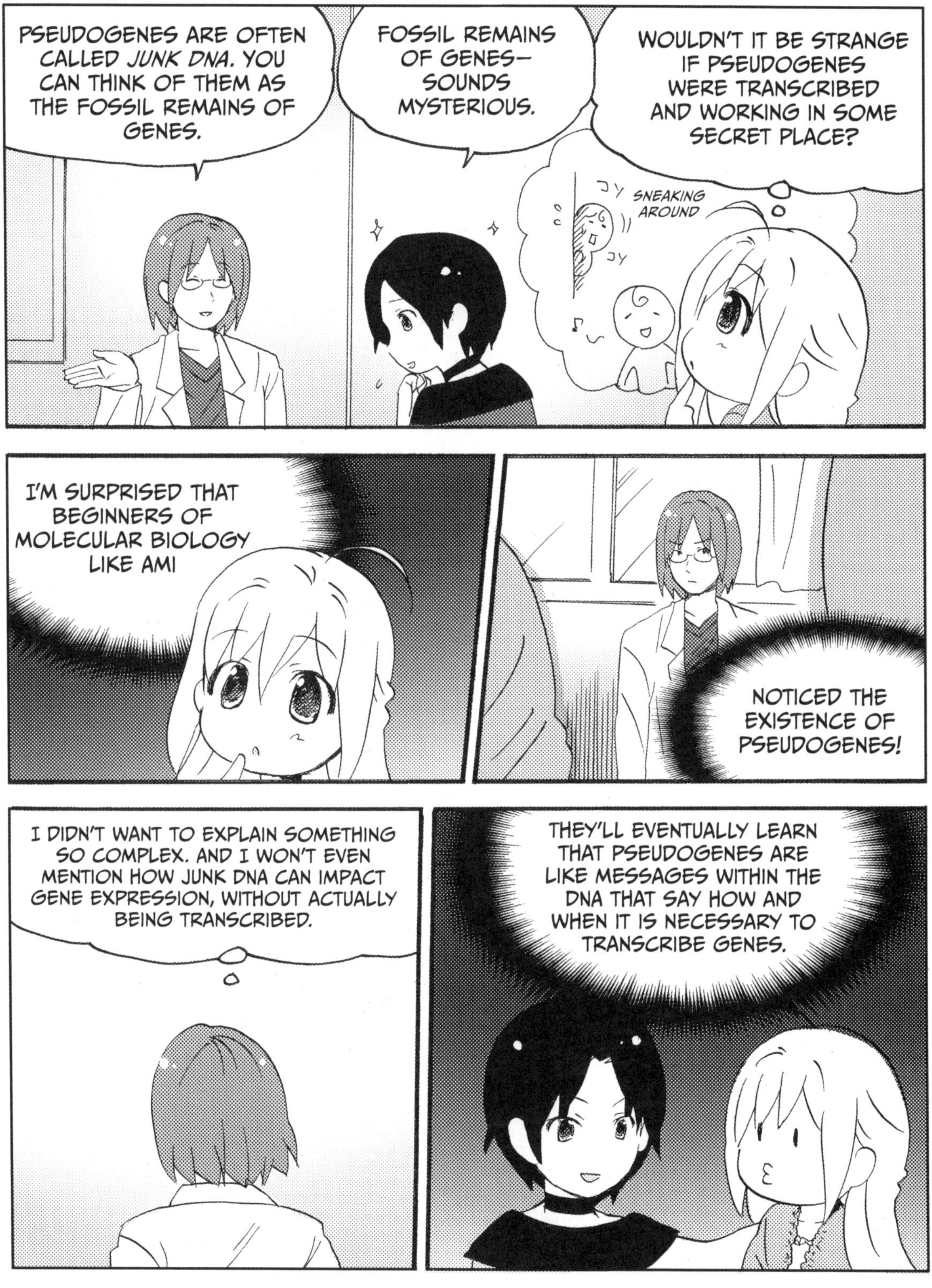
PSEUDOGENES ARE OFTEN CALLED *JUNK DNA*. YOU CAN THINK OF THEM AS THE FOSSIL REMAINS OF GENES.
FOSSIL REMAINS OF GENES— SOUNDS MYSTERIOUS.
WOULDN'T IT BE STRANGE IF PSEUDOGENES WERE TRANSCRIBED AND WORKING IN SOME SECRET PLACE?
SNEAKING AROUND
I'M SURPRISED THAT BEGINNERS OF MOLECULAR BIOLOGY LIKE AMI
NOTICED THE EXISTENCE OF PSEUDOGENES!
I DIDN'T WANT TO EXPLAIN SOMETHING SO COMPLEX. AND I WON'T EVEN MENTION HOW JUNK DNA CAN IMPACT GENE EXPRESSION, WITHOUT ACTUALLY BEING TRANSCRIBED.
THEY'LL EVENTUALLY LEARN THAT PSEUDOGENES ARE LIKE MESSAGES WITHIN THE DNA THAT SAY HOW AND WHEN IT IS NECESSARY TO TRANSCRIBE GENES.

CHROMATIN AND TRANSCRIPTION

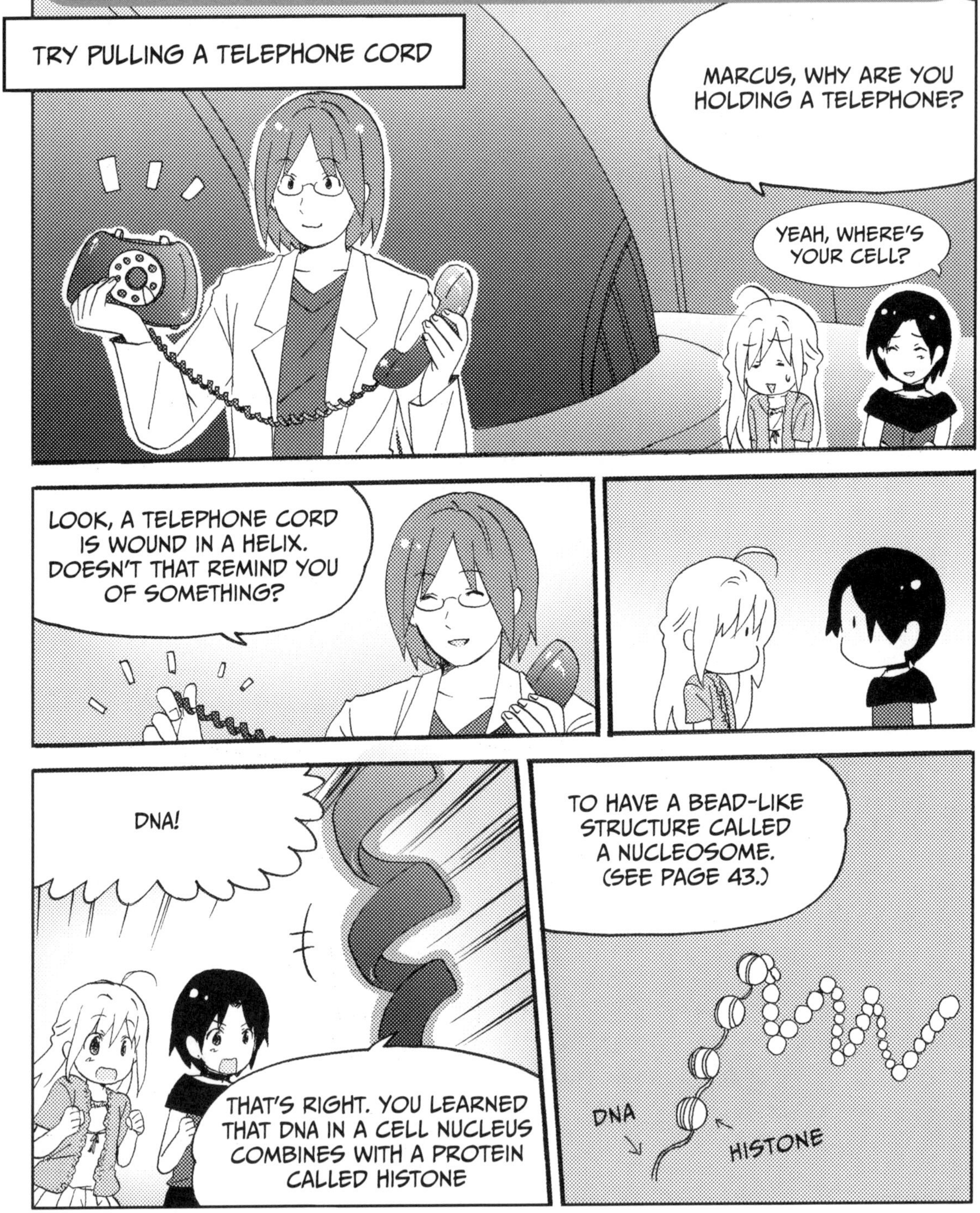

TRY PULLING A TELEPHONE CORD
MARCUS, WHY ARE YOU HOLDING A TELEPHONE?
YEAH, WHERE'S YOUR CELL?
LOOK, A TELEPHONE CORD IS WOUND IN A HELIX. DOESN'T THAT REMIND YOU OF SOMETHING?
DNA!
THAT'S RIGHT. YOU LEARNED THAT DNA IN A CELL NUCLEUS COMBINES WITH A PROTEIN CALLED HISTONE
TO HAVE A BEAD-LIKE STRUCTURE CALLED A NUCLEOSOME. (SEE PAGE 43.)
DNA
HISTONE

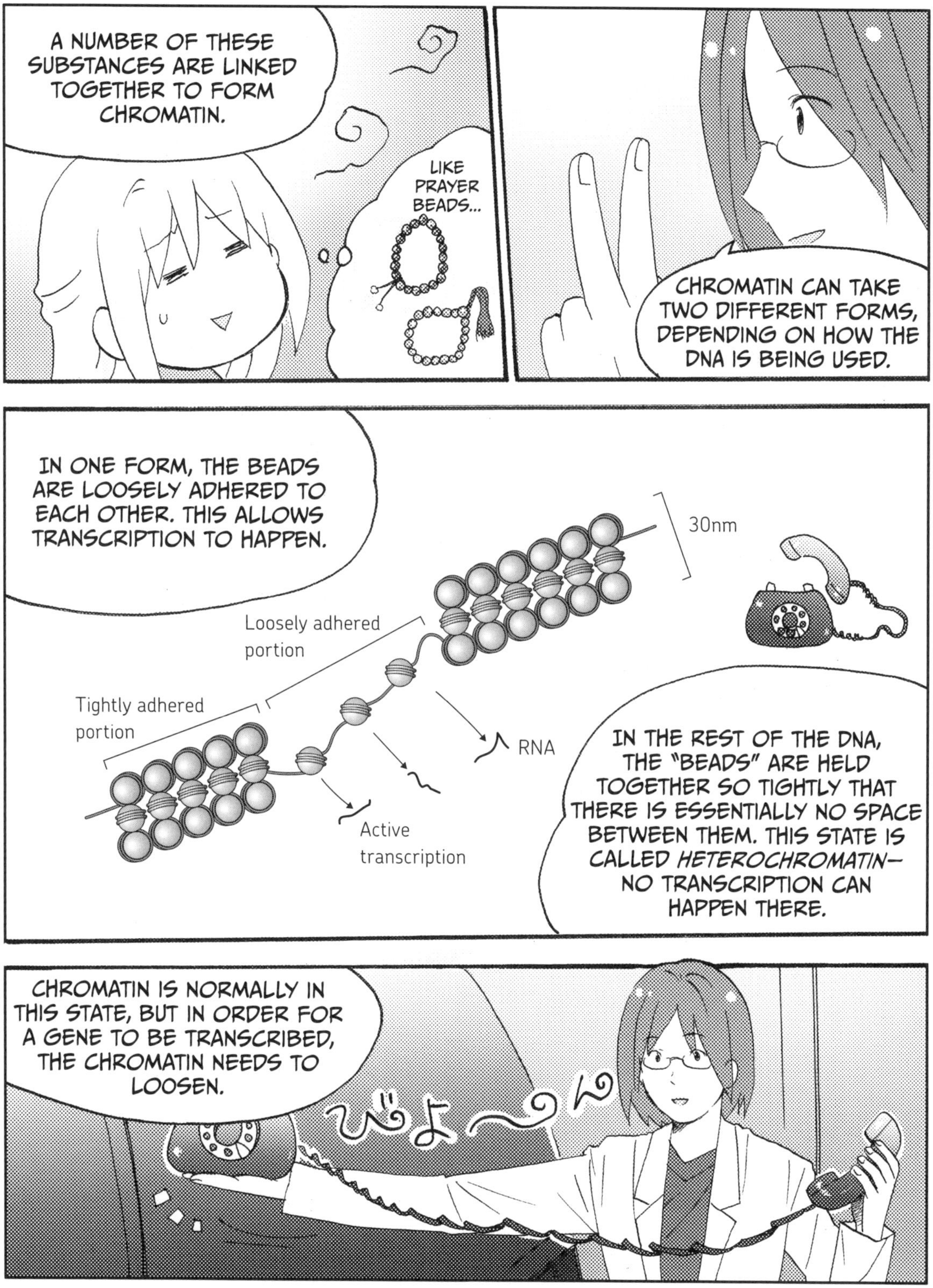
A NUMBER OF THESE SUBSTANCES ARE LINKED TOGETHER TO FORM CHROMATIN.
LIKE PRAYER BEADS...
CHROMATIN CAN TAKE TWO DIFFERENT FORMS, DEPENDING ON HOW THE DNA IS BEING USED.
IN ONE FORM, THE BEADS ARE LOOSELY ADHERED TO EACH OTHER. THIS ALLOWS TRANSCRIPTION TO HAPPEN.
30nm
Loosely adhered portion
Tightly adhered portion
RNA
Active transcription
IN THE REST OF THE DNA, THE "BEADS" ARE HELD TOGETHER SO TIGHTLY THAT THERE IS ESSENTIALLY NO SPACE BETWEEN THEM. THIS STATE IS CALLED *HETEROCHROMATIN*— NO TRANSCRIPTION CAN HAPPEN THERE.
CHROMATIN IS NORMALLY IN THIS STATE, BUT IN ORDER FOR A GENE TO BE TRANSCRIBED, THE CHROMATIN NEEDS TO LOOSEN.
びよ〜ん

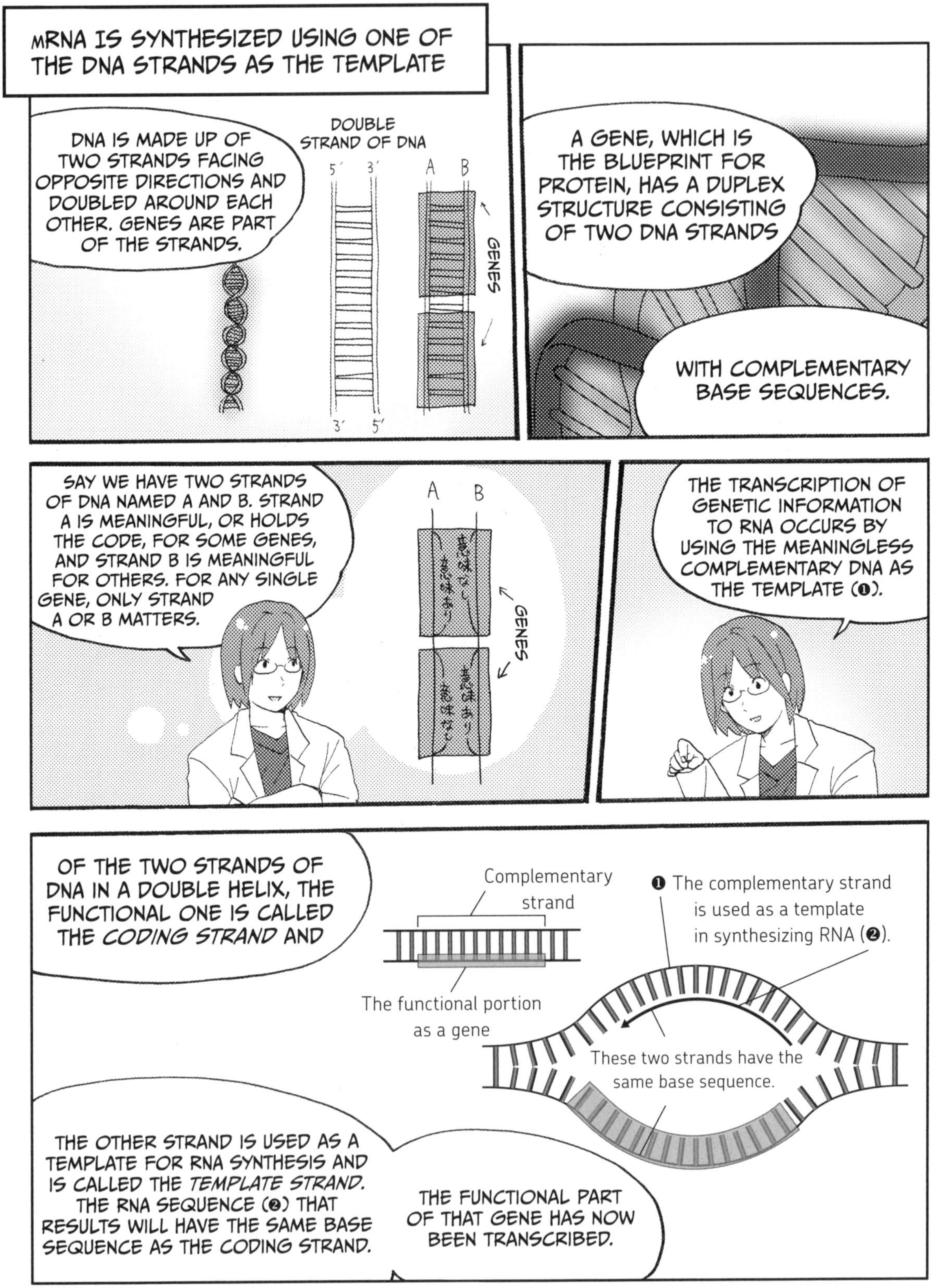
mRNA IS SYNTHESIZED USING ONE OF THE DNA STRANDS AS THE TEMPLATE
DNA IS MADE UP OF TWO STRANDS FACING OPPOSITE DIRECTIONS AND DOUBLED AROUND EACH OTHER. GENES ARE PART OF THE STRANDS.
DOUBLE STRAND OF DNA
5′ 3′
A B
GENES
3′ 5′
A GENE, WHICH IS THE BLUEPRINT FOR PROTEIN, HAS A DUPLEX STRUCTURE CONSISTING OF TWO DNA STRANDS
WITH COMPLEMENTARY BASE SEQUENCES.
SAY WE HAVE TWO STRANDS OF DNA NAMED A AND B. STRAND A IS MEANINGFUL, OR HOLDS THE CODE, FOR SOME GENES, AND STRAND B IS MEANINGFUL FOR OTHERS. FOR ANY SINGLE GENE, ONLY STRAND A OR B MATTERS.
A B
意味なし
意味あり
意味あり
意味なし
GENES
THE TRANSCRIPTION OF GENETIC INFORMATION TO RNA OCCURS BY USING THE MEANINGLESS COMPLEMENTARY DNA AS THE TEMPLATE (❶).
OF THE TWO STRANDS OF DNA IN A DOUBLE HELIX, THE FUNCTIONAL ONE IS CALLED THE *CODING STRAND* AND
Complementary strand
The functional portion as a gene
❶ The complementary strand is used as a template in synthesizing RNA (❷).
These two strands have the same base sequence.
THE OTHER STRAND IS USED AS A TEMPLATE FOR RNA SYNTHESIS AND IS CALLED THE *TEMPLATE STRAND*. THE RNA SEQUENCE (❷) THAT RESULTS WILL HAVE THE SAME BASE SEQUENCE AS THE CODING STRAND.
THE FUNCTIONAL PART OF THAT GENE HAS NOW BEEN TRANSCRIBED.

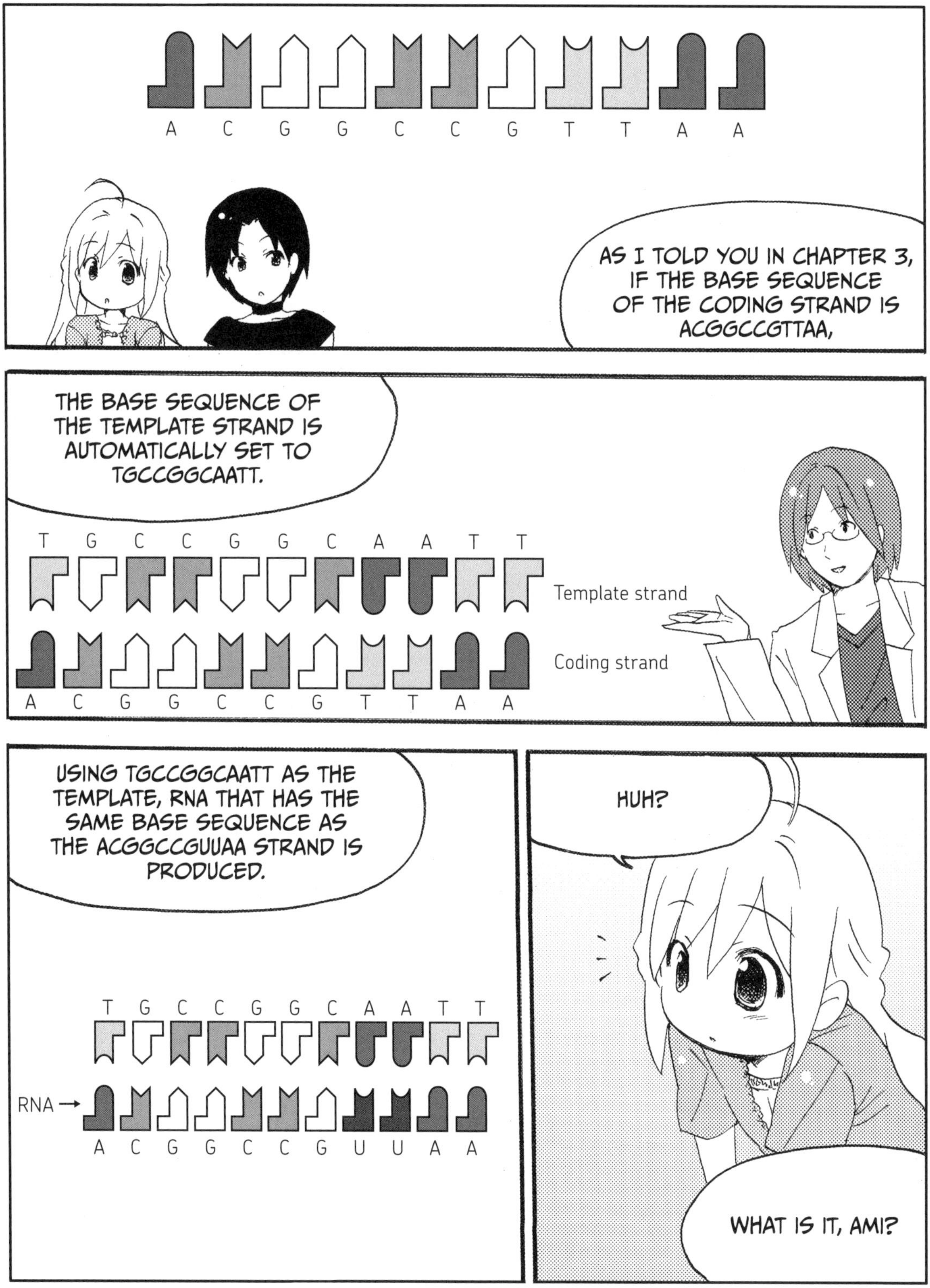
A C G G C C G T T A A
AS I TOLD YOU IN CHAPTER 3, IF THE BASE SEQUENCE OF THE CODING STRAND IS ACGGCCGTTAA,
THE BASE SEQUENCE OF THE TEMPLATE STRAND IS AUTOMATICALLY SET TO TGCCGGCAATT.
T G C C G G C A A T T
Template strand
Coding strand
A C G G C C G T T A A
USING TGCCGGCAATT AS THE TEMPLATE, RNA THAT HAS THE SAME BASE SEQUENCE AS THE ACGGCCGUUAA STRAND IS PRODUCED.
T G C C G G C A A T T
RNA →
A C G G C C G U U A A
HUH?
WHAT IS IT, AMI?

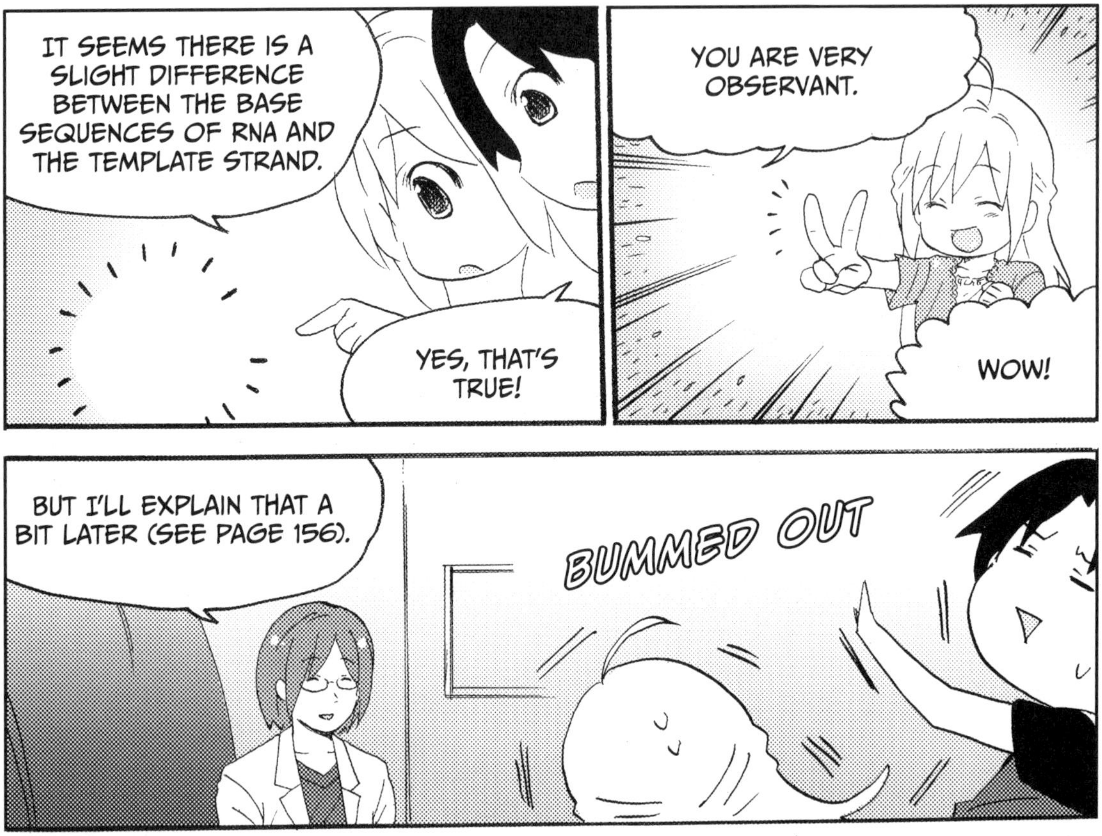
IT SEEMS THERE IS A SLIGHT DIFFERENCE BETWEEN THE BASE SEQUENCES OF RNA AND THE TEMPLATE STRAND.
YES, THAT'S TRUE!
YOU ARE VERY OBSERVANT.
WOW!
BUT I'LL EXPLAIN THAT A BIT LATER (SEE PAGE 156).
BUMMED OUT

RNA POLYMERASE COPIES GENETIC INFORMATION

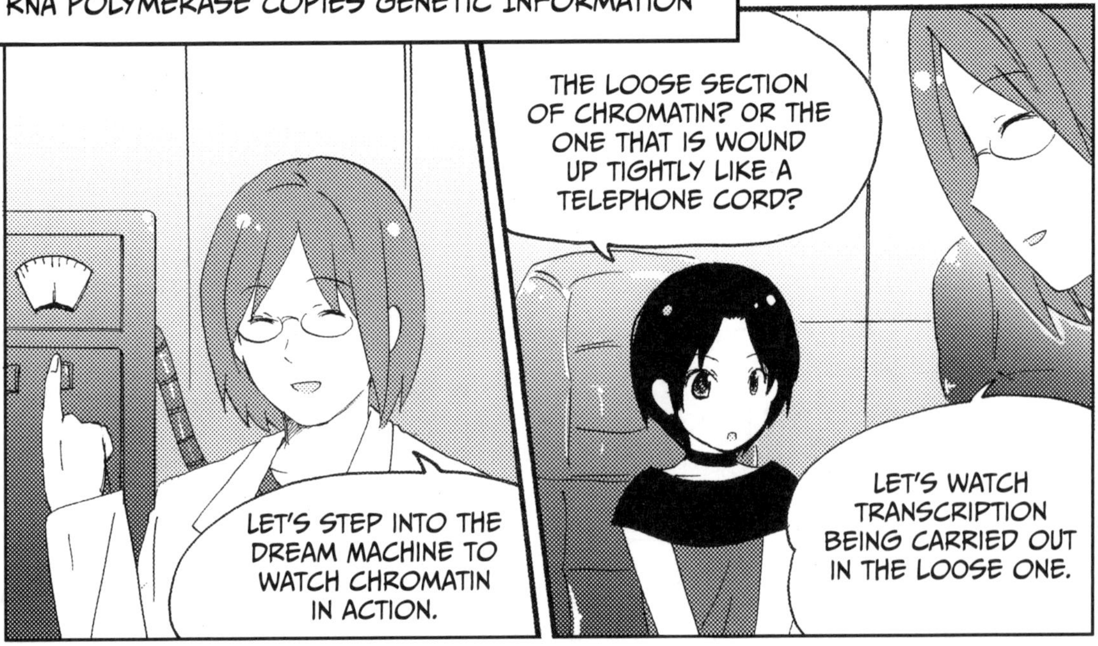
LET'S STEP INTO THE DREAM MACHINE TO WATCH CHROMATIN IN ACTION.
THE LOOSE SECTION OF CHROMATIN? OR THE ONE THAT IS WOUND UP TIGHTLY LIKE A TELEPHONE CORD?
LET'S WATCH TRANSCRIPTION BEING CARRIED OUT IN THE LOOSE ONE.

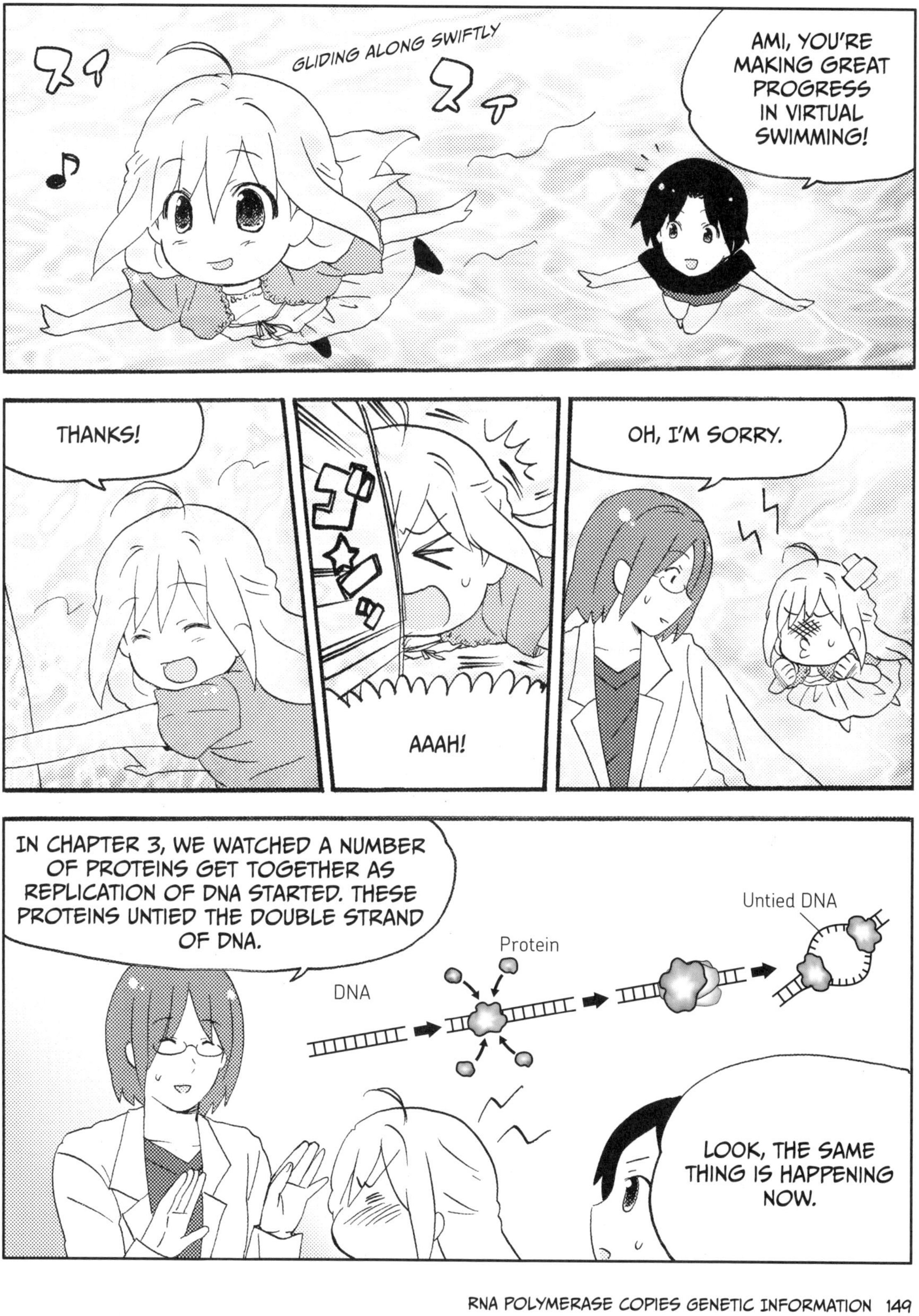

スイ
GLIDING ALONG SWIFTLY
スイ
AMI, YOU'RE MAKING GREAT PROGRESS IN VIRTUAL SWIMMING!
THANKS!
AAAH!
OH, I'M SORRY.
IN CHAPTER 3, WE WATCHED A NUMBER OF PROTEINS GET TOGETHER AS REPLICATION OF DNA STARTED. THESE PROTEINS UNTIED THE DOUBLE STRAND OF DNA.
DNA
Protein
Untied DNA
LOOK, THE SAME THING IS HAPPENING NOW.

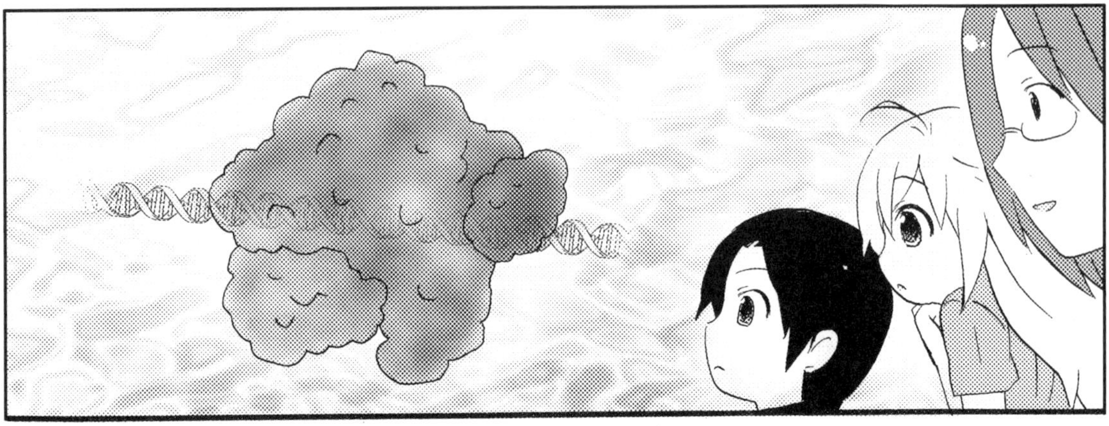

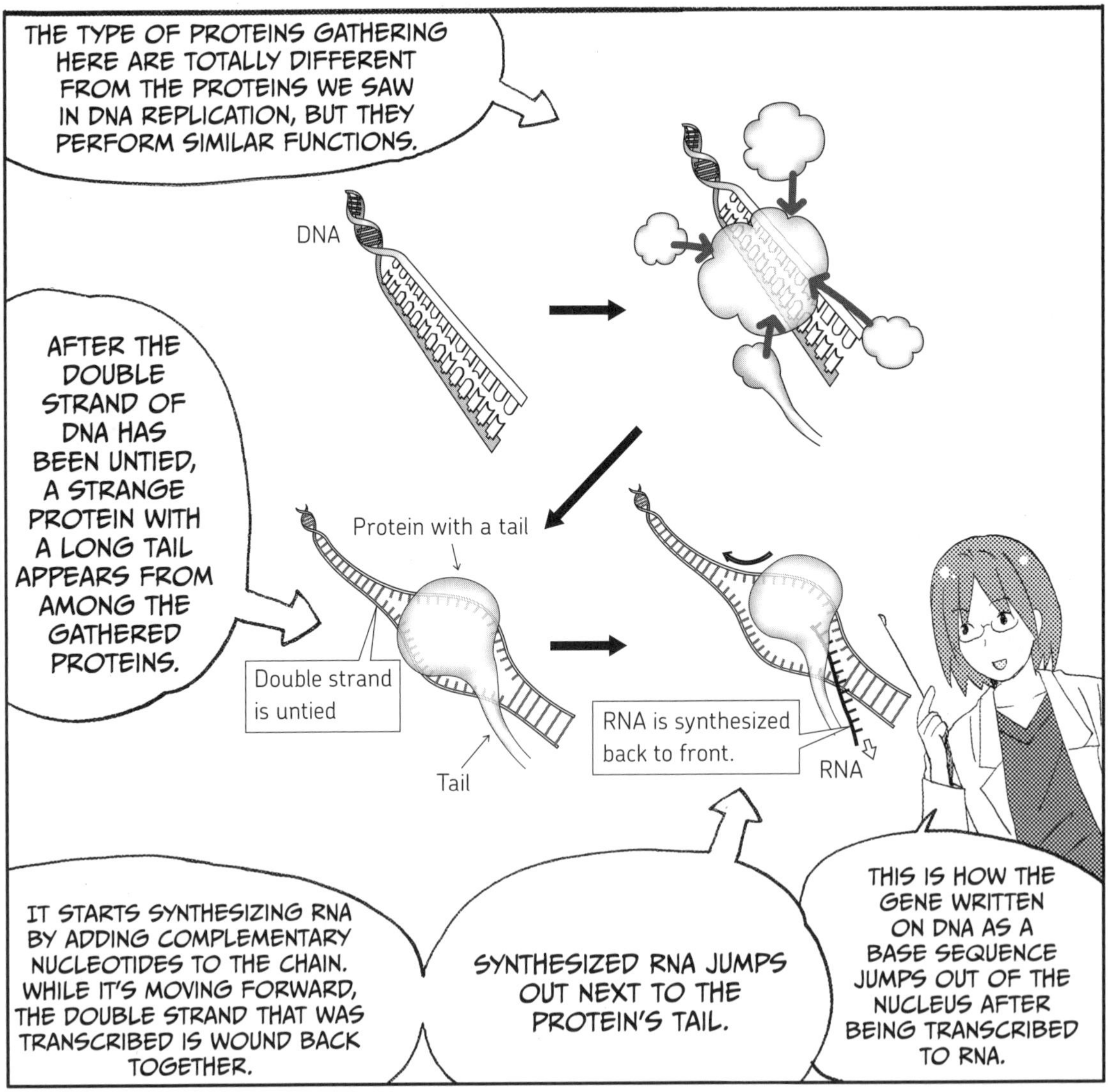
THE TYPE OF PROTEINS GATHERING HERE ARE TOTALLY DIFFERENT FROM THE PROTEINS WE SAW IN DNA REPLICATION, BUT THEY PERFORM SIMILAR FUNCTIONS.
DNA
AFTER THE DOUBLE STRAND OF DNA HAS BEEN UNTIED, A STRANGE PROTEIN WITH A LONG TAIL APPEARS FROM AMONG THE GATHERED PROTEINS.
Protein with a tail
Double strand is untied
Tail
RNA is synthesized back to front.
RNA
IT STARTS SYNTHESIZING RNA BY ADDING COMPLEMENTARY NUCLEOTIDES TO THE CHAIN. WHILE IT'S MOVING FORWARD, THE DOUBLE STRAND THAT WAS TRANSCRIBED IS WOUND BACK TOGETHER.
SYNTHESIZED RNA JUMPS OUT NEXT TO THE PROTEIN'S TAIL.
THIS IS HOW THE GENE WRITTEN ON DNA AS A BASE SEQUENCE JUMPS OUT OF THE NUCLEUS AFTER BEING TRANSCRIBED TO RNA.

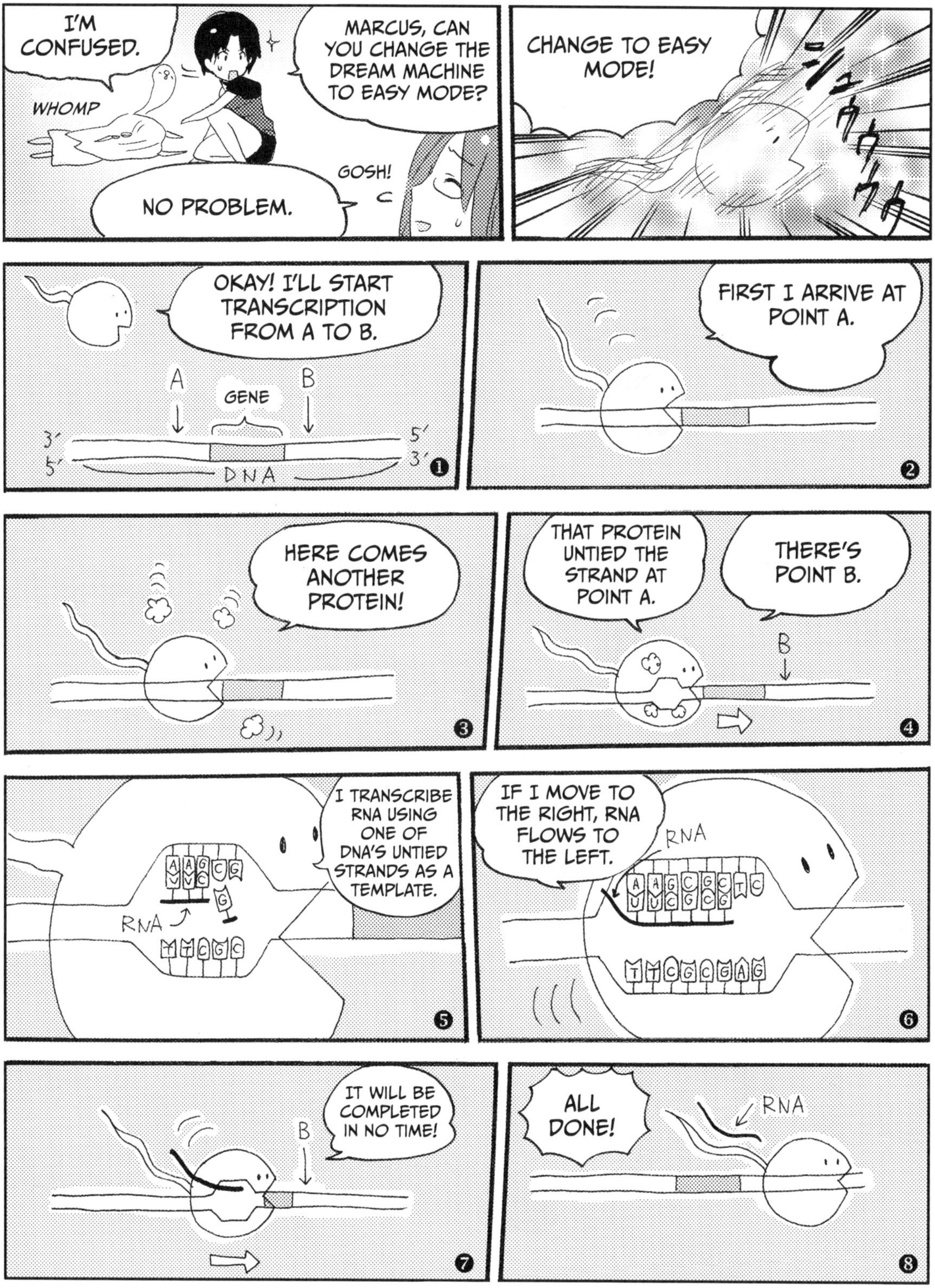
I'M CONFUSED.
WHOMP
MARCUS, CAN YOU CHANGE THE DREAM MACHINE TO EASY MODE?
NO PROBLEM.
GOSH!
CHANGE TO EASY MODE!
OKAY! I'LL START TRANSCRIPTION FROM A TO B.
A
GENE
B
3′
5′
5′
3′
DNA
FIRST I ARRIVE AT POINT A.
HERE COMES ANOTHER PROTEIN!
THAT PROTEIN UNTIED THE STRAND AT POINT A.
THERE'S POINT B.
B
I TRANSCRIBE RNA USING ONE OF DNA'S UNTIED STRANDS AS A TEMPLATE.
RNA
IF I MOVE TO THE RIGHT, RNA FLOWS TO THE LEFT.
RNA
IT WILL BE COMPLETED IN NO TIME!
B
ALL DONE!
RNA

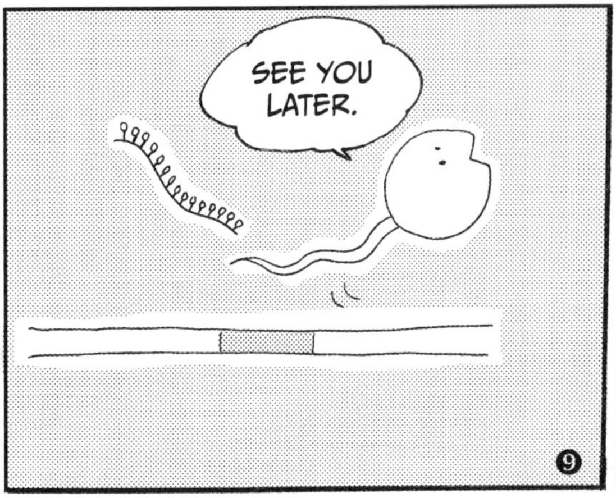
SEE YOU LATER.

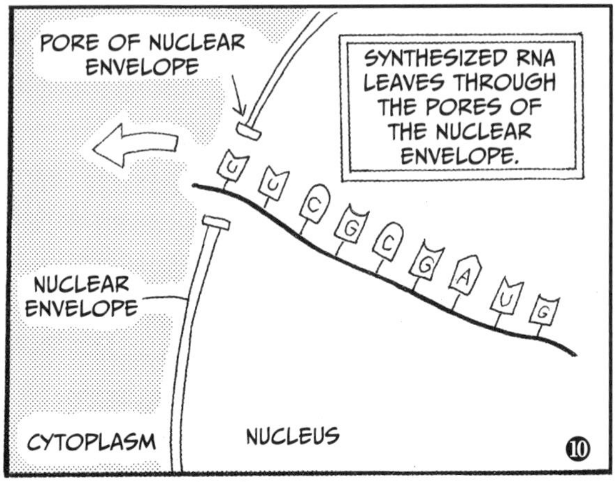
PORE OF NUCLEAR ENVELOPE
SYNTHESIZED RNA LEAVES THROUGH THE PORES OF THE NUCLEAR ENVELOPE.
U U C G C G A U G
NUCLEAR ENVELOPE
CYTOPLASM
NUCLEUS

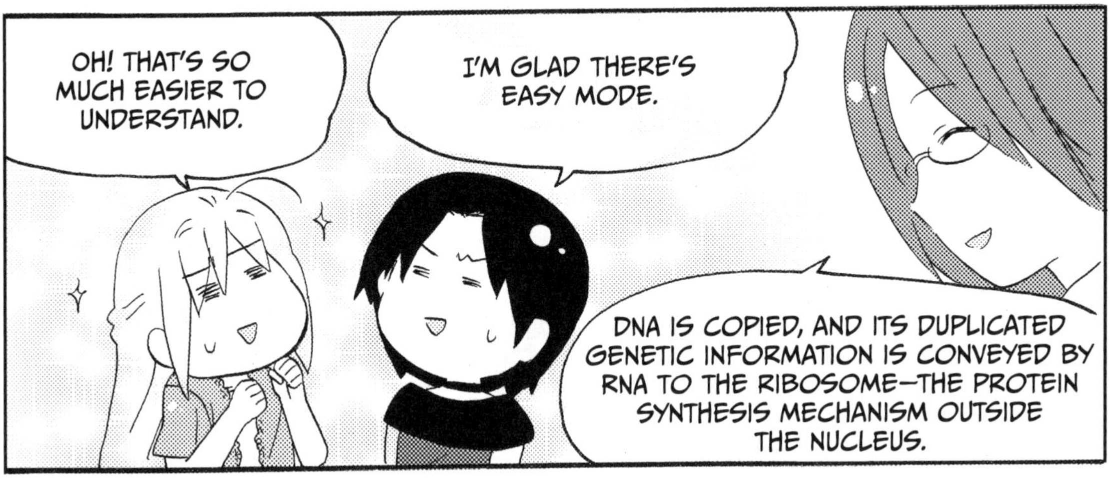
OH! THAT'S SO MUCH EASIER TO UNDERSTAND.
I'M GLAD THERE'S EASY MODE.
DNA IS COPIED, AND ITS DUPLICATED GENETIC INFORMATION IS CONVEYED BY RNA TO THE RIBOSOME—THE PROTEIN SYNTHESIS MECHANISM OUTSIDE THE NUCLEUS.

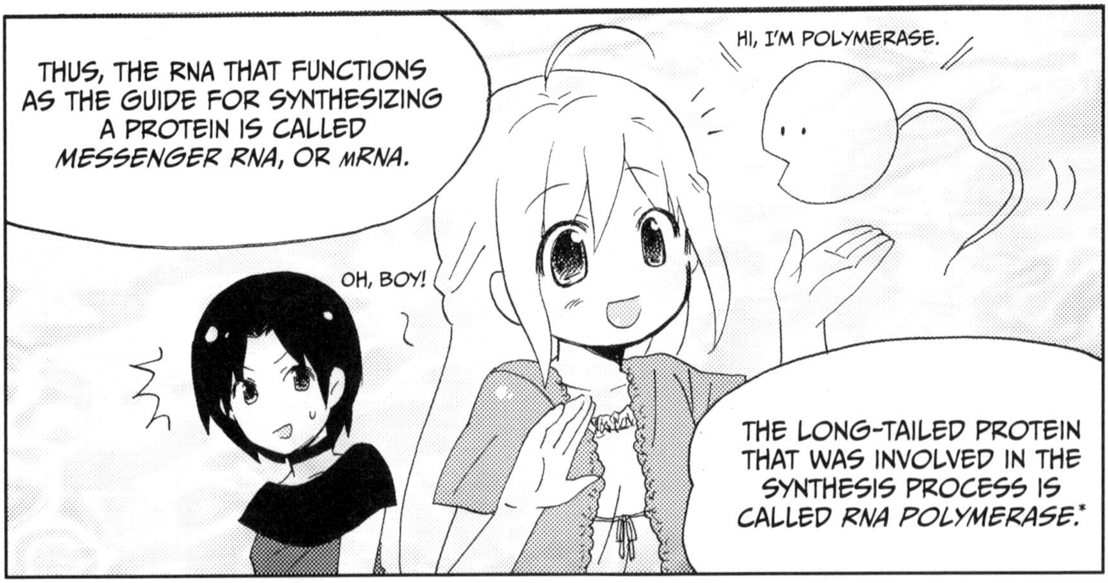
THUS, THE RNA THAT FUNCTIONS AS THE GUIDE FOR SYNTHESIZING A PROTEIN IS CALLED *MESSENGER RNA*, OR *mRNA*.
HI, I'M POLYMERASE.
OH, BOY!
THE LONG-TAILED PROTEIN THAT WAS INVOLVED IN THE SYNTHESIS PROCESS IS CALLED *RNA POLYMERASE*.*

* TO BE EXACT, THE PROTEIN IN EUKARYOTIC ORGANISMS IS CALLED *RNA POLYMERASE II*.

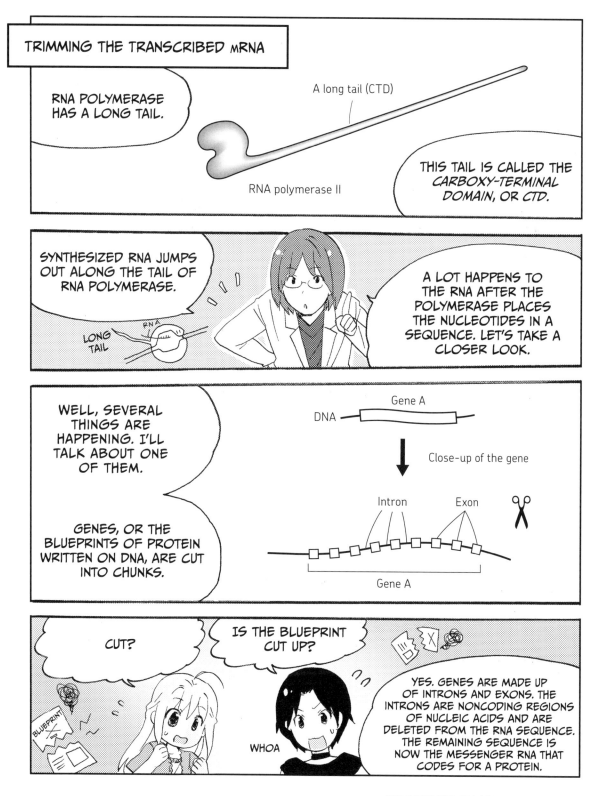

TRIMMING THE TRANSCRIBED mRNA

RNA POLYMERASE HAS A LONG TAIL.

A long tail (CTD)

RNA polymerase II

THIS TAIL IS CALLED THE *CARBOXY-TERMINAL DOMAIN, OR CTD.*

SYNTHESIZED RNA JUMPS OUT ALONG THE TAIL OF RNA POLYMERASE.

LONG TAIL

RNA

A LOT HAPPENS TO THE RNA AFTER THE POLYMERASE PLACES THE NUCLEOTIDES IN A SEQUENCE. LET'S TAKE A CLOSER LOOK.

WELL, SEVERAL THINGS ARE HAPPENING. I'LL TALK ABOUT ONE OF THEM.

GENES, OR THE BLUEPRINTS OF PROTEIN WRITTEN ON DNA, ARE CUT INTO CHUNKS.

Gene A

DNA

Close-up of the gene

Intron

Exon

Gene A

CUT?

IS THE BLUEPRINT CUT UP?

BLUEPRINT

WHOA

YES. GENES ARE MADE UP OF INTRONS AND EXONS. THE INTRONS ARE NONCODING REGIONS OF NUCLEIC ACIDS AND ARE DELETED FROM THE RNA SEQUENCE. THE REMAINING SEQUENCE IS NOW THE MESSENGER RNA THAT CODES FOR A PROTEIN.

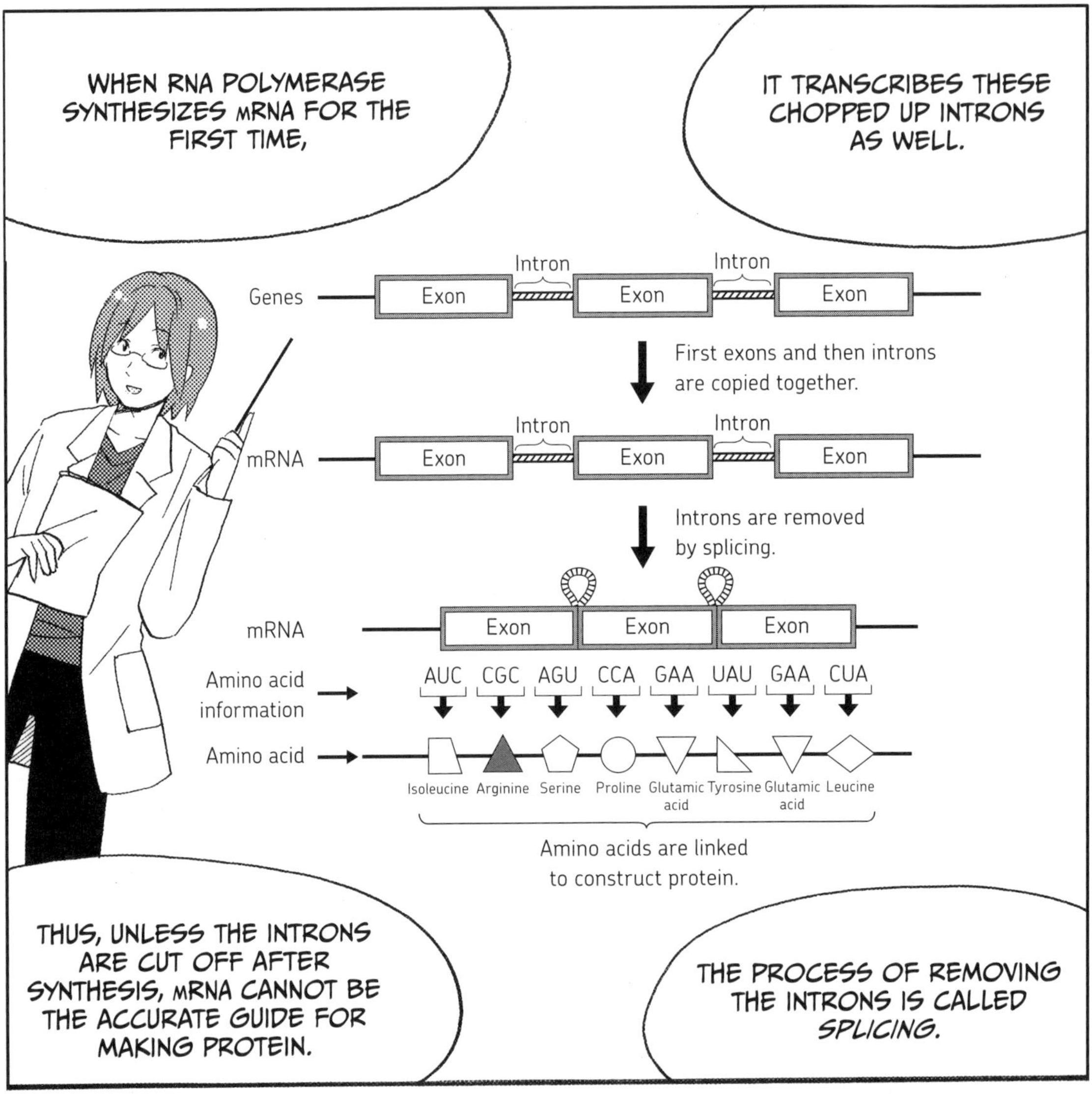
WHEN RNA POLYMERASE SYNTHESIZES mRNA FOR THE FIRST TIME,
IT TRANSCRIBES THESE CHOPPED UP INTRONS AS WELL.
Genes
Exon
Intron
Exon
Intron
Exon
First exons and then introns are copied together.
mRNA
Exon
Intron
Exon
Intron
Exon
Introns are removed by splicing.
mRNA
Exon
Exon
Exon
Amino acid information
AUC CGC AGU CCA GAA UAU GAA CUA
Amino acid
Isoleucine Arginine Serine Proline Glutamic acid Tyrosine Glutamic acid Leucine
Amino acids are linked to construct protein.
THUS, UNLESS THE INTRONS ARE CUT OFF AFTER SYNTHESIS, mRNA CANNOT BE THE ACCURATE GUIDE FOR MAKING PROTEIN.
THE PROCESS OF REMOVING THE INTRONS IS CALLED *SPLICING*.

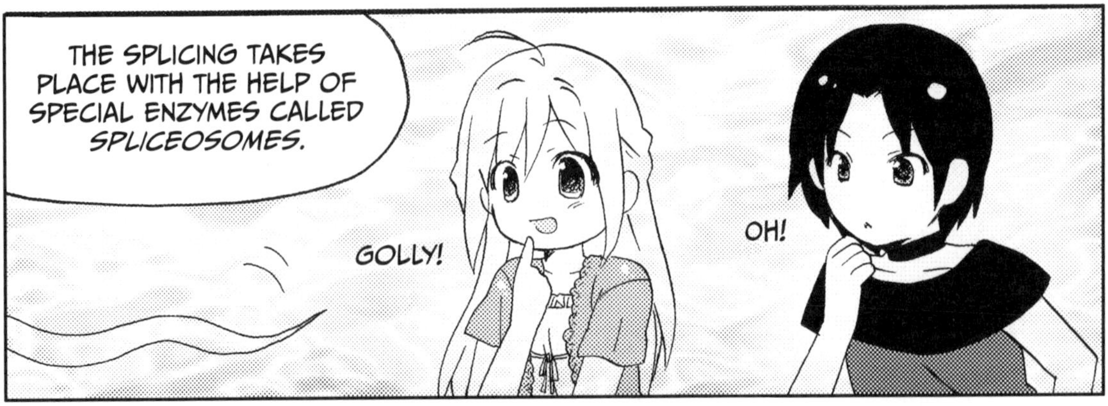
THE SPLICING TAKES PLACE WITH THE HELP OF SPECIAL ENZYMES CALLED *SPLICEOSOMES*.
GOLLY!
OH!

EXON SHUFFLING

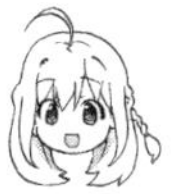

Marcus, why is a gene cut up?

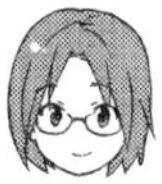

This is still being studied, and various theories have been presented.

Primitive organisms like bacteria do not have introns. Spliced genes allow easier mixing, which probably facilitated evolution.

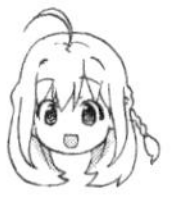

Are genes mixed for evolution?

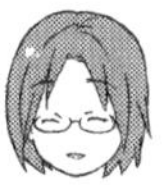

Imagine that a gene consists of a set of playing cards. If you can imagine that the cards from the ace to the king of spades make up a gene, and the cards from the ace to the king of hearts make up another gene, and that each card corresponds to an exon, the genes of hearts and spades consist of 13 exons each. (The space between each card is filled with introns.)

During the evolutionary process, the 4, 5, and 6 of spades could be replaced with the 4, 5, and 6 of hearts.

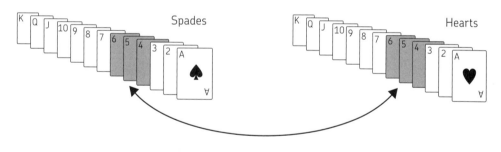

Spades
K Q J 10 9 8 7 6 5 4 3 2 A
Hearts
K Q J 10 9 8 7 6 5 4 3 2 A
Replacement

Each exon is made up of several base pairs. The introns between the cards allow the exons to switch without interrupting the sequence of any individual exon. This is what creates diversity among members of the same species.

When the replacement of exons occurs between two genes, a new gene is produced. In human genes, very similar exons exist in two totally irrelevant genes. This indicates that genes each with different functions have been produced through the mixing of exons.

This process, called *exon shuffling*, is thought to have played a part in the evolution of living organisms through the diversification of genes.

Uuuuuhhhhh...

Sorry if that was a little difficult to understand.

WHAT IS RNA?

CHARACTERS OF RNA

Do you understand that RNA is produced through DNA transcription?

Yes. A copy of DNA is RNA, right? Does that mean DNA and RNA are totally identical?

No, they're not. Ami, you noticed a little while ago that RNA is slightly different from the template strand of DNA.

Right!

Template strand
A C G G C C G T T A A
RNA →
A C G G C C G U U A A

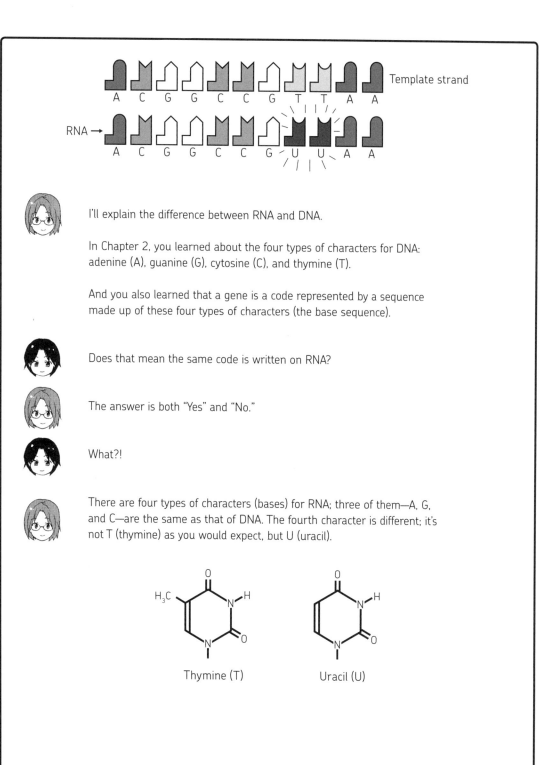

I'll explain the difference between RNA and DNA.

In Chapter 2, you learned about the four types of characters for DNA: adenine (A), guanine (G), cytosine (C), and thymine (T).

And you also learned that a gene is a code represented by a sequence made up of these four types of characters (the base sequence).

Does that mean the same code is written on RNA?

The answer is both "Yes" and "No."

What?!

There are four types of characters (bases) for RNA; three of them—A, G, and C—are the same as that of DNA. The fourth character is different; it's not T (thymine) as you would expect, but U (uracil).

O
H_3C
N
H
N
O

Thymine (T)

O
N
H
N
O

Uracil (U)

Why is that one different?

This is currently being studied, but the following hypothesis is supported by many researchers.

During DNA replication, DNA polymerase checks for mutations in the DNA. U (uracil) can be produced through the mutation of C (cytosine). When DNA encounters a uracil during this check for mutations, it would be "confused." DNA polymerase would have no way of knowing if it were a mutated cytosine that needs to be corrected, or if it is really supposed to be uracil. It's possible that the wrong kind of repair could be conducted.

That's why DNA has developed T (thymine), which is easier to identify.

Yes, but this is still a hypothesis. There are many more things waiting to be discovered in the world of molecular biology.

DNA AND RNA USE DIFFERENT SUGARS

In Chapter 2, you learned that a nucleotide (deoxyribonucleotide), the material that makes up DNA, consists of phosphoric acid, deoxyribose (a type of sugar), and a base.

Base
H
P
P
P
H
O
Phosphoric acid
H
H
H
H
OH
H
Deoxyribose

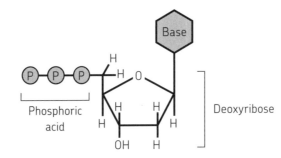

RNA is made up of a special kind of nucleotide, called a *ribonucleotide*. It is also made up of phosphoric acid, sugar, and a base. This sugar is not deoxyribose but simply *ribose*.

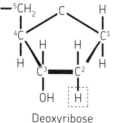

Deoxyribose
(Material of DNA)

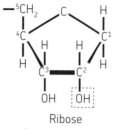

Ribose
(Material of RNA)

 In addition to the difference in one of the characters, thymine or uracil, there is another variation between DNA and RNA. DNA uses deoxyribose as a sugar, and RNA uses ribose.

 What's the difference between deoxyribose and ribose?

 They differ only in one point, whether hydrogen (H) or hydroxyl (OH) becomes the second carbon (C). Hydrogen binds to the second carbon with deoxyribose, and hydroxyl binds to the second carbon with ribose. This is the only difference, but it causes big changes in the properties of DNA and RNA in molecular form.

 How do their properties differ?

 The reactivity of RNA containing hydroxyl is much higher than that of DNA. This is because the oxygen atom (O) in hydroxyl is more reactive with other atoms.

RNA IS FLEXIBLE

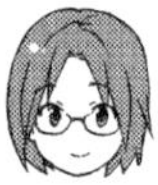

As described earlier, we can list two differences that cause chemical differences between RNA molecules and DNA molecules. They are the differences in the base, namely T (thymine) or U (uracil), and the differences in the sugar of the nucleotide. There is another point that contributes significantly to the differences between them. I've explained that DNA forms double strands, but most RNA moves around as a single strand.

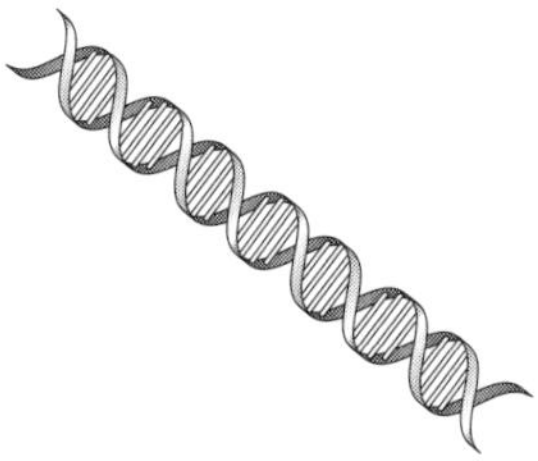

DNA has a double strand.

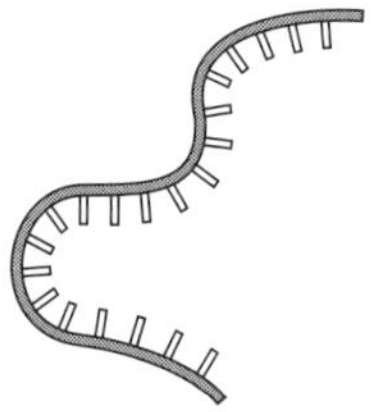

RNA has a single strand.

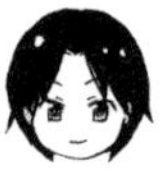

Really? Why? Double strands are more useful, aren't they?

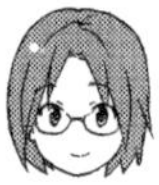

This is definitely the case with DNA. But it is advantageous for RNA to be a single strand. A single strand allows RNA to be flexible without being bound by the fixed structure of the double helix. RNA can thus transform into various shapes within a single molecule.

Suppose there is a base sequence AGGCCC and another base sequence GGGCCU somewhere on a single RNA.

Since A and U, and G and C can find and create pairs together, this portion alone in the molecule has a double-strand structure. Then, RNA will be shaped like this.

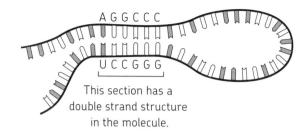

A G G C C C

U C C G G G

This section has a
double strand structure
in the molecule.

RNA is not just a string-like molecule; it can transform into various shapes
just by changing the base sequence. Since RNA can take various forms, it
is capable of assuming various roles in addition to merely functioning as a
duplicate molecule. This is very important.

THERE ARE MANY TYPES OF RNA

Various roles—like what?

Messenger RNA (mRNA) is not the only kind of RNA you will find. RNA
can take the form of *transfer RNA (tRNA)* and *ribosomal RNA (rRNA)* as
well. Both play a very important role when constructing protein from the
genetic information copied to mRNA.

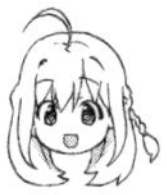

It sounds very mysterious. I have heard the word DNA but I didn't know anything about RNA. Actually, RNA looks like it does a whole lot more than DNA.

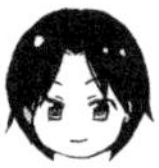

Yes, you're right. Since RNA is more flexible...

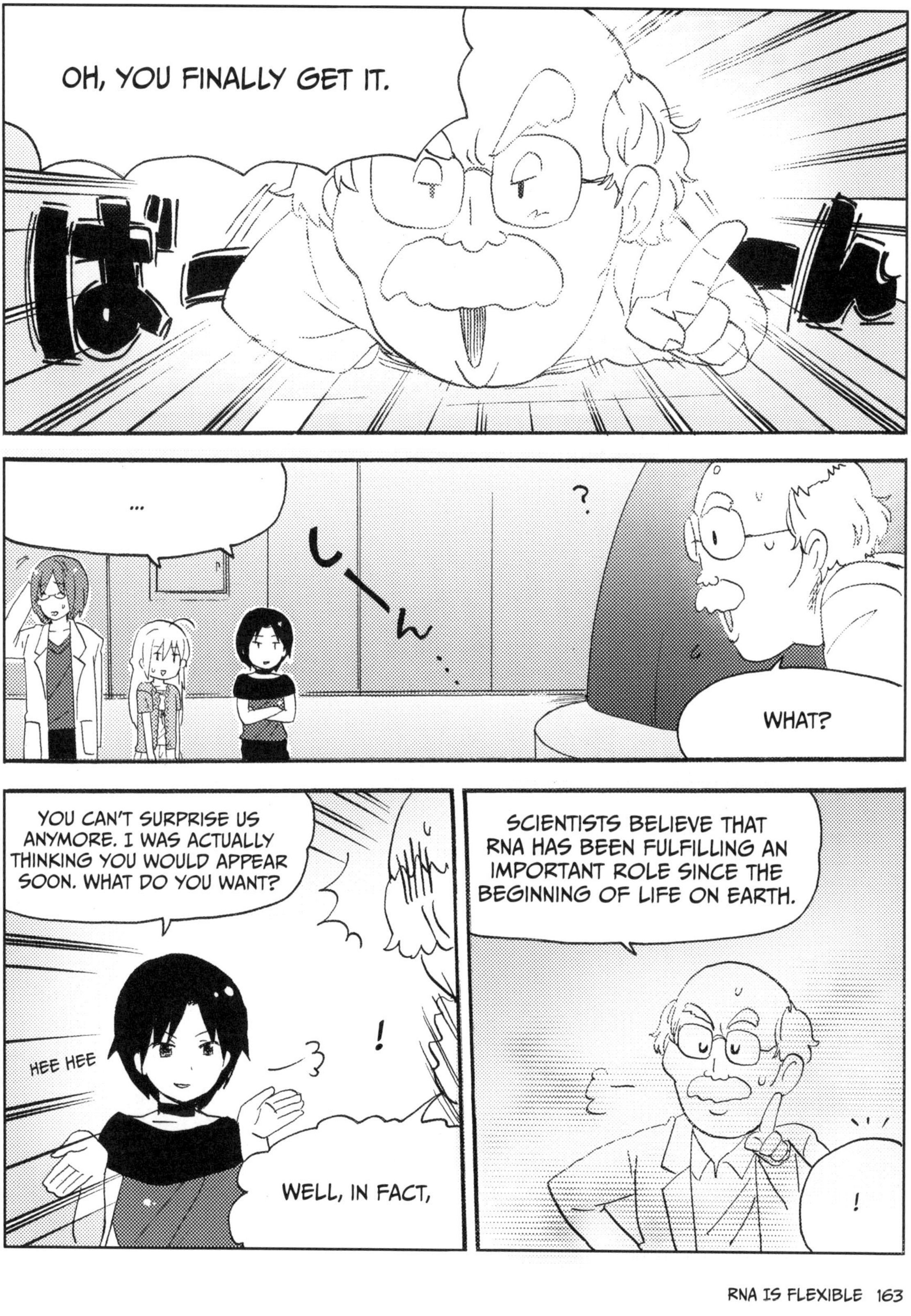
OH, YOU FINALLY GET IT.
ばーーん
...
?
しーん
WHAT?
YOU CAN'T SURPRISE US ANYMORE. I WAS ACTUALLY THINKING YOU WOULD APPEAR SOON. WHAT DO YOU WANT?
HEE HEE
!
WELL, IN FACT,
SCIENTISTS BELIEVE THAT RNA HAS BEEN FULFILLING AN IMPORTANT ROLE SINCE THE BEGINNING OF LIFE ON EARTH.
!

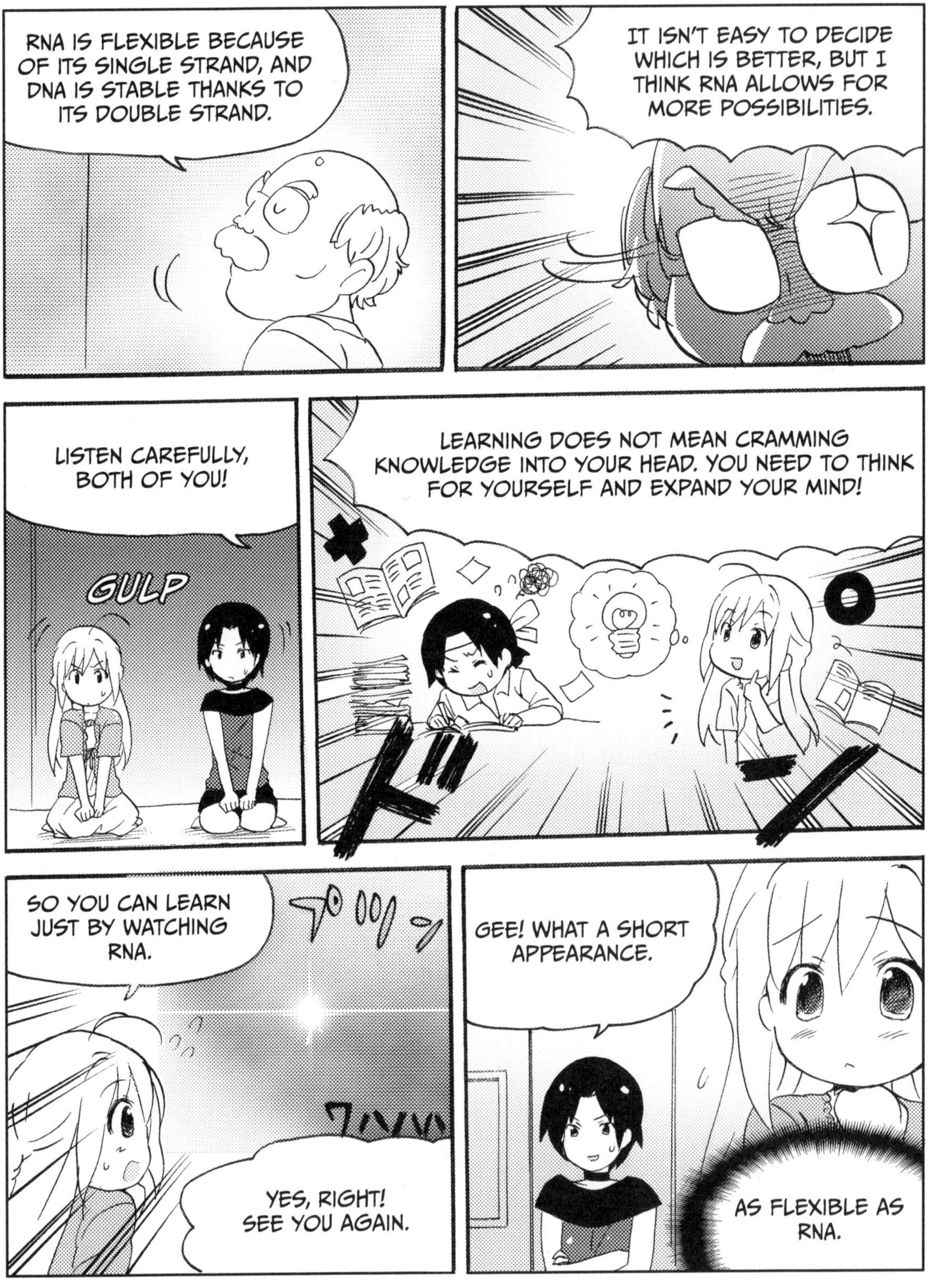
RNA IS FLEXIBLE BECAUSE OF ITS SINGLE STRAND, AND DNA IS STABLE THANKS TO ITS DOUBLE STRAND.
IT ISN'T EASY TO DECIDE WHICH IS BETTER, BUT I THINK RNA ALLOWS FOR MORE POSSIBILITIES.
LISTEN CAREFULLY, BOTH OF YOU!
GULP
LEARNING DOES NOT MEAN CRAMMING KNOWLEDGE INTO YOUR HEAD. YOU NEED TO THINK FOR YOURSELF AND EXPAND YOUR MIND!
SO YOU CAN LEARN JUST BY WATCHING RNA.
YES, RIGHT! SEE YOU AGAIN.
GEE! WHAT A SHORT APPEARANCE.
AS FLEXIBLE AS RNA.

TRANSFER RNA

RIBOSOME: THE PROTEIN SYNTHESIS MECHANISM

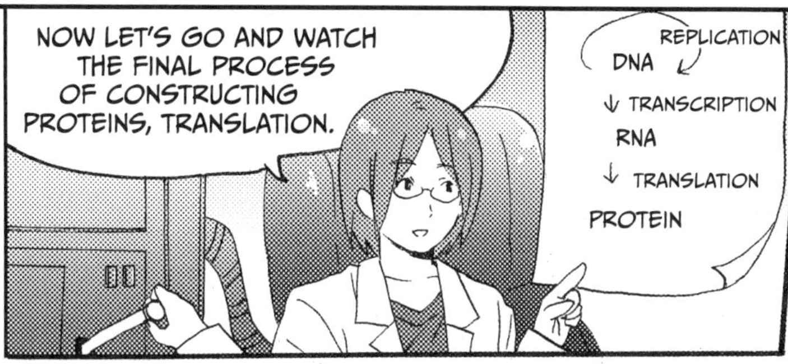
NOW LET'S GO AND WATCH THE FINAL PROCESS OF CONSTRUCTING PROTEINS, TRANSLATION.
REPLICATION
DNA
TRANSCRIPTION
RNA
TRANSLATION
PROTEIN

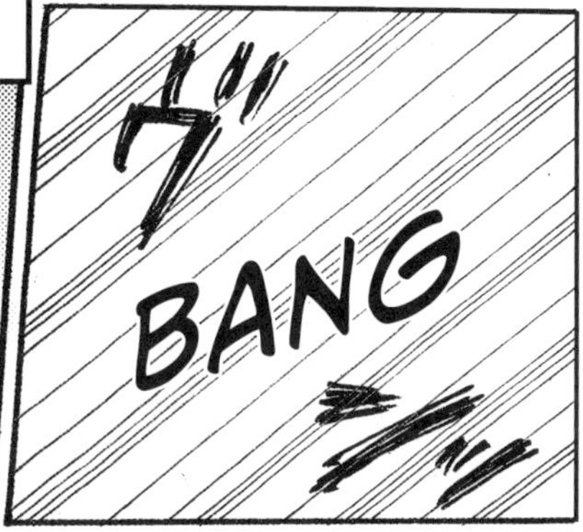
BANG

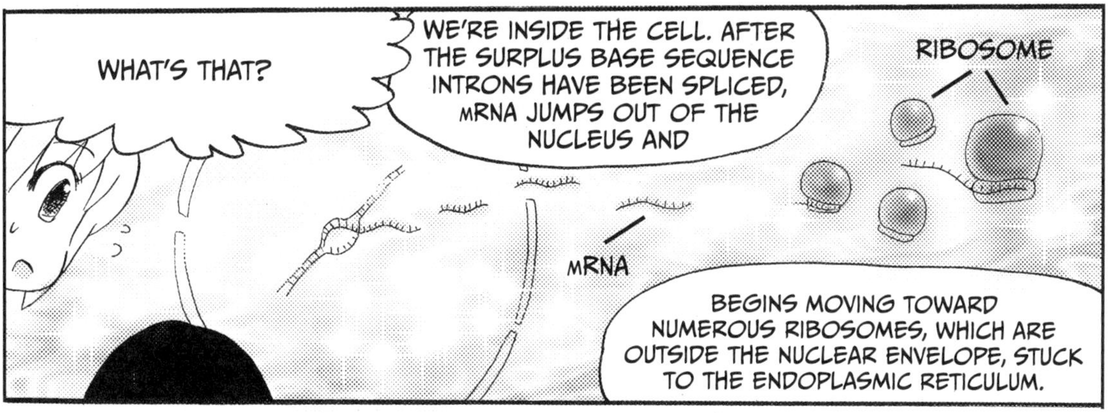
WHAT'S THAT?
WE'RE INSIDE THE CELL. AFTER THE SURPLUS BASE SEQUENCE INTRONS HAVE BEEN SPLICED, mRNA JUMPS OUT OF THE NUCLEUS AND
RIBOSOME
mRNA
BEGINS MOVING TOWARD NUMEROUS RIBOSOMES, WHICH ARE OUTSIDE THE NUCLEAR ENVELOPE, STUCK TO THE ENDOPLASMIC RETICULUM.

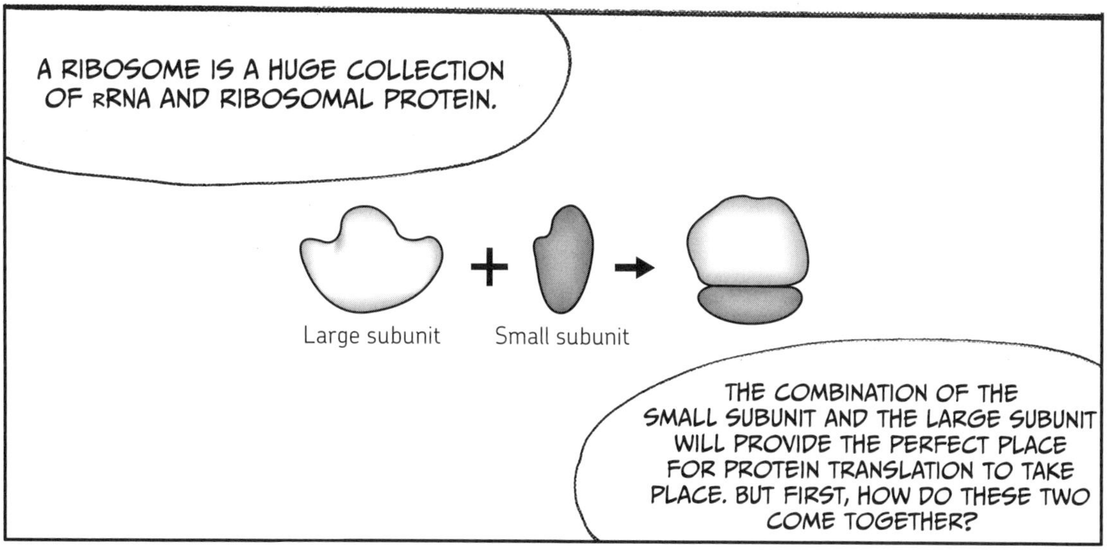
A RIBOSOME IS A HUGE COLLECTION OF rRNA AND RIBOSOMAL PROTEIN.
Large subunit
Small subunit
THE COMBINATION OF THE SMALL SUBUNIT AND THE LARGE SUBUNIT WILL PROVIDE THE PERFECT PLACE FOR PROTEIN TRANSLATION TO TAKE PLACE. BUT FIRST, HOW DO THESE TWO COME TOGETHER?

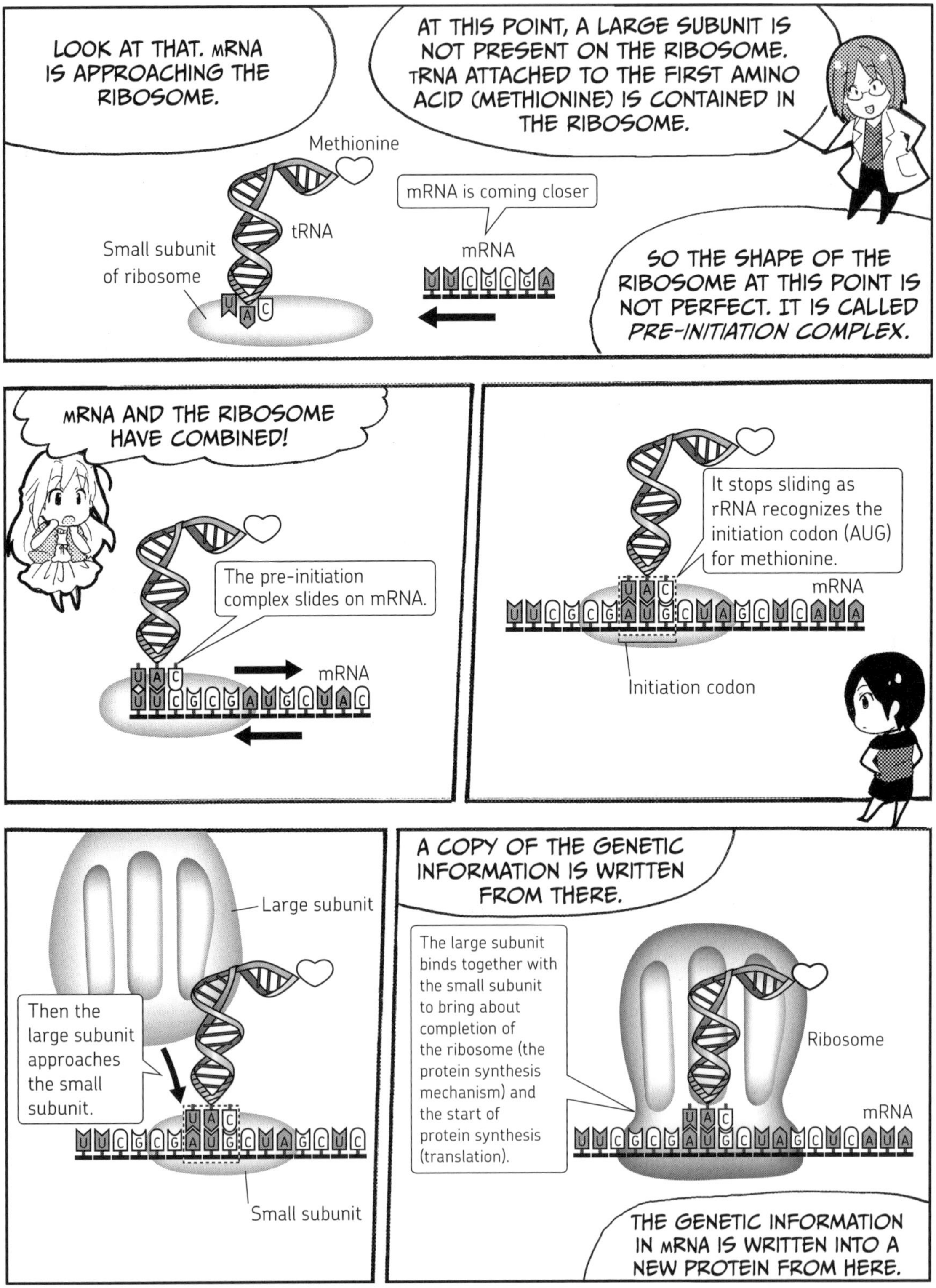
LOOK AT THAT. mRNA IS APPROACHING THE RIBOSOME.
AT THIS POINT, A LARGE SUBUNIT IS NOT PRESENT ON THE RIBOSOME. tRNA ATTACHED TO THE FIRST AMINO ACID (METHIONINE) IS CONTAINED IN THE RIBOSOME.
Methionine
tRNA
Small subunit of ribosome
mRNA is coming closer
mRNA
SO THE SHAPE OF THE RIBOSOME AT THIS POINT IS NOT PERFECT. IT IS CALLED *PRE-INITIATION COMPLEX.*
mRNA AND THE RIBOSOME HAVE COMBINED!
The pre-initiation complex slides on mRNA.
mRNA
It stops sliding as rRNA recognizes the initiation codon (AUG) for methionine.
mRNA
Initiation codon
Large subunit
Then the large subunit approaches the small subunit.
Small subunit
A COPY OF THE GENETIC INFORMATION IS WRITTEN FROM THERE.
The large subunit binds together with the small subunit to bring about completion of the ribosome (the protein synthesis mechanism) and the start of protein synthesis (translation).
Ribosome
mRNA
THE GENETIC INFORMATION IN mRNA IS WRITTEN INTO A NEW PROTEIN FROM HERE.

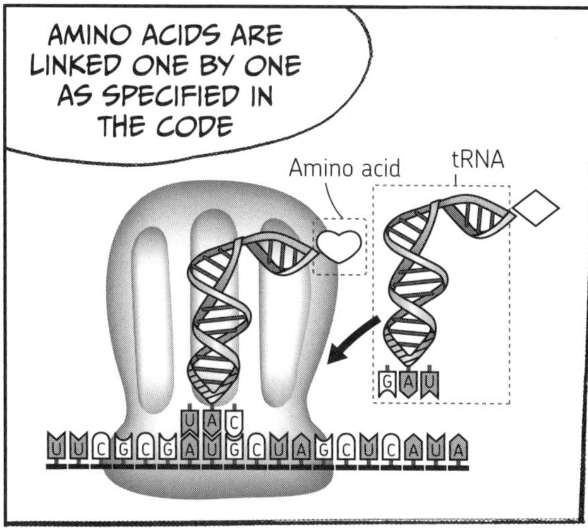
AMINO ACIDS ARE LINKED ONE BY ONE AS SPECIFIED IN THE CODE
Amino acid
tRNA

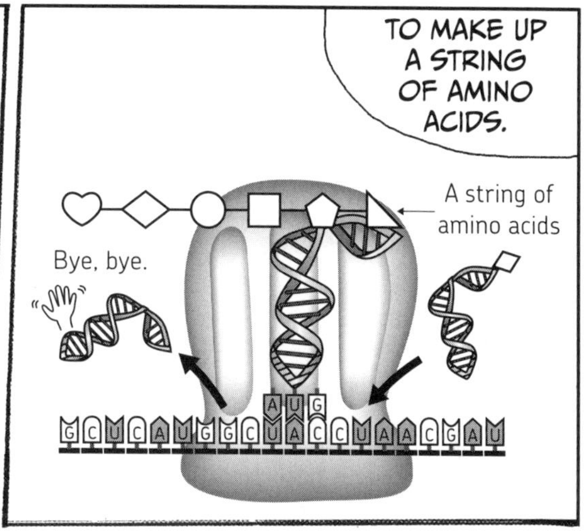
TO MAKE UP A STRING OF AMINO ACIDS.
A string of amino acids
Bye, bye.

MECHANICS OF THE GENETIC CODE

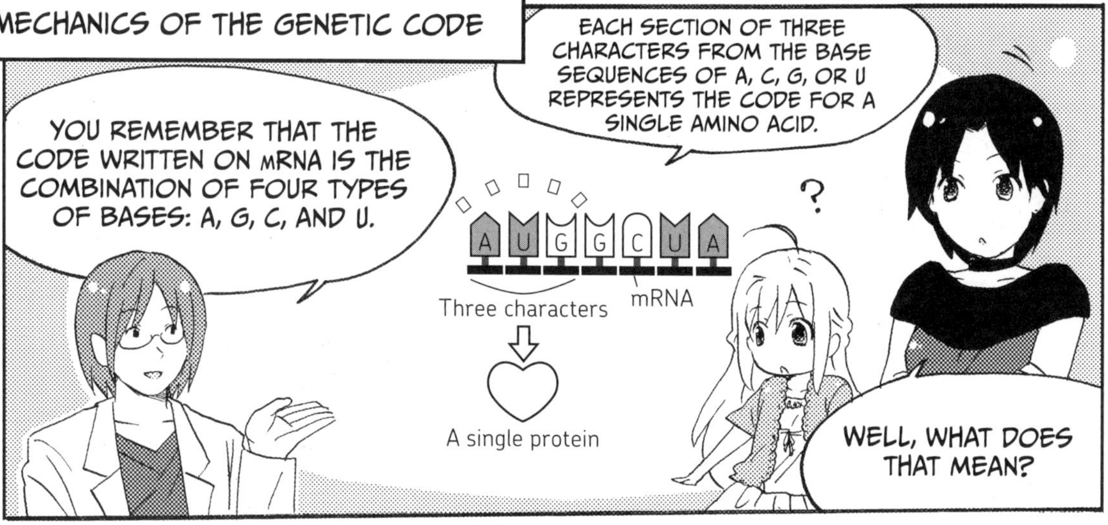
YOU REMEMBER THAT THE CODE WRITTEN ON mRNA IS THE COMBINATION OF FOUR TYPES OF BASES: A, G, C, AND U.
EACH SECTION OF THREE CHARACTERS FROM THE BASE SEQUENCES OF A, C, G, OR U REPRESENTS THE CODE FOR A SINGLE AMINO ACID.
A U G G C U A
Three characters
mRNA
A single protein
?
WELL, WHAT DOES THAT MEAN?

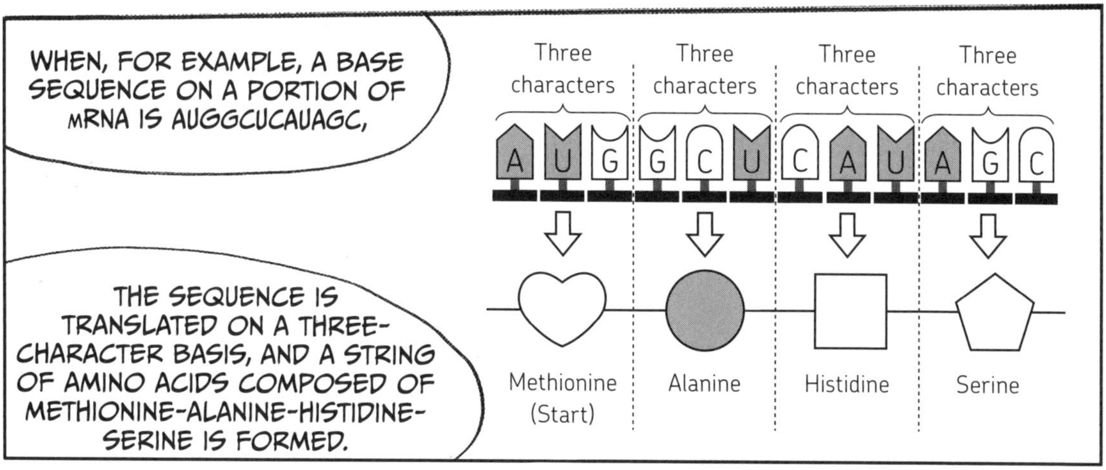
WHEN, FOR EXAMPLE, A BASE SEQUENCE ON A PORTION OF mRNA IS AUGGCUCAUAGC,
Three characters
Three characters
Three characters
Three characters
A U G G C U C A U A G C
Methionine (Start)
Alanine
Histidine
Serine
THE SEQUENCE IS TRANSLATED ON A THREE-CHARACTER BASIS, AND A STRING OF AMINO ACIDS COMPOSED OF METHIONINE-ALANINE-HISTIDINE-SERINE IS FORMED.

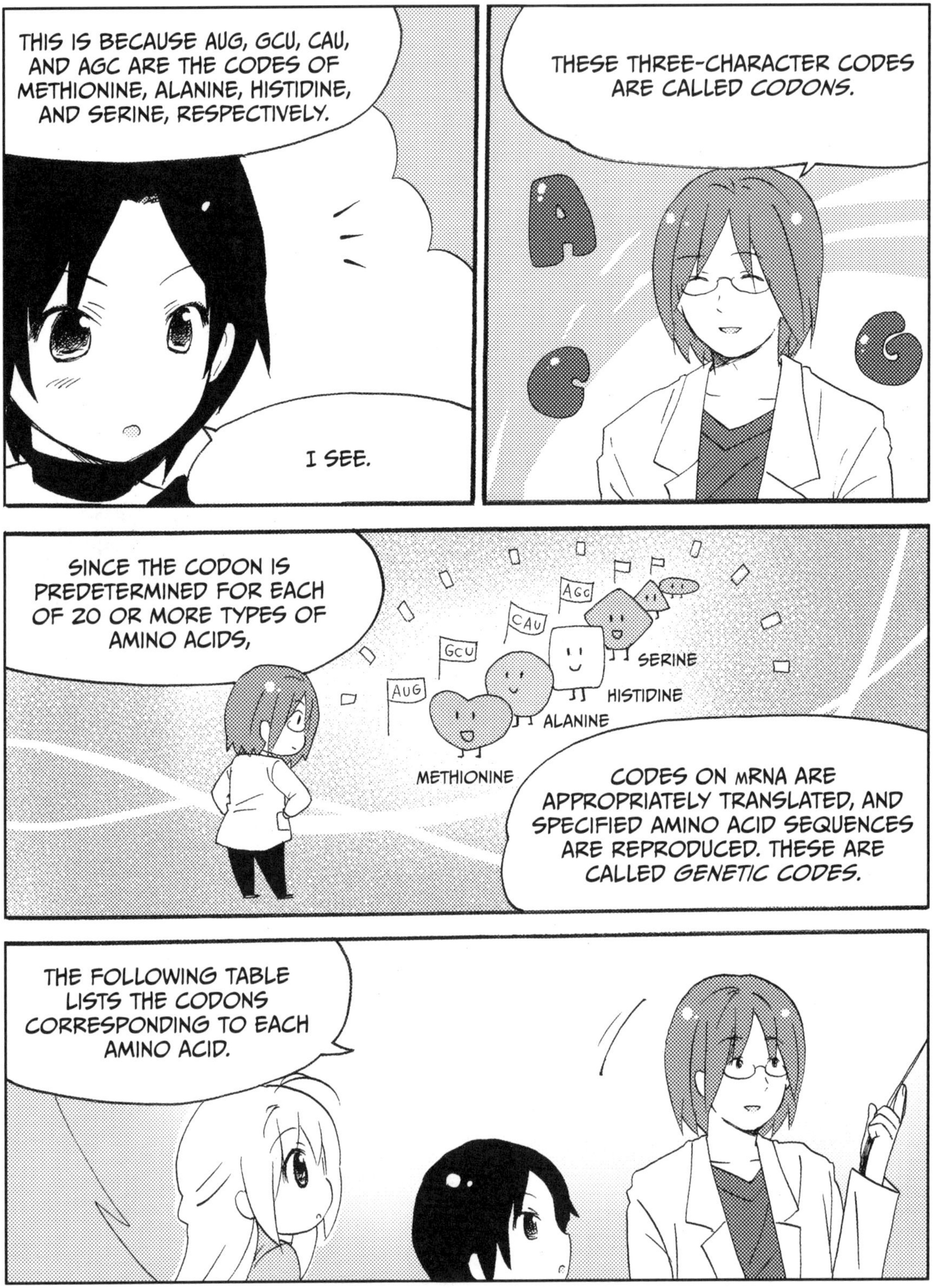
THIS IS BECAUSE AUG, GCU, CAU, AND AGC ARE THE CODES OF METHIONINE, ALANINE, HISTIDINE, AND SERINE, RESPECTIVELY.
I SEE.
THESE THREE-CHARACTER CODES ARE CALLED *CODONS*.
SINCE THE CODON IS PREDETERMINED FOR EACH OF 20 OR MORE TYPES OF AMINO ACIDS,
AUG
GCU
CAU
AGC
METHIONINE
ALANINE
HISTIDINE
SERINE
CODES ON mRNA ARE APPROPRIATELY TRANSLATED, AND SPECIFIED AMINO ACID SEQUENCES ARE REPRODUCED. THESE ARE CALLED *GENETIC CODES*.
THE FOLLOWING TABLE LISTS THE CODONS CORRESPONDING TO EACH AMINO ACID.

First character	Second character: U	Second character: C	Second character: A	Second character: G	Third character
U	(UUU) Phenylalanine	(UCU) Serine	(UAU) Tyrosine	(UGU) Cysteine	U
	(UUC) Phenylalanine	(UCC) Serine	(UAC) Tyrosine	(UGC) Cysteine	C
	(UUA) Leucine	(UCA) Serine	(UAA) Stop codon	(UGA) Stop codon	A
	(UUG) Leucine	(UCG) Serine	(UAG) Stop codon	(UGG) Tryptophane	G
C	(CUU) Leucine	(CCU) Proline	(CAU) Histidine	(CGU) Arginine	U
	(CUC) Leucine	(CCC) Proline	(CAC) Histidine	(CGC) Arginine	C
	(CUA) Leucine	(CCA) Proline	(CAA) Glutamine	(CGA) Arginine	A
	(CUG) Leucine	(CCG) Proline	(CAG) Glutamine	(CGG) Arginine	G
A	(AUU) Isoleucine	(ACU) Threonine	(AAU) Asparagine	(AGU) Serine	U
	(AUC) Isoleucine	(ACC) Threonine	(AAC) Asparagine	(AGC) Serine	C
	(AUA) Isoleucine	(ACA) Threonine	(AAA) Lycine	(AGA) Arginine	A
	(AUG) Methionine (start codon)	(ACG) Threonine	(AAG) Lycine	(AGG) Arginine	G
G	(GUU) Valine	(GCU) Alanine	(GAU) Asparagic acid	(GGU) Glycine	U
	(GUC) Valine	(GCC) Alanine	(GAC) Asparagic acid	(GGC) Glycine	C
	(GUA) Valine	(GCA) Alanine	(GAA) Glutamic acid	(GGA) Glycine	A
	(GUG) Valine	(GCG) Alanine	(GAG) Glutamic acid	(GGG) Glycine	G

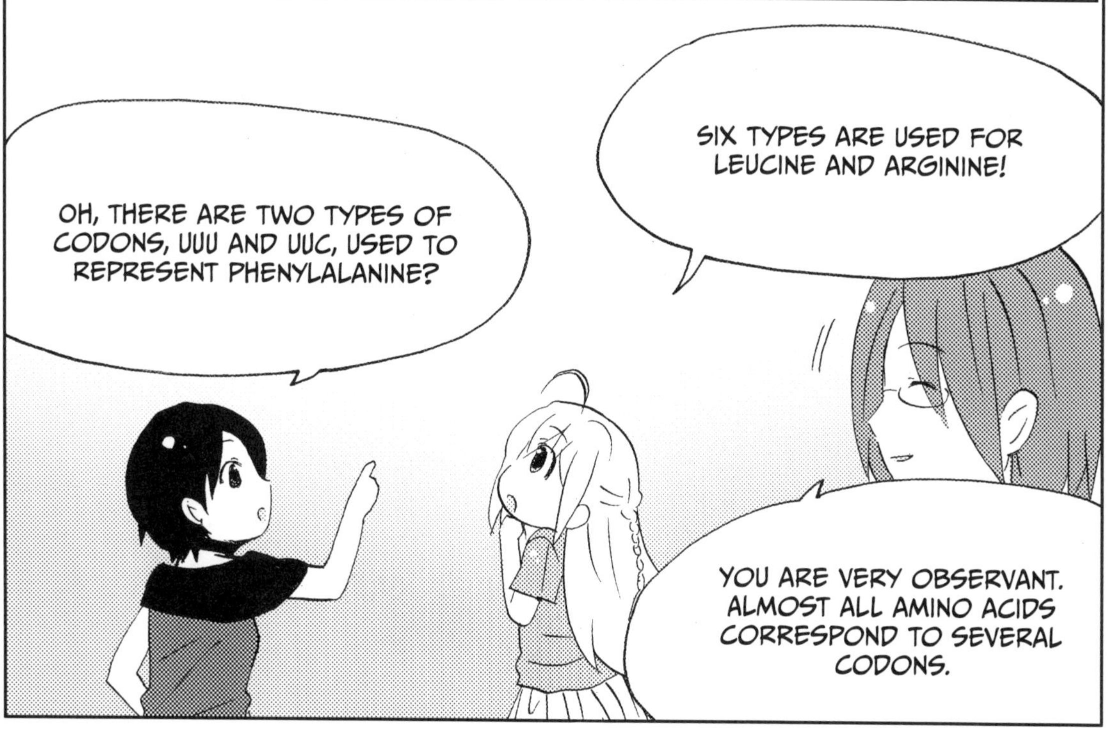
OH, THERE ARE TWO TYPES OF CODONS, UUU AND UUC, USED TO REPRESENT PHENYLALANINE?
SIX TYPES ARE USED FOR LEUCINE AND ARGININE!
YOU ARE VERY OBSERVANT. ALMOST ALL AMINO ACIDS CORRESPOND TO SEVERAL CODONS.

tRNA TRANSFERS AMINO ACIDS

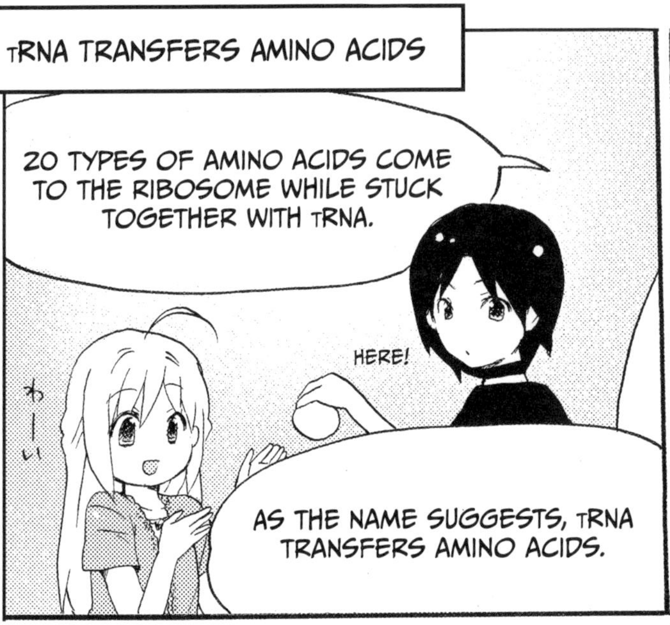
20 TYPES OF AMINO ACIDS COME TO THE RIBOSOME WHILE STUCK TOGETHER WITH tRNA.
HERE!
わーい
AS THE NAME SUGGESTS, tRNA TRANSFERS AMINO ACIDS.

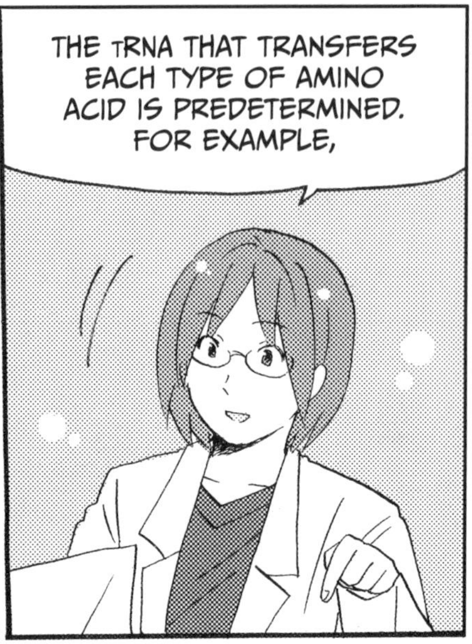
THE tRNA THAT TRANSFERS EACH TYPE OF AMINO ACID IS PREDETERMINED. FOR EXAMPLE,

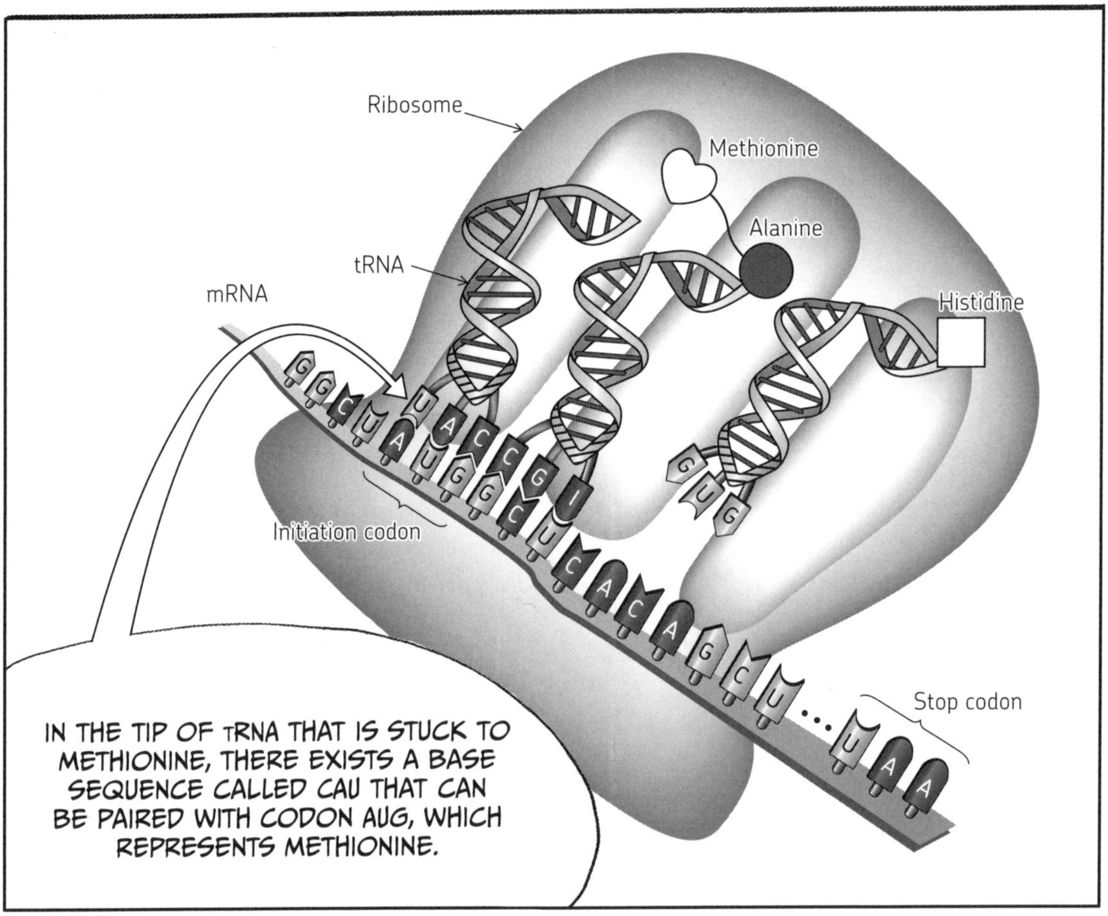
Ribosome
Methionine
Alanine
tRNA
mRNA
Histidine
Initiation codon
Stop codon
IN THE TIP OF tRNA THAT IS STUCK TO METHIONINE, THERE EXISTS A BASE SEQUENCE CALLED CAU THAT CAN BE PAIRED WITH CODON AUG, WHICH REPRESENTS METHIONINE.

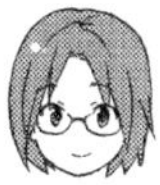

The three base sequences on tRNA that can be paired with this codon make up an *anticodon*.

Likewise, every tRNA that is bound to the amino acid alanine has a sequence (or anticodon) IGC that can be paired with GCU, the code for alanine from mRNA. For the first character of an anticodon, a strange character, *I* for inosine, is sometimes used.

Alanine
← tRNA
←Ribosome
←Anticodon
←Codon
mRNA
C
G
I
G
C
U

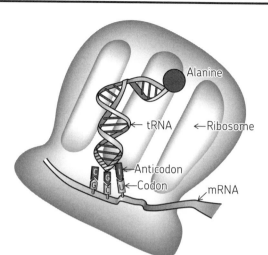

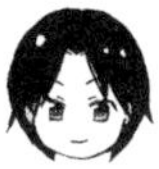

Does "sometimes" mean it is a substitute for another base?

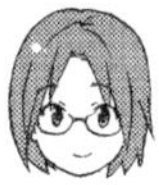

Well, inosine is a special base that can stick together with two or three types of bases. As I just described, the power of the third base of a codon for making a pair with the first base of an anticodon is weak. As a result, such a base can make a pair with other bases, too. Such base pairing is called *wobble base pairing*.

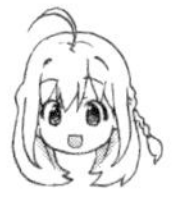

It's like the joker in a stack of playing cards!

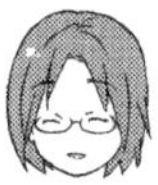

Yes, I agree. You could say it's a wildcard.

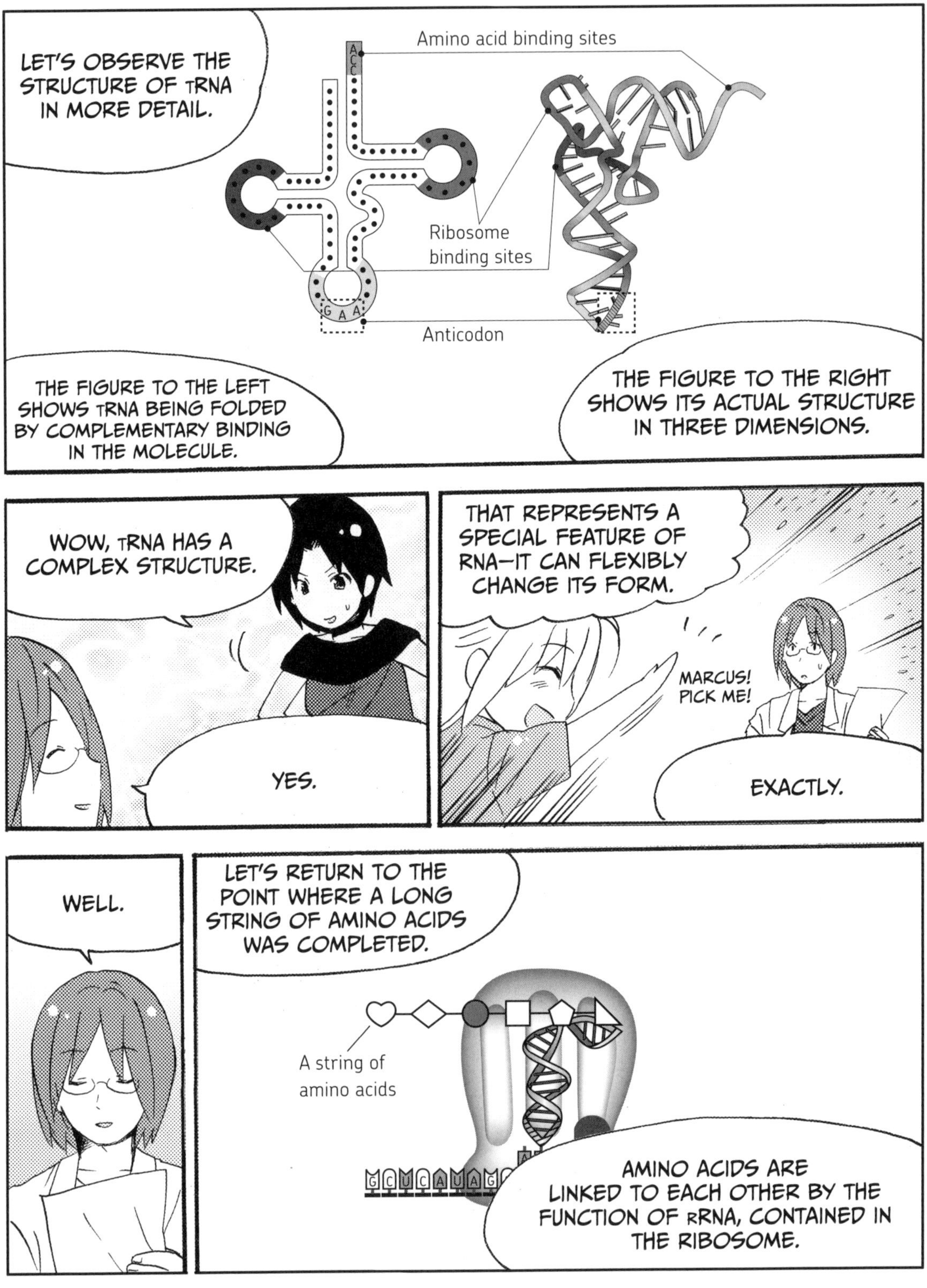
LET'S OBSERVE THE STRUCTURE OF tRNA IN MORE DETAIL.
Amino acid binding sites
Ribosome binding sites
Anticodon
THE FIGURE TO THE LEFT SHOWS tRNA BEING FOLDED BY COMPLEMENTARY BINDING IN THE MOLECULE.
THE FIGURE TO THE RIGHT SHOWS ITS ACTUAL STRUCTURE IN THREE DIMENSIONS.
WOW, tRNA HAS A COMPLEX STRUCTURE.
YES.
THAT REPRESENTS A SPECIAL FEATURE OF RNA—IT CAN FLEXIBLY CHANGE ITS FORM.
MARCUS! PICK ME!
EXACTLY.
WELL.
LET'S RETURN TO THE POINT WHERE A LONG STRING OF AMINO ACIDS WAS COMPLETED.
A string of amino acids
AMINO ACIDS ARE LINKED TO EACH OTHER BY THE FUNCTION OF rRNA, CONTAINED IN THE RIBOSOME.

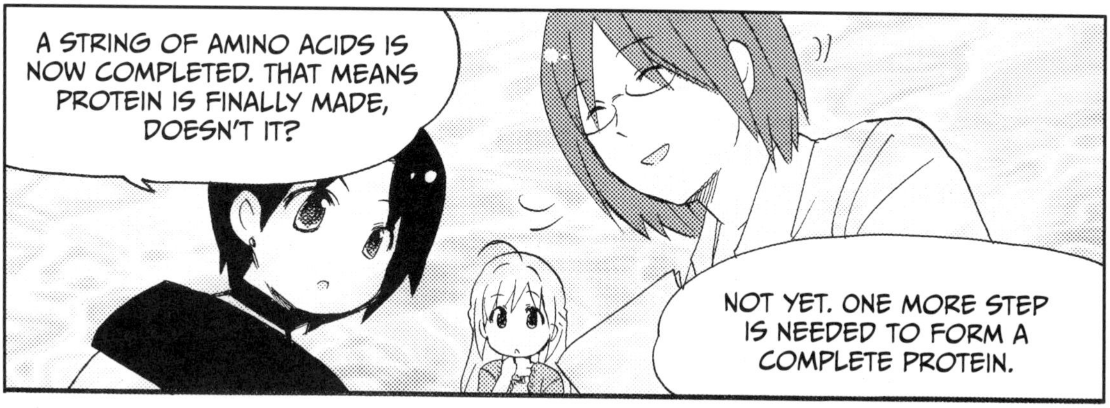
A STRING OF AMINO ACIDS IS NOW COMPLETED. THAT MEANS PROTEIN IS FINALLY MADE, DOESN'T IT?
NOT YET. ONE MORE STEP IS NEEDED TO FORM A COMPLETE PROTEIN.

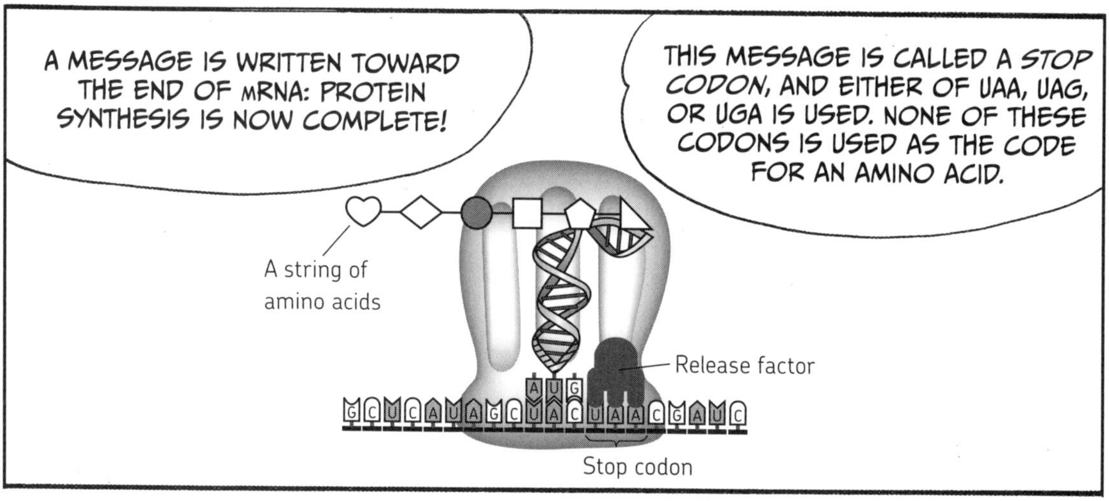
A MESSAGE IS WRITTEN TOWARD THE END OF mRNA: PROTEIN SYNTHESIS IS NOW COMPLETE!
THIS MESSAGE IS CALLED A *STOP CODON*, AND EITHER OF UAA, UAG, OR UGA IS USED. NONE OF THESE CODONS IS USED AS THE CODE FOR AN AMINO ACID.
A string of amino acids
Release factor
AUG
GCUCAUAGCUACUAACGAUC
Stop codon

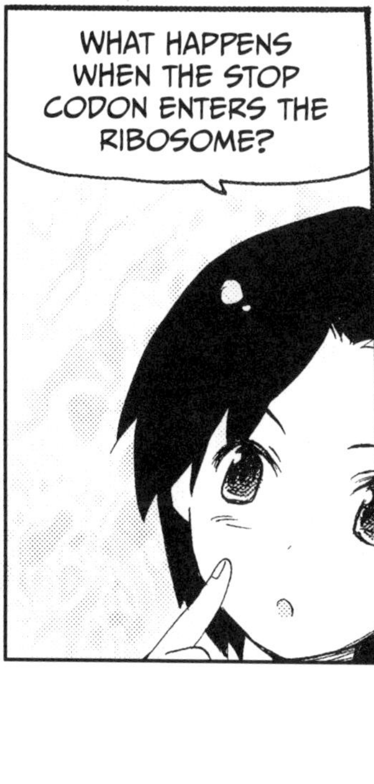
WHAT HAPPENS WHEN THE STOP CODON ENTERS THE RIBOSOME?

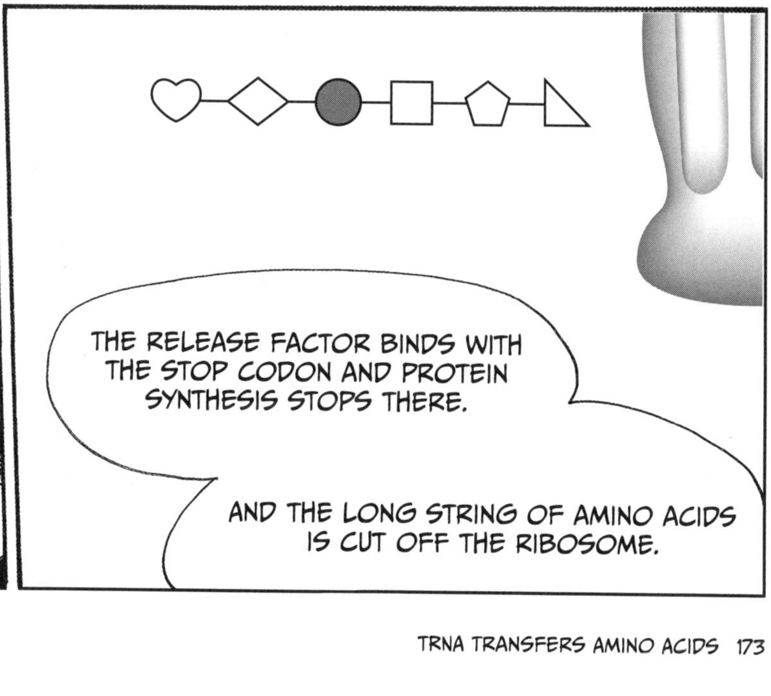
THE RELEASE FACTOR BINDS WITH THE STOP CODON AND PROTEIN SYNTHESIS STOPS THERE.
AND THE LONG STRING OF AMINO ACIDS IS CUT OFF THE RIBOSOME.

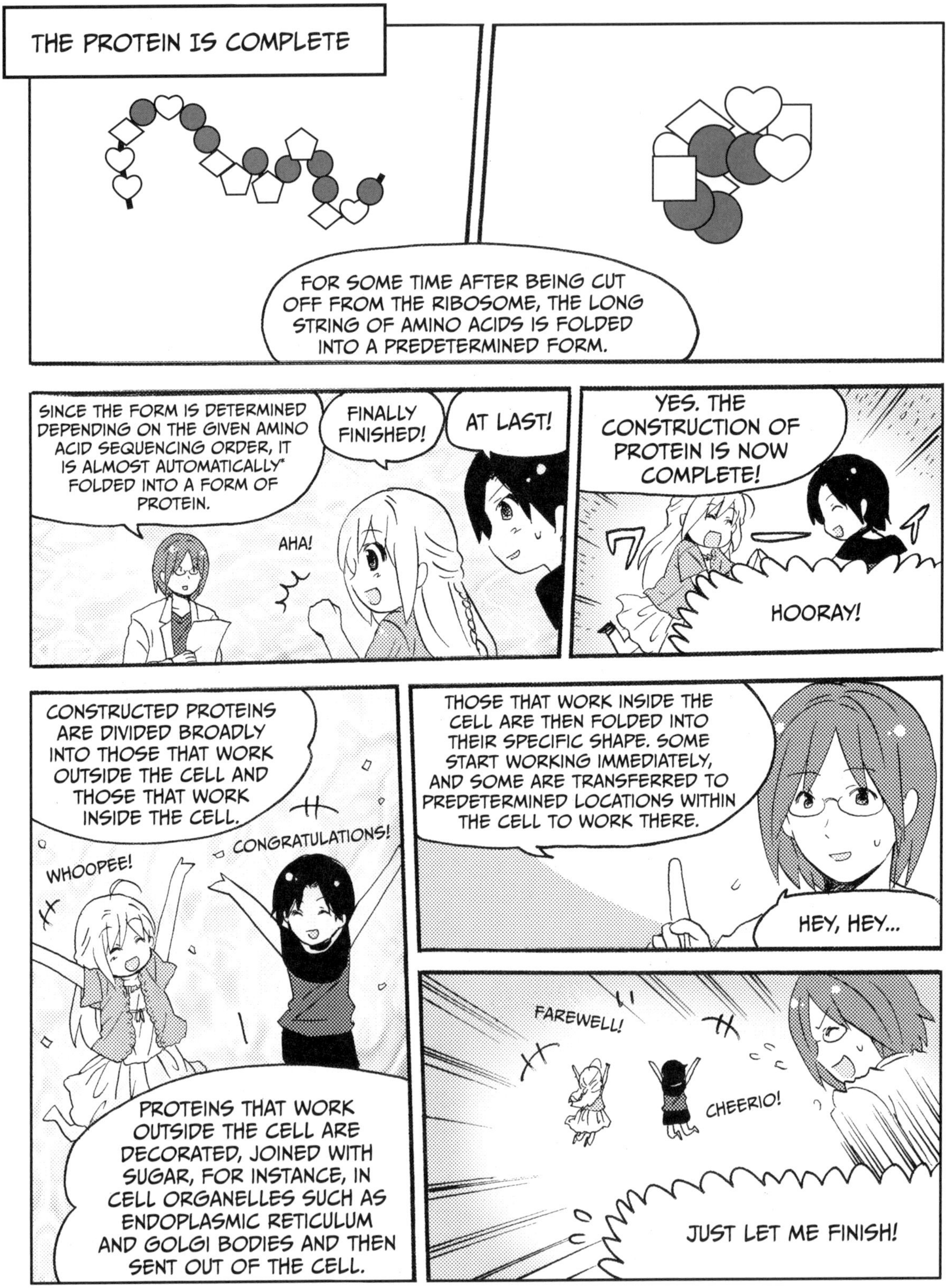
THE PROTEIN IS COMPLETE
FOR SOME TIME AFTER BEING CUT OFF FROM THE RIBOSOME, THE LONG STRING OF AMINO ACIDS IS FOLDED INTO A PREDETERMINED FORM.
SINCE THE FORM IS DETERMINED DEPENDING ON THE GIVEN AMINO ACID SEQUENCING ORDER, IT IS ALMOST AUTOMATICALLY* FOLDED INTO A FORM OF PROTEIN.
FINALLY FINISHED!
AT LAST!
AHA!
YES. THE CONSTRUCTION OF PROTEIN IS NOW COMPLETE!
HOORAY!
CONSTRUCTED PROTEINS ARE DIVIDED BROADLY INTO THOSE THAT WORK OUTSIDE THE CELL AND THOSE THAT WORK INSIDE THE CELL.
WHOOPEE!
CONGRATULATIONS!
PROTEINS THAT WORK OUTSIDE THE CELL ARE DECORATED, JOINED WITH SUGAR, FOR INSTANCE, IN CELL ORGANELLES SUCH AS ENDOPLASMIC RETICULUM AND GOLGI BODIES AND THEN SENT OUT OF THE CELL.
THOSE THAT WORK INSIDE THE CELL ARE THEN FOLDED INTO THEIR SPECIFIC SHAPE. SOME START WORKING IMMEDIATELY, AND SOME ARE TRANSFERRED TO PREDETERMINED LOCATIONS WITHIN THE CELL TO WORK THERE.
HEY, HEY...
FAREWELL!
CHEERIO!
JUST LET ME FINISH!

* TO BE EXACT, A PROTEIN IS NOT AUTOMATICALLY FOLDED. FOLDING IS CARRIED OUT WITH THE HELP OF OTHER PROTEINS.

5
GENETIC TECHNOLOGY AND RESEARCH

WHAT IS GENETIC RECOMBINATION TECHNOLOGY?

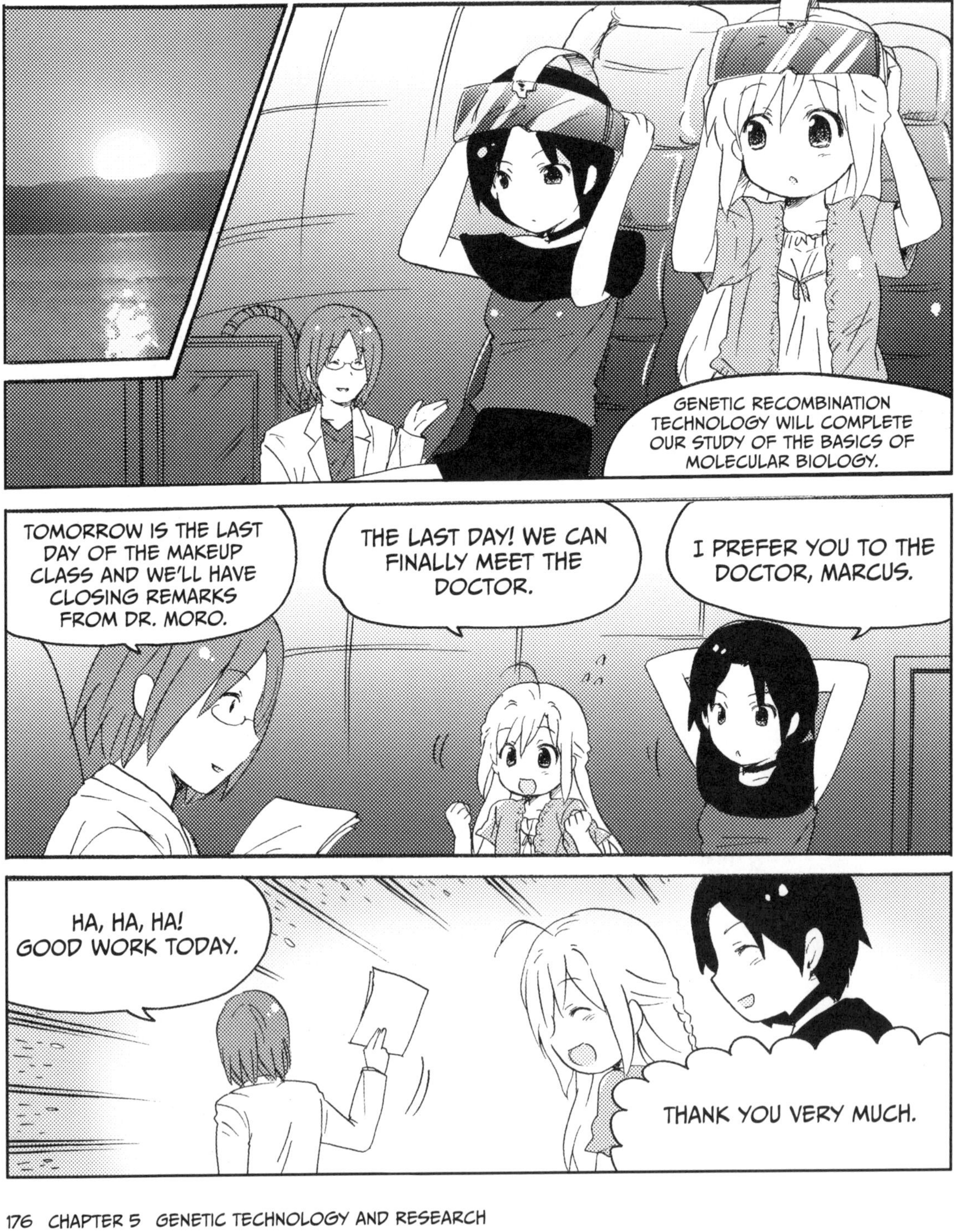
GENETIC RECOMBINATION TECHNOLOGY WILL COMPLETE OUR STUDY OF THE BASICS OF MOLECULAR BIOLOGY.
TOMORROW IS THE LAST DAY OF THE MAKEUP CLASS AND WE'LL HAVE CLOSING REMARKS FROM DR. MORO.
THE LAST DAY! WE CAN FINALLY MEET THE DOCTOR.
I PREFER YOU TO THE DOCTOR, MARCUS.
HA, HA, HA! GOOD WORK TODAY.
THANK YOU VERY MUCH.

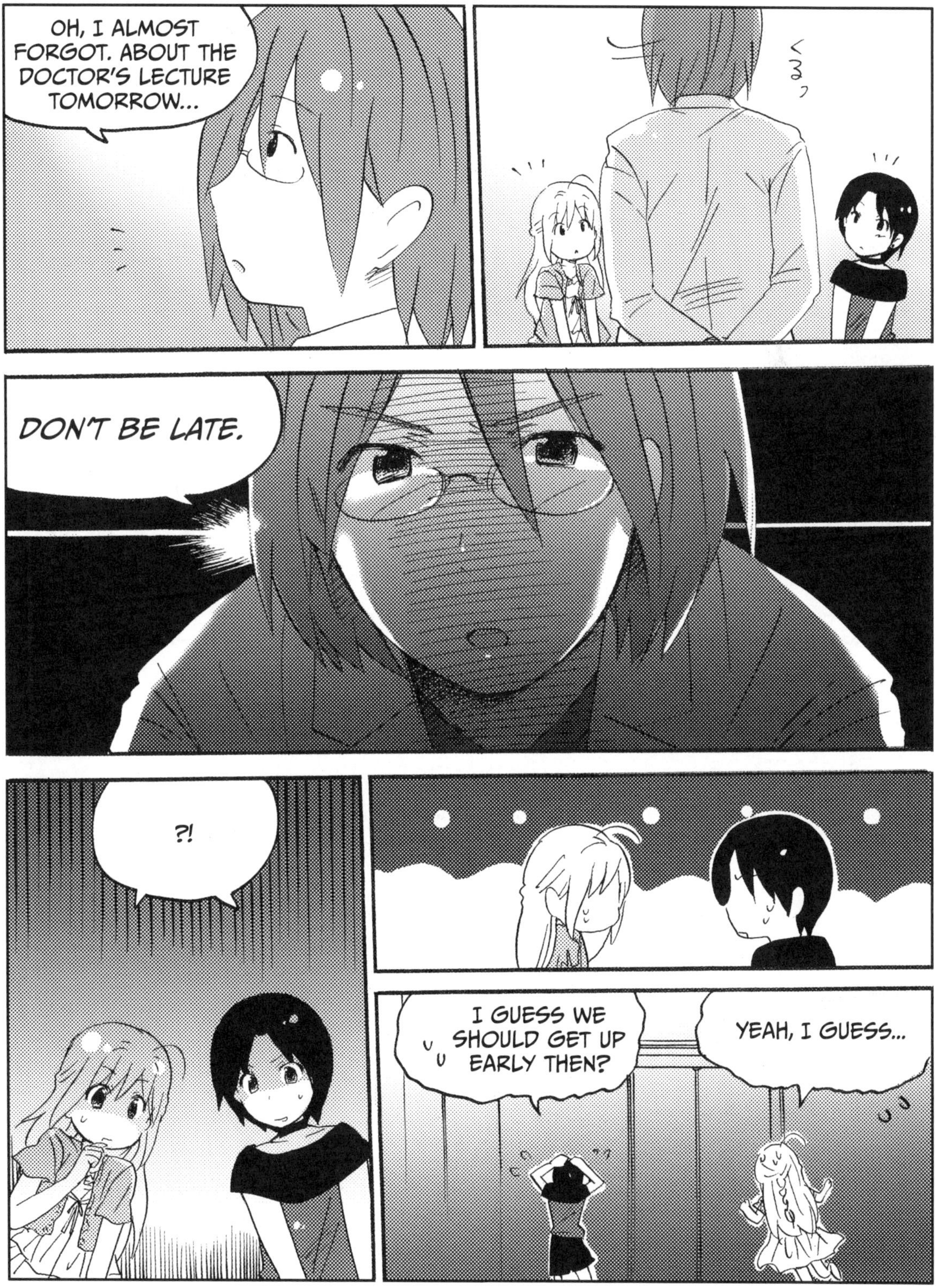
OH, I ALMOST FORGOT. ABOUT THE DOCTOR'S LECTURE TOMORROW...
くるっ
DON'T BE LATE.
?!
I GUESS WE SHOULD GET UP EARLY THEN?
YEAH, I GUESS...

THE LAST DAY OF CLASS

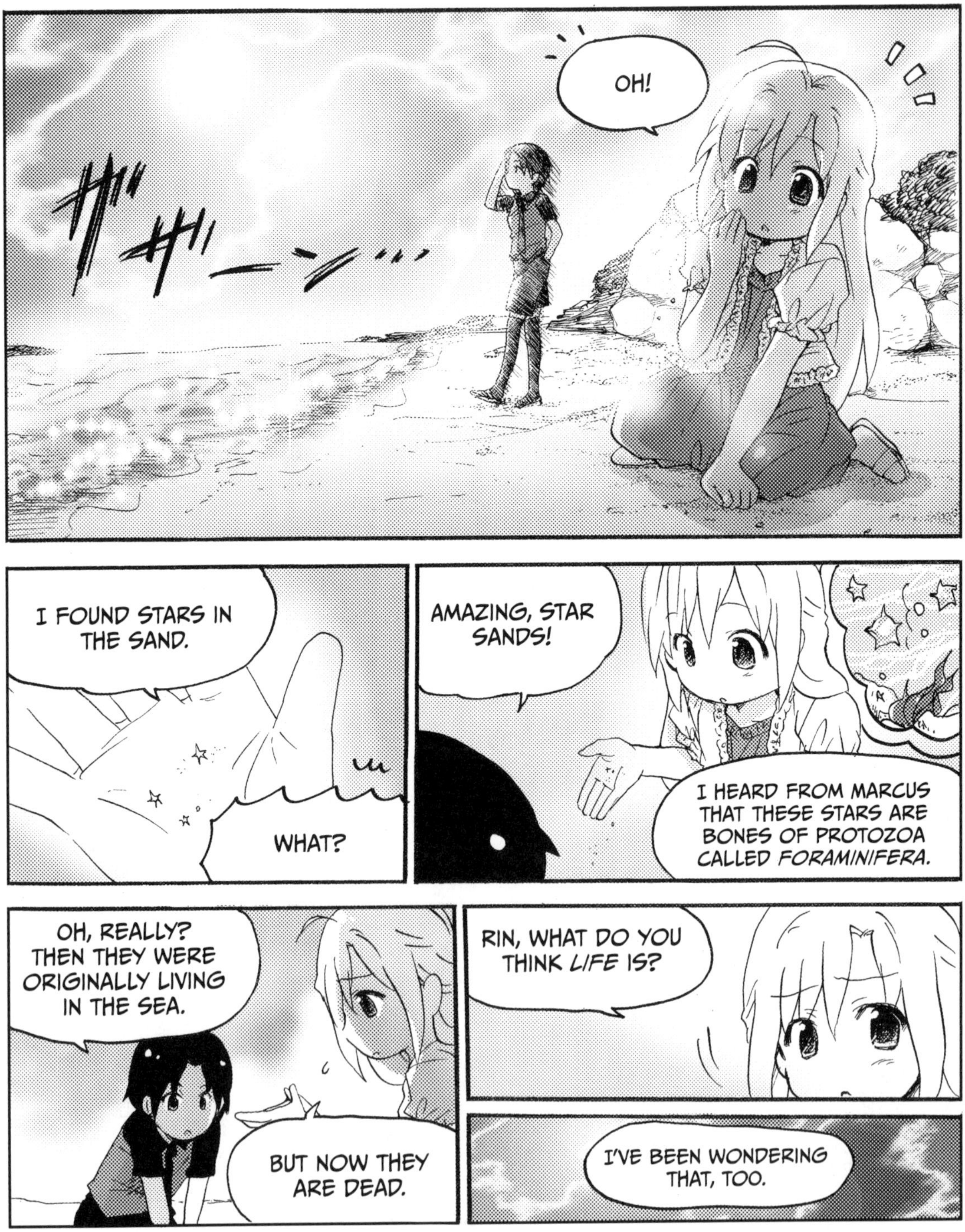

OH!
ザザーン…
I FOUND STARS IN THE SAND.
WHAT?
AMAZING, STAR SANDS!
I HEARD FROM MARCUS THAT THESE STARS ARE BONES OF PROTOZOA CALLED *FORAMINIFERA*.
OH, REALLY? THEN THEY WERE ORIGINALLY LIVING IN THE SEA.
BUT NOW THEY ARE DEAD.
RIN, WHAT DO YOU THINK *LIFE* IS?
I'VE BEEN WONDERING THAT, TOO.

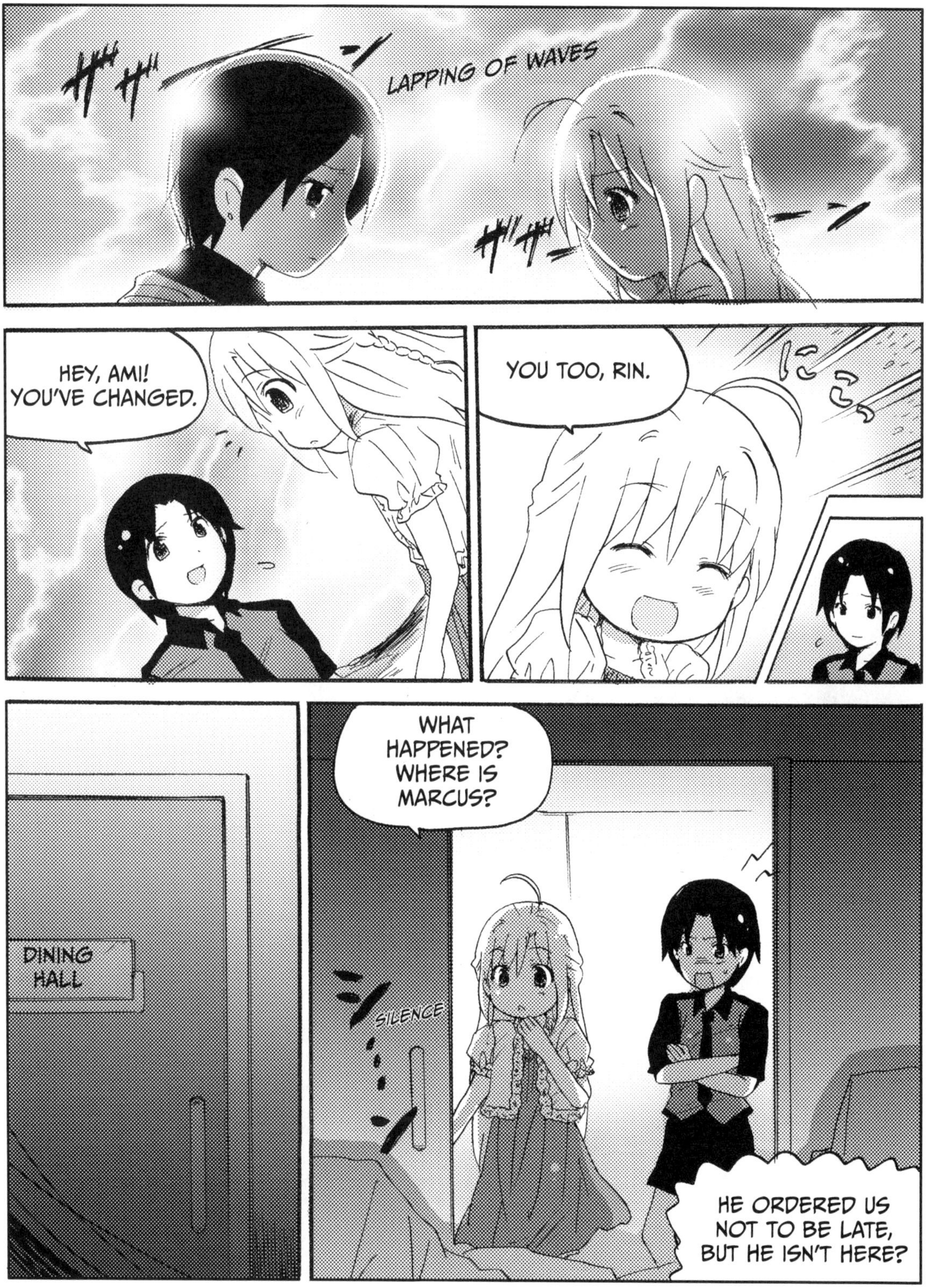
LAPPING OF WAVES
HEY, AMI! YOU'VE CHANGED.
YOU TOO, RIN.
DINING HALL
WHAT HAPPENED? WHERE IS MARCUS?
SILENCE
HE ORDERED US NOT TO BE LATE, BUT HE ISN'T HERE?

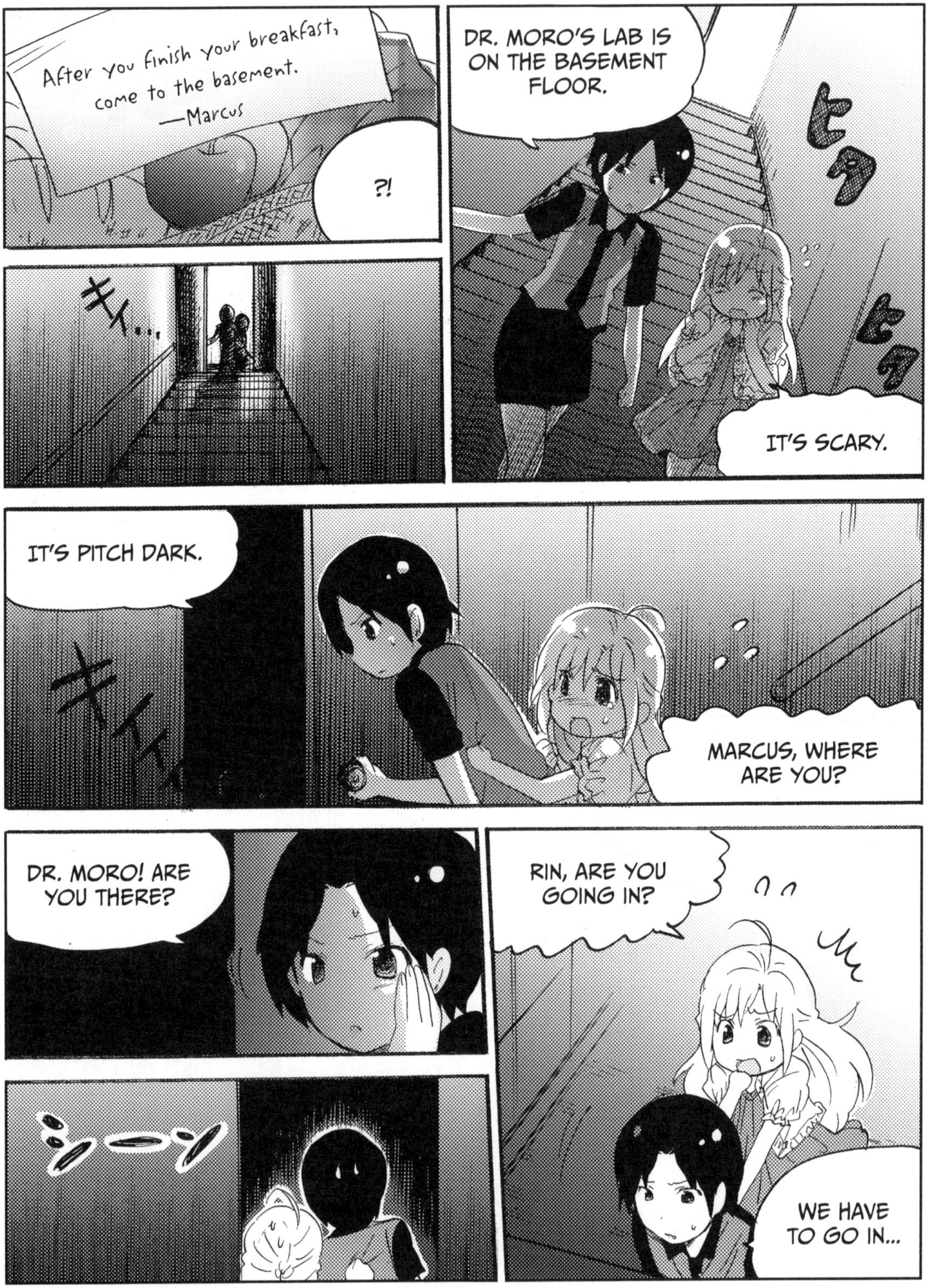
After you finish your breakfast, come to the basement.
—Marcus
?!
キイ...
DR. MORO'S LAB IS ON THE BASEMENT FLOOR.
ヒタ ヒタ
IT'S SCARY.
IT'S PITCH DARK.
キイイ
MARCUS, WHERE ARE YOU?
DR. MORO! ARE YOU THERE?
シーン
RIN, ARE YOU GOING IN?
WE HAVE TO GO IN...

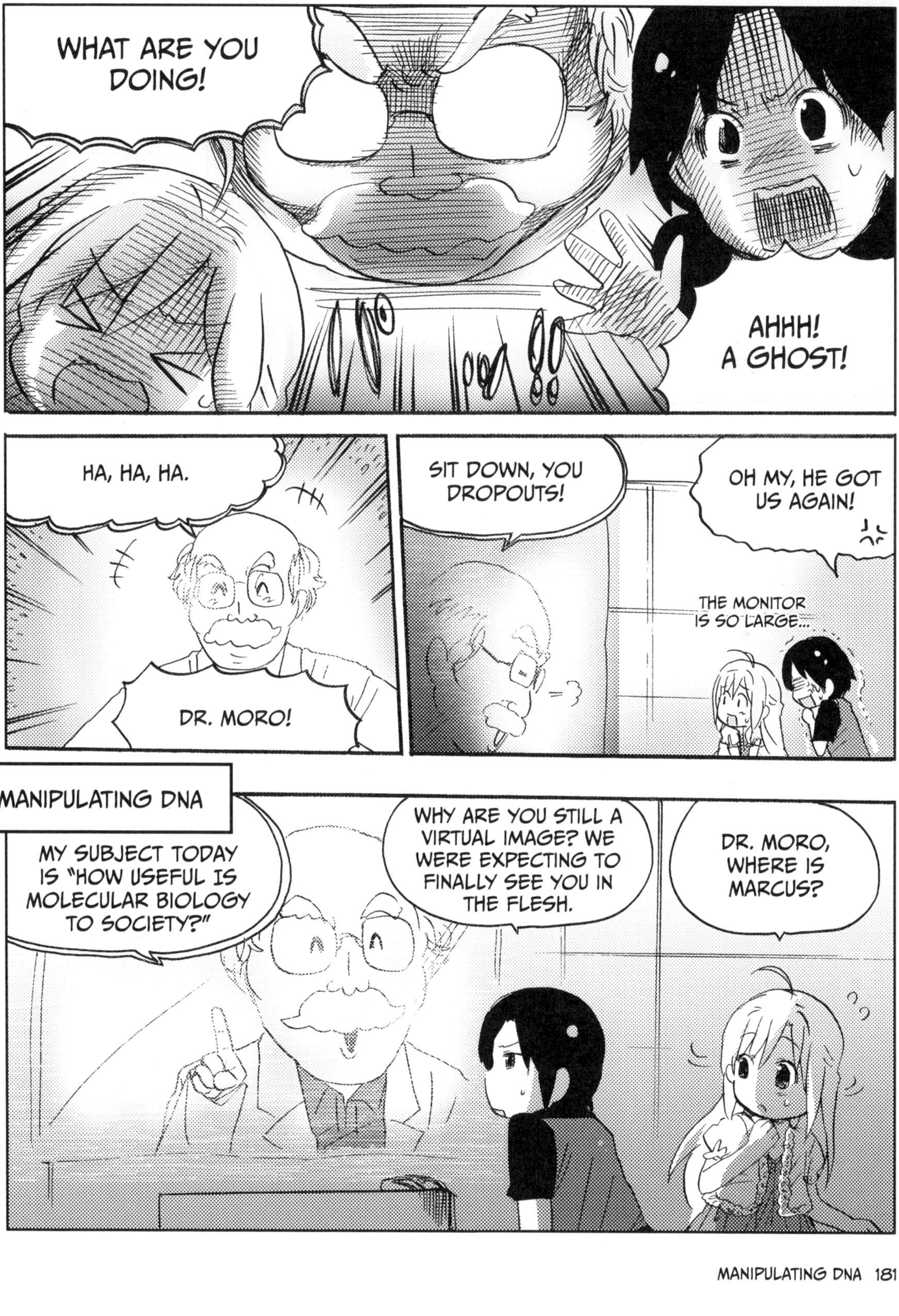
WHAT ARE YOU DOING!
AHHH! A GHOST!
HA, HA, HA.
DR. MORO!
SIT DOWN, YOU DROPOUTS!
OH MY, HE GOT US AGAIN!
THE MONITOR IS SO LARGE...
MANIPULATING DNA
MY SUBJECT TODAY IS "HOW USEFUL IS MOLECULAR BIOLOGY TO SOCIETY?"
WHY ARE YOU STILL A VIRTUAL IMAGE? WE WERE EXPECTING TO FINALLY SEE YOU IN THE FLESH.
DR. MORO, WHERE IS MARCUS?

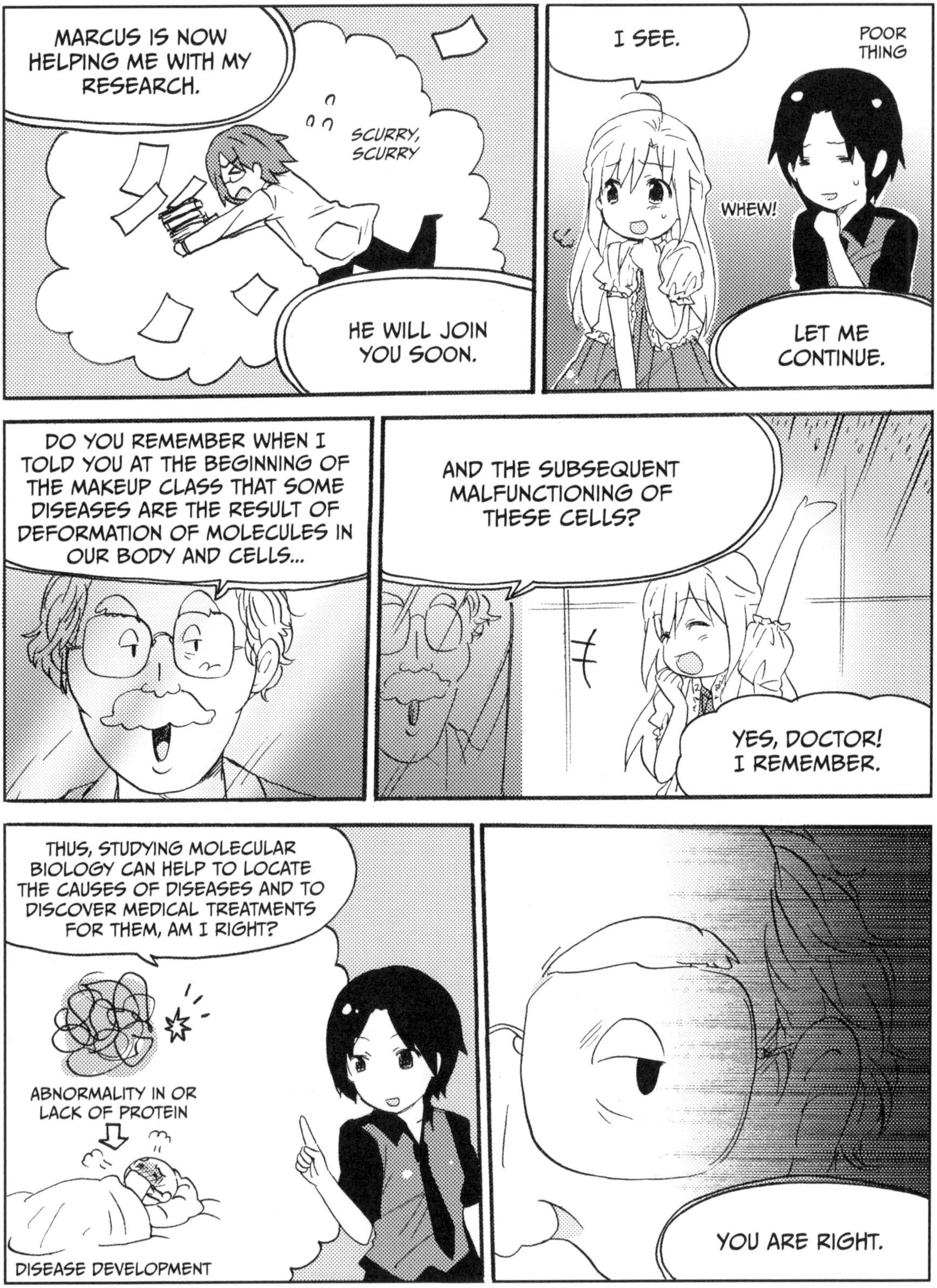
MARCUS IS NOW HELPING ME WITH MY RESEARCH.
SCURRY, SCURRY
HE WILL JOIN YOU SOON.
I SEE.
POOR THING
WHEW!
LET ME CONTINUE.
DO YOU REMEMBER WHEN I TOLD YOU AT THE BEGINNING OF THE MAKEUP CLASS THAT SOME DISEASES ARE THE RESULT OF DEFORMATION OF MOLECULES IN OUR BODY AND CELLS...
AND THE SUBSEQUENT MALFUNCTIONING OF THESE CELLS?
YES, DOCTOR! I REMEMBER.
THUS, STUDYING MOLECULAR BIOLOGY CAN HELP TO LOCATE THE CAUSES OF DISEASES AND TO DISCOVER MEDICAL TREATMENTS FOR THEM, AM I RIGHT?
ABNORMALITY IN OR LACK OF PROTEIN
DISEASE DEVELOPMENT
YOU ARE RIGHT.

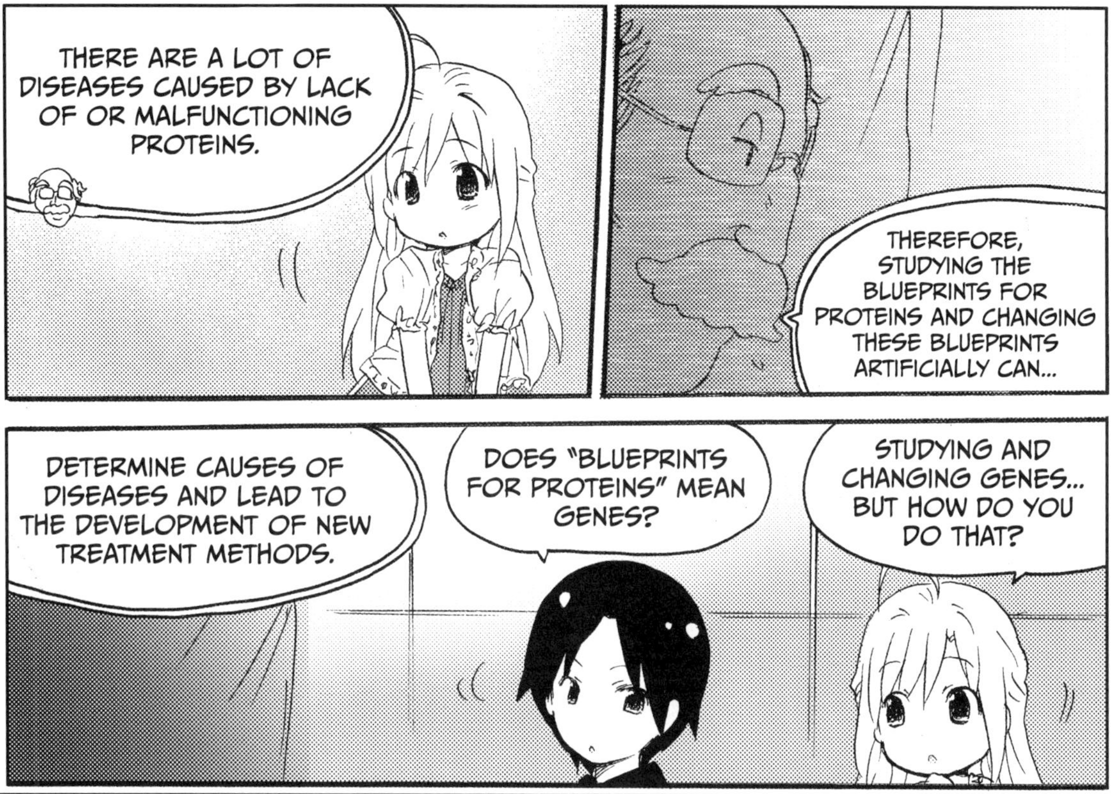
THERE ARE A LOT OF DISEASES CAUSED BY LACK OF OR MALFUNCTIONING PROTEINS.
THEREFORE, STUDYING THE BLUEPRINTS FOR PROTEINS AND CHANGING THESE BLUEPRINTS ARTIFICIALLY CAN...
DETERMINE CAUSES OF DISEASES AND LEAD TO THE DEVELOPMENT OF NEW TREATMENT METHODS.
DOES "BLUEPRINTS FOR PROTEINS" MEAN GENES?
STUDYING AND CHANGING GENES... BUT HOW DO YOU DO THAT?

BREED IMPROVEMENT AND GENETIC RECOMBINATION TECHNOLOGY

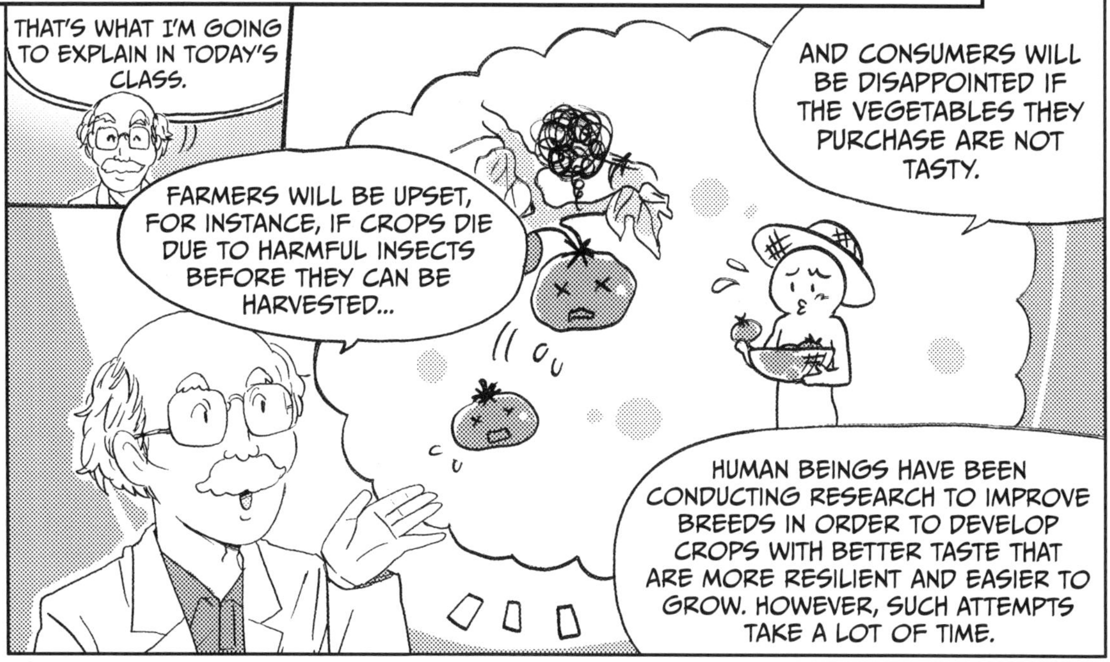
THAT'S WHAT I'M GOING TO EXPLAIN IN TODAY'S CLASS.
FARMERS WILL BE UPSET, FOR INSTANCE, IF CROPS DIE DUE TO HARMFUL INSECTS BEFORE THEY CAN BE HARVESTED...
AND CONSUMERS WILL BE DISAPPOINTED IF THE VEGETABLES THEY PURCHASE ARE NOT TASTY.
HUMAN BEINGS HAVE BEEN CONDUCTING RESEARCH TO IMPROVE BREEDS IN ORDER TO DEVELOP CROPS WITH BETTER TASTE THAT ARE MORE RESILIENT AND EASIER TO GROW. HOWEVER, SUCH ATTEMPTS TAKE A LOT OF TIME.

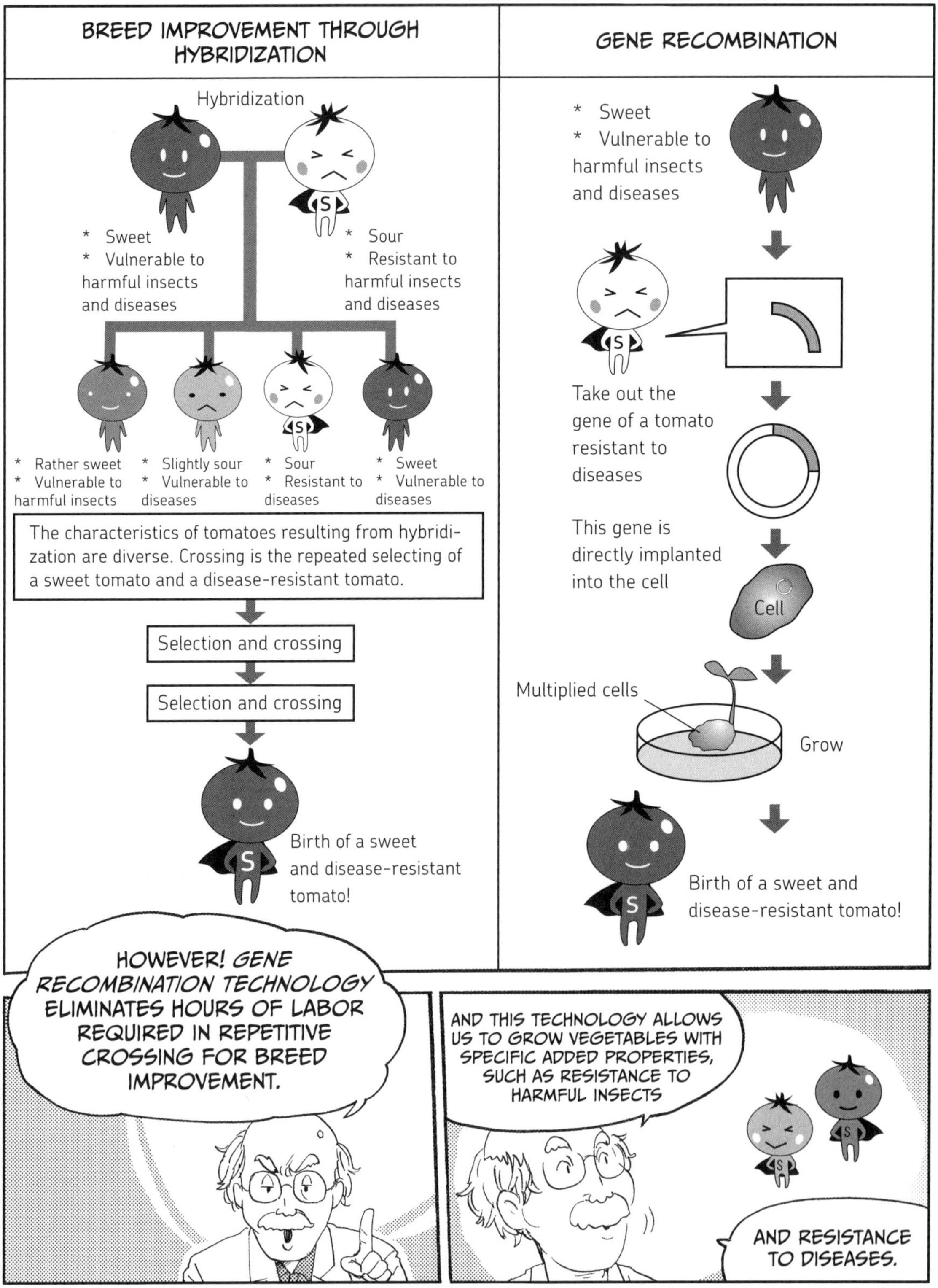
BREED IMPROVEMENT THROUGH HYBRIDIZATION
Hybridization
* Sweet
* Vulnerable to harmful insects and diseases
* Sour
* Resistant to harmful insects and diseases
* Rather sweet
* Vulnerable to harmful insects
* Slightly sour
* Vulnerable to diseases
* Sour
* Resistant to diseases
* Sweet
* Vulnerable to diseases
The characteristics of tomatoes resulting from hybridization are diverse. Crossing is the repeated selecting of a sweet tomato and a disease-resistant tomato.
Selection and crossing
Selection and crossing
Birth of a sweet and disease-resistant tomato!
GENE RECOMBINATION
* Sweet
* Vulnerable to harmful insects and diseases
Take out the gene of a tomato resistant to diseases
This gene is directly implanted into the cell
Cell
Multiplied cells
Grow
Birth of a sweet and disease-resistant tomato!
HOWEVER! GENE RECOMBINATION TECHNOLOGY ELIMINATES HOURS OF LABOR REQUIRED IN REPETITIVE CROSSING FOR BREED IMPROVEMENT.
AND THIS TECHNOLOGY ALLOWS US TO GROW VEGETABLES WITH SPECIFIC ADDED PROPERTIES, SUCH AS RESISTANCE TO HARMFUL INSECTS
AND RESISTANCE TO DISEASES.

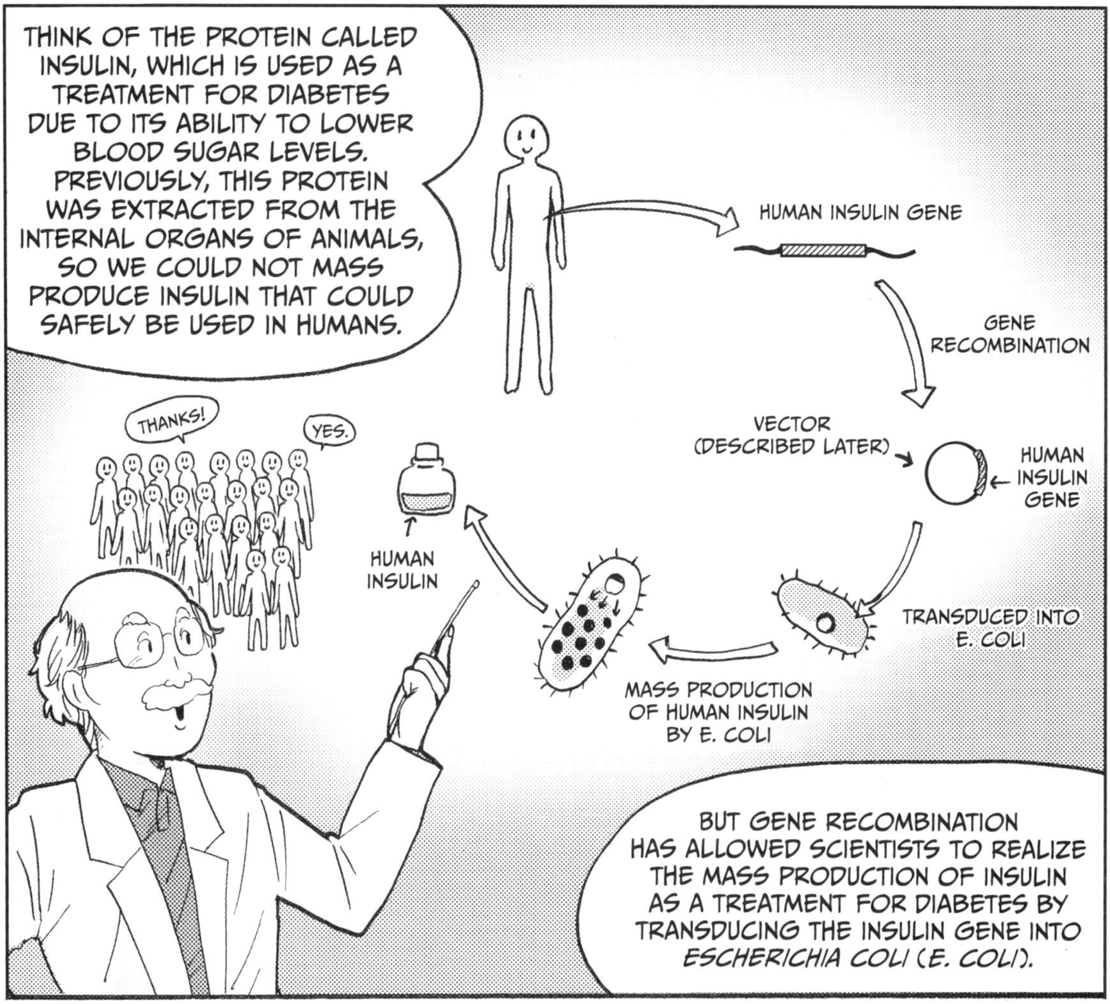
THINK OF THE PROTEIN CALLED INSULIN, WHICH IS USED AS A TREATMENT FOR DIABETES DUE TO ITS ABILITY TO LOWER BLOOD SUGAR LEVELS. PREVIOUSLY, THIS PROTEIN WAS EXTRACTED FROM THE INTERNAL ORGANS OF ANIMALS, SO WE COULD NOT MASS PRODUCE INSULIN THAT COULD SAFELY BE USED IN HUMANS.
HUMAN INSULIN GENE
GENE RECOMBINATION
VECTOR (DESCRIBED LATER)
HUMAN INSULIN GENE
TRANSDUCED INTO E. COLI
MASS PRODUCTION OF HUMAN INSULIN BY E. COLI
HUMAN INSULIN
THANKS!
YES.
BUT GENE RECOMBINATION HAS ALLOWED SCIENTISTS TO REALIZE THE MASS PRODUCTION OF INSULIN AS A TREATMENT FOR DIABETES BY TRANSDUCING THE INSULIN GENE INTO *ESCHERICHIA COLI* (*E. COLI*).

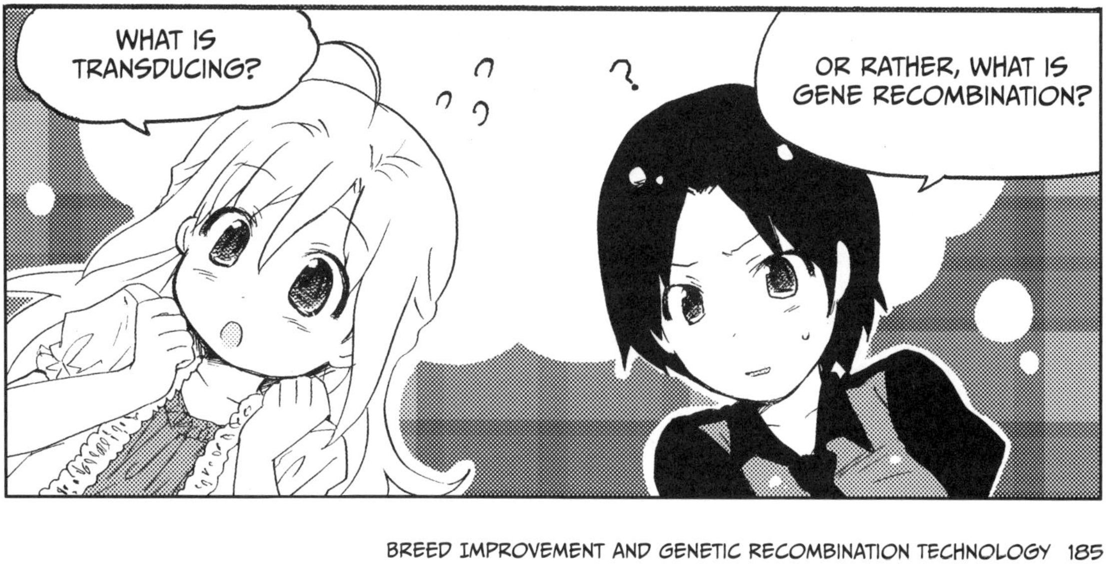
WHAT IS TRANSDUCING?
?
OR RATHER, WHAT IS GENE RECOMBINATION?

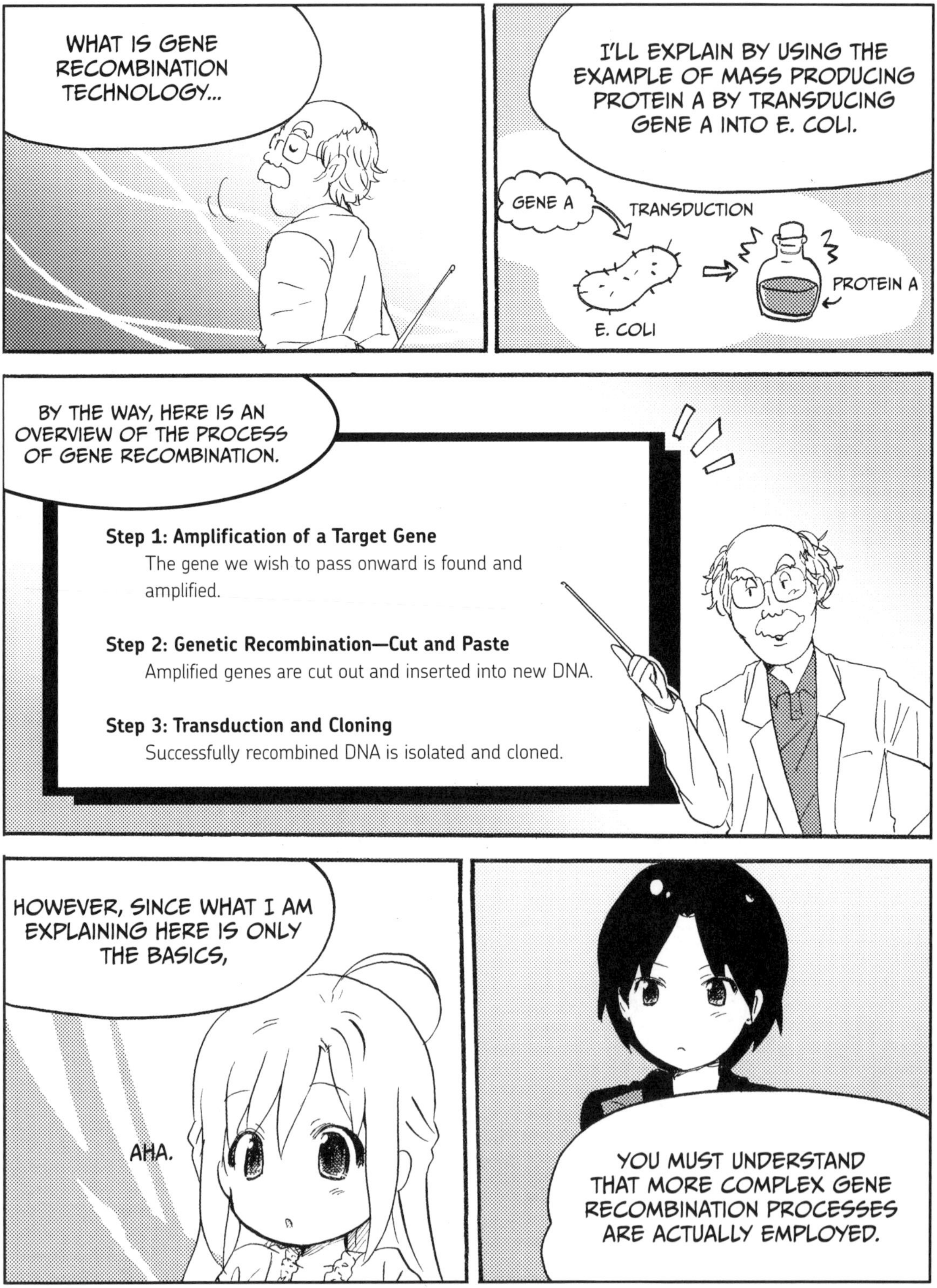
WHAT IS GENE RECOMBINATION TECHNOLOGY...
I'LL EXPLAIN BY USING THE EXAMPLE OF MASS PRODUCING PROTEIN A BY TRANSDUCING GENE A INTO E. COLI.
GENE A
TRANSDUCTION
E. COLI
PROTEIN A
BY THE WAY, HERE IS AN OVERVIEW OF THE PROCESS OF GENE RECOMBINATION.
Step 1: Amplification of a Target Gene
The gene we wish to pass onward is found and amplified.
Step 2: Genetic Recombination—Cut and Paste
Amplified genes are cut out and inserted into new DNA.
Step 3: Transduction and Cloning
Successfully recombined DNA is isolated and cloned.
HOWEVER, SINCE WHAT I AM EXPLAINING HERE IS ONLY THE BASICS,
AHA.
YOU MUST UNDERSTAND THAT MORE COMPLEX GENE RECOMBINATION PROCESSES ARE ACTUALLY EMPLOYED.

AN EXAMPLE OF GENETIC RECOMBINATION TECHNOLOGY

STEP 1: TARGET GENE IS MULTIPLIED

The width of a molecule forming the double helix of DNA is merely two nanometers (nm). One nm is one billionth of a meter (one millionth mm). DNA, the true identifier of genes, is so small that we cannot see it with our eyes.

Amino acid
DNA
Virus
Bacterium
1
10
100
1000 [nm]

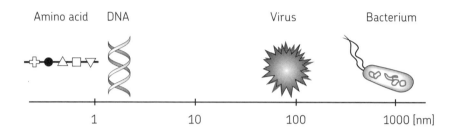

You know, it is very hard to handle invisible objects. So how can we make it visible?

We talked about water at the beginning of this book. We can't see individual water molecules. However, when a vast number of water molecules get together, we can recognize the liquid as water. This also holds true for DNA. If individual DNA is not visible, why shouldn't we multiply it until it becomes visible?

One of the technologies used for that purpose is *PCR* (*polymerase chain reaction*). PCR is a process that allows scientists to multiply a specific gene in a sample of DNA. Using this technology allows us, for instance, to make millions of copies of one gene and purify the DNA from that gene alone, or to detect the presence of the gene in a sample (for more information about PCR and how it works, see page 203).

Gene A
DNA
Amplifies gene A alone
through the use of PCR

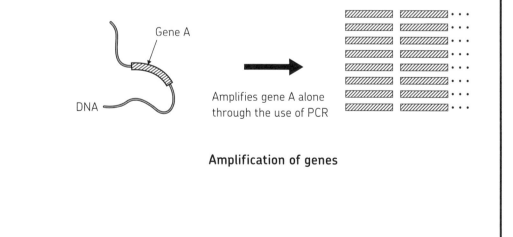

Amplification of genes

Researchers can now obtain genes from a large number of animals and plants (humans too!) using a database called a *cDNA library*. The genes are dissolved in a small volume of solution, which contains the cDNA of gene A. PCR works by using an RNA primer, a set of enzymes, and a pool of free nucleotides to make many copies of (amplify) gene A. PCR works differently than the cell's DNA replication process but uses many of the same raw materials.

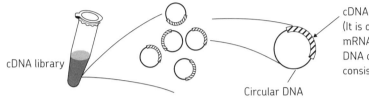

cDNA library

Circular DNA

cDNA
(It is obtained by converting mRNA to DNA. Although original DNA contains introns, cDNA consists of exons alone.)

STEP 2: GENETIC RECOMBINATION—CUT AND PASTE

Another kind of technology is required to insert the amplified gene A into a new strand of DNA. This technique makes up the major part of gene recombination technology.

Its principle is *cut and paste*. It's similar to the well known computer task of inserting a sentence or word at another location. In the first step, you cut gene A away from the rest of the DNA using a special enzyme called a *restriction enzyme*. But as gene A comes away, there is a little *overhang* or area of overlap on each side.

Overlap? What do you mean by an area of overlap?

Restriction enzymes act like scissors, cutting the strands of DNA at a given sequence of base pairs. For example, a restriction enzyme called EcoRI cuts only the portion where the base sequence is GAATTC. Look at the following figure carefully and you'll find that the base sequence of the partner of the double helix is also GAATTC when read backward. EcoRI cuts this portion in a zigzag manner, leaving the sequence AATT without any pairs, which creates an area of overlap. These overhangs are sometimes called *sticky ends*.

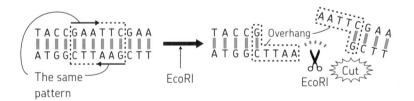

T A C C G A A T T C G A A
A T G G C T T A A G C T T

The same pattern

EcoRI

T A C C G: Overhang
A T G G C T T A A

A A T T C G A A
G C T T

Cut

EcoRI

One neat thing about PCR is that, when you are amplifying gene A, you can simply add on the DNA sequence recognized by the restriction enzyme to both ends of the gene. As gene A amplifies, this new sequence gets copied over, along with the DNA for gene A (for details, see page 203).

Using the same restriction enzyme, this technology cuts both the ends of the DNA for gene A, as well as the DNA within the target cell, and leaves the same area of overlap on each piece. In this way, you can paste gene A into the DNA in the target cell.

Then PCR mixes the two types of DNA (gene and destination DNA) and connects the overhangs with an enzyme called *DNA ligase*. This completes the pasting process.

Overhang
Overhang
+
=
Paste
Gene
Destination DNA

STEP 3: TRANSDUCTION AND CLONING

I've mentioned *destination DNA* several times. Why do we have to insert gene A into another DNA?

When inserting a gene into a living organism and having it implement its function (express the gene), you must insert it into a special carrier (made of DNA) through the use of the above cut and paste method.

This carrier is called a *vector*. It is made of circular or hoop-shaped DNA that was derived from a DNA called *plasmid* contained in the cells of bacteria such as E. coli. Researchers modified the plasmid, and a variety of vectors have been developed to serve various purposes (some vectors are derived from viruses).

Plasmids replicate arbitrarily in the cells of bacteria. Thus, if you transduce (there are several transduction methods, including the electric shock method) the plasmid-derived vector being inserted with gene A into bacteria such as *Escherichia coli* (*E. coli*)—which can easily multiply in a laboratory environment—you can easily create a vast number of them.

Location for the gene insertion is predetermined
Vector
Gene A
Cut and paste
Transduction
E. coli
Multiplication

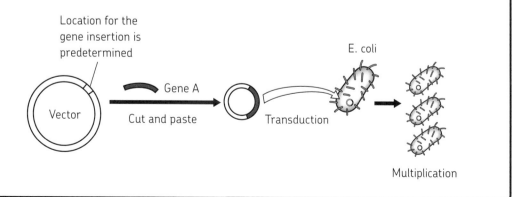

After the bacteria have multiplied many times, DNA purification allows us to obtain a massive number of copies of the plasmid (called *clones*) containing gene A. This process is called *cloning*.

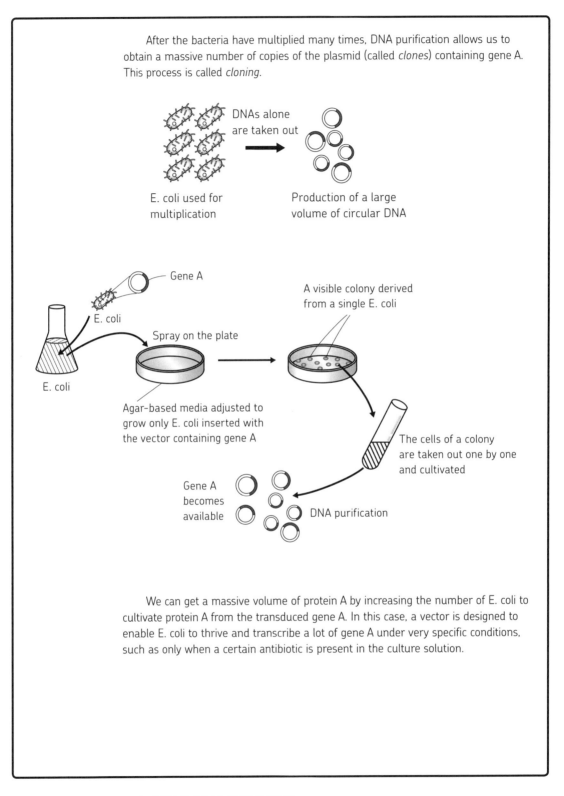

DNAs alone
are taken out
E. coli used for
multiplication
Production of a large
volume of circular DNA

Gene A
E. coli
E. coli
Spray on the plate
A visible colony derived
from a single E. coli
Agar-based media adjusted to
grow only E. coli inserted with
the vector containing gene A
The cells of a colony
are taken out one by one
and cultivated
Gene A
becomes
available
DNA purification

We can get a massive volume of protein A by increasing the number of E. coli to cultivate protein A from the transduced gene A. In this case, a vector is designed to enable E. coli to thrive and transcribe a lot of gene A under very specific conditions, such as only when a certain antibiotic is present in the culture solution.

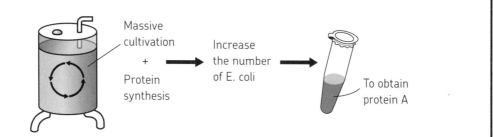

Massive cultivation + Protein synthesis → Increase the number of E. coli → To obtain protein A

Currently, it is possible to construct proteins not only from bacteria such as E. coli, but also from various types of cells, including insect cells and cells of mammals. Vectors specialized for each are being developed.

METHODS FOR DETECTING AND ISOLATING DNA

A little while ago, we learned that as chromosomes condense they become visible under the microscope. But when we are working with just one gene, how can we see it? We can visualize the products of PCR, or any mixture of DNA, using a process called *gel electrophoresis*. This process manipulates many of the chemical properties of DNA to separate, purify, and visualize the DNA that we isolate and synthesize (using PCR) in the lab.

A solution containing a mixture of DNA is poured into holes (wells) at the top of an agar-based plate. The agar medium is made of strands of protein that act like mesh. As an electric current is run through the plate, the pieces of DNA move through the agar based on their size. Longer strands of DNA get stuck in the mesh of the agar and migrate more slowly than shorter strands. Why do we use an electric current? The sugar-phosphate backbone of DNA is negatively charged; as the current is applied the negative charge of the DNA is attracted to the positive pole.

Then we can use chemicals to actually *see* the DNA. Certain molecules can bind in between the nitrogenous bases of DNA. The chemical structure of these molecules makes them fluoresce under ultraviolet light. Adding this kind of chemical (ethidium bromide, for example) allows us to see DNA by having it fluoresce under UV light. In this manner, we can use gel electrophoresis not only to separate but also to see the pieces of DNA in a mixture.

If you have run PCR for a specific gene, you can simply cut out the bands of DNA on the gel that correspond to the gene you want (that is, with the same length as your gene). A pure sample of DNA can be obtained by dissolving the agar, extracting out the DNA using alcohol, and washing the sample clean. In this way, you can obtain a large volume of *pure DNA* (not containing DNA strands with a different base sequence or length) that can be used for other purposes, such as gene recombination.

DNA
Negative
Agar is dissolved
An electric current is applied
Targeted portion is isolated
Intended DNA is obtained as pure DNA
Agar-based plate (agarose gel)
DNA strands separate into several groups
Positive

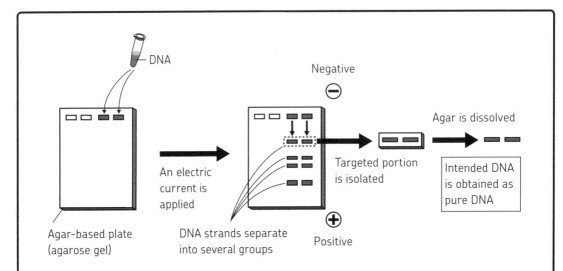

TRANSGENIC ANIMALS (KNOCKOUT MOUSE)

While gene recombination technology is applied in the improvement of agricultural crops and mass production of medicines, the number of areas it can be applied in does not stop there.

First of all, it is contributing greatly to the study of molecular biology itself. For instance, by transducing a certain gene into a cultured cell and analyzing what happens—namely, how the cell changes—

We can determine the properties of the protein constructed by the gene?

That's right! Genetic recombination technologies allow us to discover the function (or functions) of a particular protein. Moreover, with slightly more sophisticated techniques, you can study how cells behave once a gene is *deleted* to give you even more information about how a protein works in the cell. These methods are not limited to cell culture alone; they can also be used to make *transgenic animals*. Transgenic animals have an extra gene *added to* or *deleted from* their genome. They are very useful in studying how genes affect growth or disease development in a whole organism (mice are the most common, but fruit flies, rabbits, zebrafish, yeast, and mustard plants are also used).

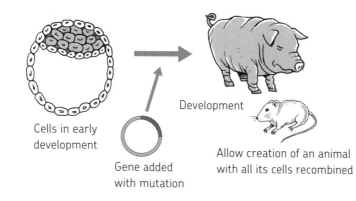

Cells in early
development

Development

Gene added
with mutation

Allow creation of an animal
with all its cells recombined

By using gene transduction methods on animal cells at a very early stage of development, you can incorporate a new gene into some, but not all, of the cells of the developing animal. This makes an animal with a mixed genome, called a *mosaic*. By breeding the offspring of mosaic mice together (some eggs and sperm will have the new gene, called a *transgene*) you can create animals that have the transgene as part of their DNA, so every cell will express that new gene.

By transducing a gene into an *ES cell* that is mutated or a gene sequence that deactivates another gene (already in the mouse's genome), you can create transgenic animals that lack a certain gene (and thus a functioning copy of that protein), called *knockout mice*.

What is an ES cell?

An ES cell is an *embryonic stem cell*. It is a very special kind of cell that is harvested from an embryo before any of the cells have started to specialize into one tissue or another. Because an ES cell has the ability to develop into a cell in any kind of tissue in the body, it is called *totipotent*. If you add in or knock out a gene in an embryonic stem cell and then put it back into the growing embryo, the mouse will become a mosaic.

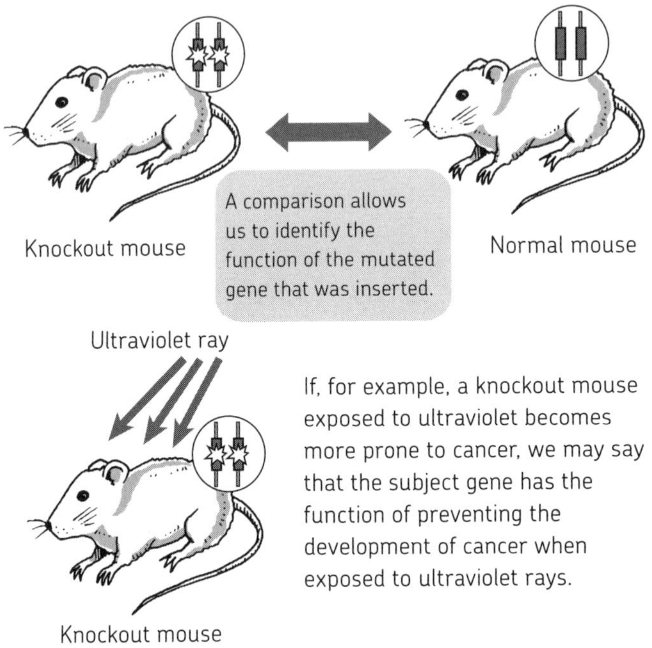

Knockout mouse

A comparison allows us to identify the function of the mutated gene that was inserted.

Normal mouse

Ultraviolet ray

If, for example, a knockout mouse exposed to ultraviolet becomes more prone to cancer, we may say that the subject gene has the function of preventing the development of cancer when exposed to ultraviolet rays.

Knockout mouse

Knockout mice are particularly useful. We can determine the original function of a gene by studying the differences between a knockout mouse and a normal mouse. For example, if a knockout mouse exposed to ultraviolet light develops more skin cancer than a normal (*wild-type*) mouse, we may say that the target gene likely helps prevent the development of skin cancer caused by UV rays. Note that we would need further studies to show *how* the gene prevents skin cancer. The 2007 Nobel Prize in Physiology and Medicine was awarded to the three scientists who discovered the method for producing knockout mice using ES cells.

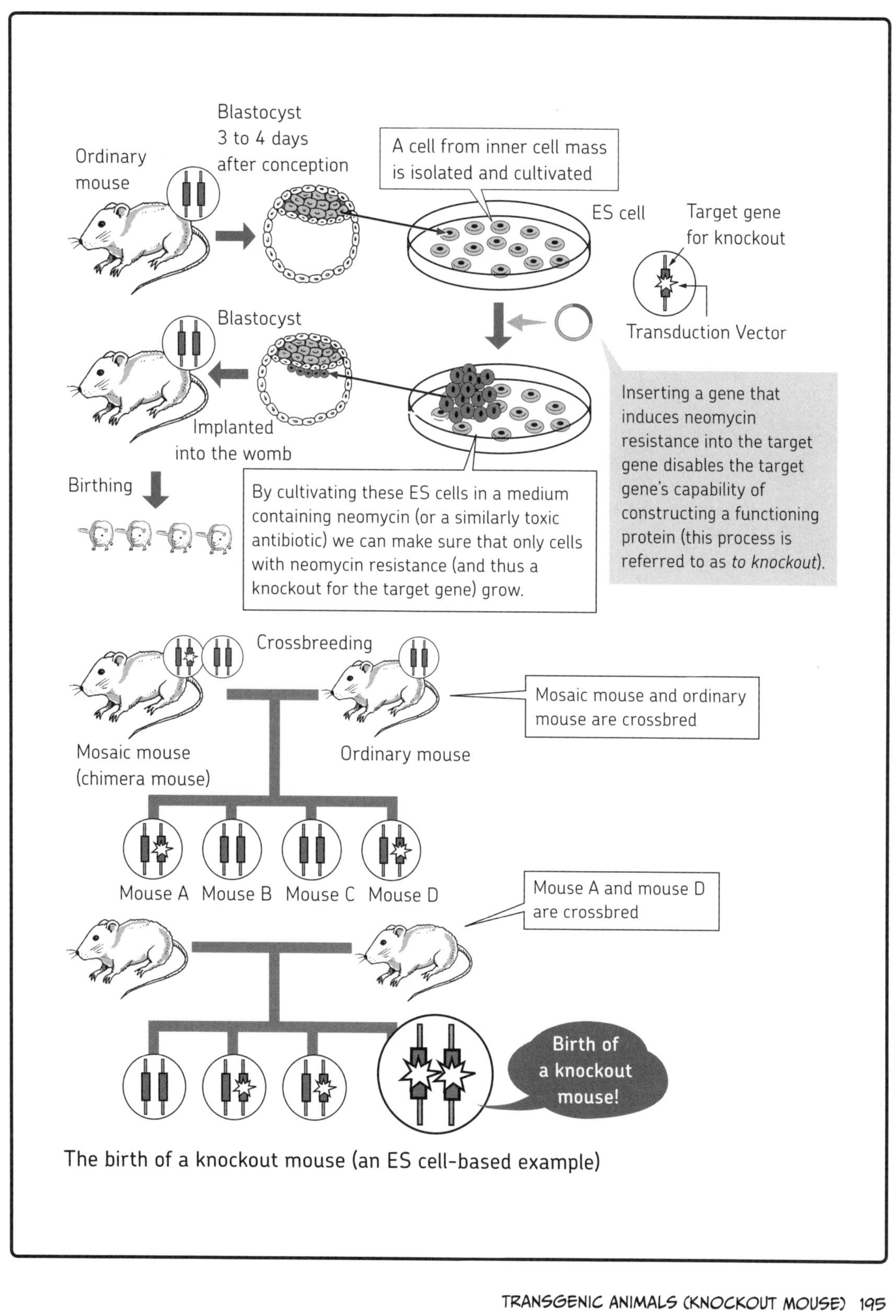
Ordinary mouse
Blastocyst 3 to 4 days after conception
A cell from inner cell mass is isolated and cultivated
ES cell
Target gene for knockout
Transduction Vector
Blastocyst
Implanted into the womb
Inserting a gene that induces neomycin resistance into the target gene disables the target gene's capability of constructing a functioning protein (this process is referred to as *to knockout*).
Birthing
By cultivating these ES cells in a medium containing neomycin (or a similarly toxic antibiotic) we can make sure that only cells with neomycin resistance (and thus a knockout for the target gene) grow.
Crossbreeding
Mosaic mouse and ordinary mouse are crossbred
Mosaic mouse (chimera mouse)
Ordinary mouse
Mouse A
Mouse B
Mouse C
Mouse D
Mouse A and mouse D are crossbred
Birth of a knockout mouse!

The birth of a knockout mouse (an ES cell-based example)

PERSONALIZED MEDICINE AND GENE THERAPY: ARE GENETICS THE FUTURE OF DISEASE PREVENTION?

 Have you ever heard the term *metabolic syndrome*?

 My father has metabolic syndrome.

 Wait, wait, are you saying your father has the syndrome just because his waistline is large? That is wrong.

 Oh, why?

 Metabolic syndrome refers to the state where someone with visceral fat obesity has developed two or more of the following: a high blood sugar level, high blood pressure, or dyslipidemia. Diseases like high blood pressure and dyslipidemia are called *lifestyle-related diseases* and are considered to be the result of lifestyle, including dietary habits and exercise. Lifestyle-related diseases include serious ailments such as diabetes, cardiac infarction, brain infarction, and cancer of the large intestine, which in some cases can result in death.

 I think inadequate exercise, as well as excessive eating and drinking, is the cause.

 I agree, but there is evidence that some of these lifestyle-related diseases are also caused by genetic factors.

 Genetic factors?

 Demographic studies, that is, studies looking at a whole population, have shown that mutations in certain genes make a person much more likely to develop lifestyle-related diseases independent of their diet and exercise level.

I see, so for conditions like metabolic syndrome it's not just someone's lifestyle that is the cause.

That's right. And a change in the gene often involves the replacement of only one base in the DNA sequence with another. The interesting thing is that the proteins still work close to normal, but slight changes in the structure can change how the protein works in a subtle way. There is now evidence that such small replacements, over the course of a lifetime, can increase the risk of suffering a heart attack or getting certain types of cancer.

A G G T T C G A T G C
A G G T T C G G T G C
Prone to suffering from a specific disease
Mere replacement of A with G

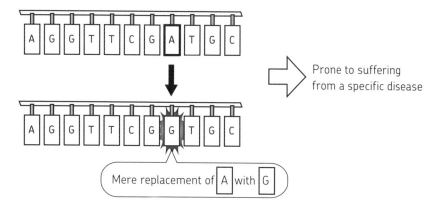

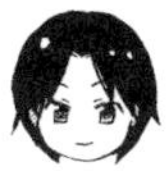

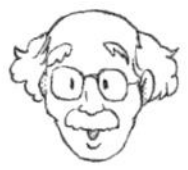

Note: This is a conceptual diagram and does not indicate that this base sequence tends to cause a specific disease.

Thanks to recent research, our knowledge of how genes can contribute to disease is growing. In the future, you may be able to have your own genome sequenced to look for diseases you might be at risk of developing. Physicians call this concept *personalized medicine*, tailoring your medical treatment to your known risk of diseases based on your genes.

Knowing which diseases I am at risk for—that's scary. I would rather not know.

I would want to know! Because then I would be able to take all necessary measures.

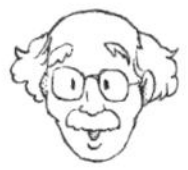

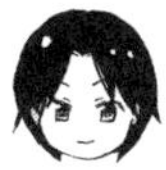

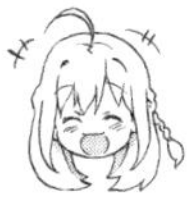

Necessary measures?

You know, just so I could prepare mentally.

In addition to mental preparedness, you would be able to make changes to your lifestyle. By knowing ahead of time what diseases you are at risk for, you can tailor your diet, how much you exercise, and what activities you do to help delay the onset of these preventable diseases. Personalized medicine may one day be a cornerstone of preventative health care.

That's right, change my diet and exercise more. That's what I meant!

I guess that makes sense . . .

GENE THERAPY

There is another frontier in medical treatment that you should know about. This new treatment is called *gene therapy*.

I've heard of gene therapy. I think it was on the news.

What kind of treatment is it?

Imagine that a baby is born with an abnormal gene and that gene is important for life. With a serious genetic disease, the baby could die at birth or before it ever gets a chance to grow.

In order to save the lives of people with serious genetic diseases, contemporary medicine sometimes employs an approach called gene therapy. Normal, healthy, genes are inserted into specialized vectors that are artificially transduced into some of the cells of the sick person. Using this method, the normal version of the gene is added back into the body so that cells can make a functioning, normal version of the missing or nonfunctional protein.

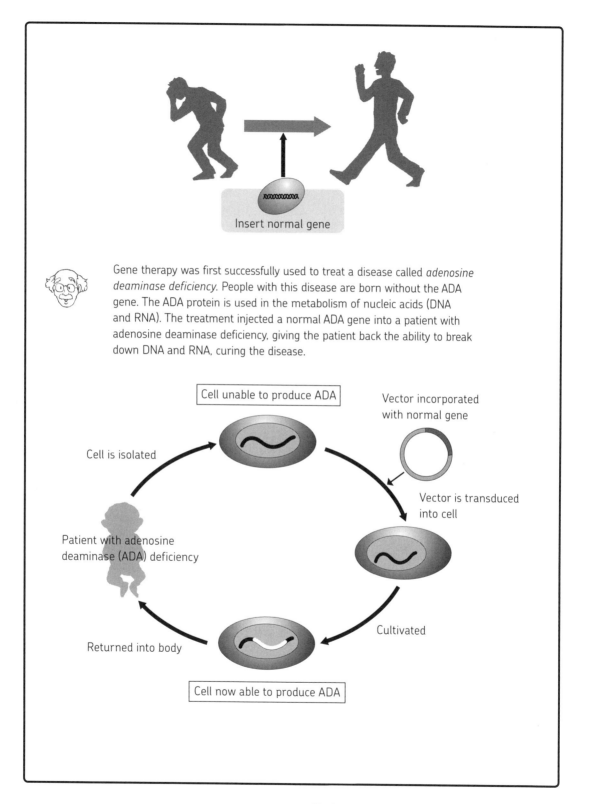

Insert normal gene

Gene therapy was first successfully used to treat a disease called *adenosine deaminase deficiency*. People with this disease are born without the ADA gene. The ADA protein is used in the metabolism of nucleic acids (DNA and RNA). The treatment injected a normal ADA gene into a patient with adenosine deaminase deficiency, giving the patient back the ability to break down DNA and RNA, curing the disease.

Cell unable to produce ADA

Vector incorporated with normal gene

Cell is isolated

Vector is transduced into cell

Patient with adenosine deaminase (ADA) deficiency

Cultivated

Returned into body

Cell now able to produce ADA

It seems like gene therapy is a dream come true for people with genetic diseases.

 Sadly, things are not so simple.

 Why not?

 Gene therapy is very difficult technically, often dangerous, and still experimental. Genetic diseases where gene therapy could work well, like ADA deficiency, are rare. While some trials of gene therapy have been used on patients with brain tumors and breast cancer, the trials have had limited success. Additionally, there are many ethical complications that have stopped gene therapy from coming into widespread use. There are restrictions on the handling of genetic material and on what tissues are candidates for gene insertion. There are other ethical considerations involved in testing experimental therapies on children.

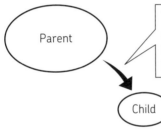

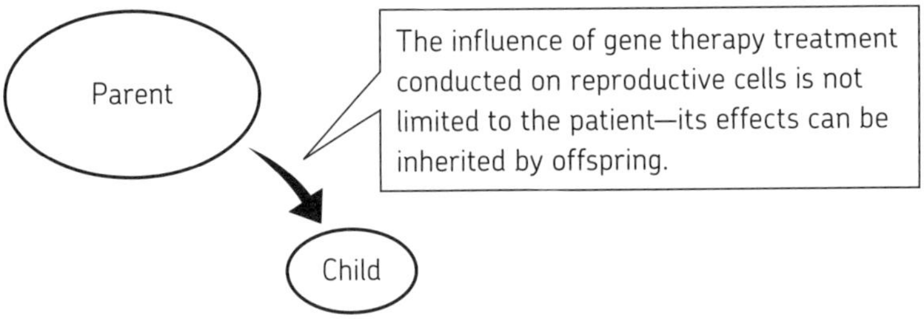

The influence of gene therapy treatment conducted on reproductive cells is not limited to the patient—its effects can be inherited by offspring.

 Because developing gene therapy treatments requires a lot of money and manpower, its usage is mostly limited to diseases that are otherwise incurable.

 But . . .

 I know what you're going to say. How can we forget about all the patients who would benefit from further study of gene therapy? We aren't ignoring them. It's just that gene therapy, like any other new technology, has limitations.

 It seems like a difficult problem.

THE RNA RENAISSANCE

Scientists believe that, long ago, RNA was the first and, at that time, *only* nucleic acid. Well before human life existed on earth, RNA carried out the role of encoding genes and passing them on to future generations. At some point, RNA's role was replaced by DNA, a much more stable molecule for storing genetic information. As researchers began learning about molecular biology, they thought that protein synthesis was the only important function that RNA had.

But recent discoveries suggest that RNA has many different and important functions. On top of that, scientists are becoming experts at manipulating RNA to do even more jobs within the cell. In the current world of molecular biology, studying and manipulating RNA is more in fashion than studying DNA. Some researchers refer to our expanding knowledge and experimentation with RNA as the *RNA renaissance*. RNA, which until recently was thought of as just a copy, has captured the imagination of scientists all over the world.

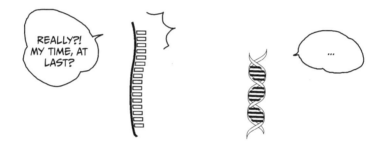

RNA INTERFERENCE: USING RNA TO ALTER GENE EXPRESSION

The 2006 Nobel Prize in Physiology and Medicine was awarded to two American molecular biologists—Andrew Z. Fire and Craig C. Mello—who discovered a phenomenon called *RNA interference*. A very short, complementary strand of RNA can bind to a molecule of mRNA in the cell. This binding creates a double-stranded molecule of RNA, which the body recognizes as foreign. Enzymes in the cytoplasm then attack and break down the mRNA and interfering RNA, thus stopping that mRNA from being made into a protein.

Why on earth was the Nobel Prize was awarded for discovering such a strange and destructive process?

Although this interference is destructive to mRNA, it is convenient from the viewpoint of the entire cell. Breaking down mRNA is one way of suppressing that gene's expression, which helps maintain the delicate balance of proteins in the cell. It was originally thought that the interfering short RNA strands were a mechanism to defend the cell against viruses, but recent studies have determined that too much

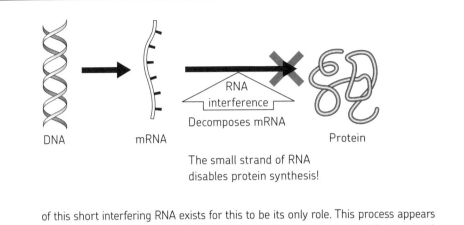

DNA
mRNA
RNA
interference
Decomposes mRNA
Protein
The small strand of RNA
disables protein synthesis!

of this short interfering RNA exists for this to be its only role. This process appears to have emerged from the cells' defense mechanism against viruses. Viruses are the only organisms that still use double-stranded RNA to encode their genes, and the cell has many enzymes to break down double-stranded RNA. Studies have shown that a vast number of such "interfering" short-strand RNAs exist in our cells, apart from any viruses.

Any functioning society must have balance. Similarly, your cells must also have balance. This happens through the control of gene expression. RNA, a nucleic acid, is one factor that contributes to the balance of gene expression.

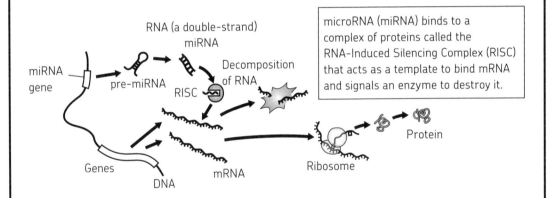

RNA (a double-strand)
miRNA
miRNA
gene
pre-miRNA
RISC
Decomposition
of RNA
microRNA (miRNA) binds to a
complex of proteins called the
RNA-Induced Silencing Complex (RISC)
that acts as a template to bind mRNA
and signals an enzyme to destroy it.
Protein
Ribosome
Genes
DNA
mRNA

As you can see, RNA has many functions. We had already learned about mRNA, tRNA, and ribosomal RNA used in making proteins. Now we also know about another function of RNA. RNA can be used to alter gene expression using *short interfering RNA* (*siRNA*) and *microRNA* (*miRNA*). Other kinds of RNA act on their own as enzymes, called *ribozymes*. It has been discovered that over 70 percent of the mouse genome can be used to transcribe RNA. Substantial portions of the human genome are also used to synthesize RNA. While scientists have made great strides in learning about the functions of RNA, there are still many kinds of RNA molecules whose functions remain a mystery.

CAN RNA CURE DISEASES?

With all of the advances in RNA research, scientists have begun looking at how RNA might be used to treat diseases. Currently, many companies are working on drug discovery projects using siRNA. RNA has the following two features that make it amenable for drug development: RNA can be easily synthesized with any combination of base sequence (A, G, C, U), allowing for a large variety of shapes and sizes; and the cell can easily break down RNA, allowing for quick metabolism of any RNA-based drug.

Certain molecules of RNA have the ability to bind to proteins. RNA medications work by binding to abnormal, disease-causing proteins. The ability for a specific sequence of RNA to bind to a protein is analyzed and optimized using a procedure called the *SELEX method* (*Systematic Evaluation of Ligands by Exponential Enrichment*). Using this method, sets of RNA molecules that are very similar (but not the same) in structure are analyzed to see which best bind to a specific protein. RNA that is targeted to bind to a protein is called an *RNA aptamer*.

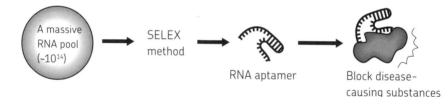

A massive RNA pool (~10^{14}) → SELEX method → RNA aptamer → Block disease-causing substances

The first RNA-based medicine to be approved by the US FDA was *Pegapatnib*, an RNA aptamer that binds VEGF 165 and is used to treat age-related macular degeneration, a crippling eye disease. Because RNA and RNA-based medications are easily broken down, these medications are believed to have few side effects. Whether RNA aptamers will become miracle-drugs has yet to be seen and depends on the result of much more future research.

HOW EXACTLY DOES PCR WORK?

The method for multiplying, or amplifying, genes for use in the laboratory is called *PCR* (*Polymerase chain reaction*; see page 187). Let's examine more details of this method here. Remember from Chapter 3 that the protein (enzyme) that replicates DNA is called *DNA polymerase*. Using DNA polymerase, PCR multiplies a copy of a specific gene in an exponential manner.

This process happens in a very controlled manner in the body, but for use in the lab, PCR was designed to create continuous replication of a gene by altering the enzymes and the temperature of the reaction. A large pool of nucleotides and two RNA primers for the gene of interest (one for each strand, forward and backward) are added to a buffer containing a small sample of DNA. An increase in temperature melts the DNA into two single strands; as the solution cools, the RNA primer can bind to the DNA and DNA polymerase makes a copy of the gene using the free nucleotides in the solution.

DNA polymerase
Untying
DNA
❶
❷
Primer
❸
92°C
56°C
72°C

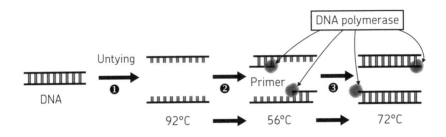

❶ 92°C: DNA is separated to a single strand at 92°C.
❷ 56°C: When cooled to 50°C, mating primer is bonded with each strand.
❸ 72°C: At 72°C, DNA polymerase starts DNA synthesis.

When the temperature is increased to 92°C again, two completed DNA strands separate from each other. Then, the steps are repeated over and over again.

Usually, proteins break down, or *denature*, at high temperatures (40-50°C or 100-125°F). A very special version of the DNA polymerase enzyme is used in PCR, one that can withstand very high temperatures. This enzyme was isolated from a bacterium called *Thermus aquaticus*. This bacteria is found only in hot springs and was discovered in the geyser pools of Yellowstone National Park, living in near-boiling (165°F) water.

Karry Mullis developed PCR in 1993 and won the Nobel Prize in Chemistry that same year. Since then, compact machines have been developed which can be programmed to control the temperature and time of each step of the PCR reaction to mass produce copies of a gene. As was mentioned earlier (page 188), adding additional base pairs to the primer incorporates this new sequence into the copied gene. It is in this manner that PCR can be used to create genes that can be cut and pasted into a plasmid with restriction enzymes.

Newer methods of PCR have been developed that allow you to turn RNA into DNA to be copied via PCR (*RT-PCR*) and to count the number of copies of DNA produced so accurately that you can use equations to figure out how many copies of a gene were present in a sample (*quantitative PCR*). These methods working together can help you figure out how much mRNA a cell is making of a particular gene (directly measure gene expression) and also make up the basis of DNA testing in forensic science.

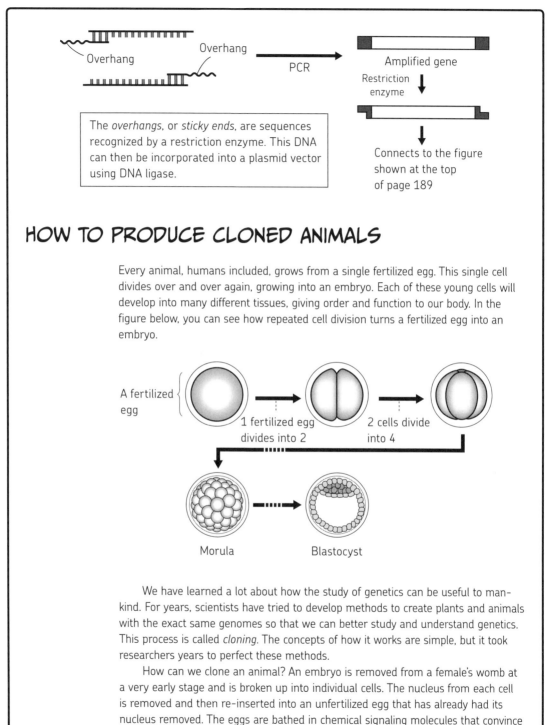

Overhang

Overhang

PCR

Amplified gene

Restriction
enzyme

The *overhangs*, or *sticky ends*, are sequences recognized by a restriction enzyme. This DNA can then be incorporated into a plasmid vector using DNA ligase.

Connects to the figure shown at the top of page 189

HOW TO PRODUCE CLONED ANIMALS

Every animal, humans included, grows from a single fertilized egg. This single cell divides over and over again, growing into an embryo. Each of these young cells will develop into many different tissues, giving order and function to our body. In the figure below, you can see how repeated cell division turns a fertilized egg into an embryo.

A fertilized egg

1 fertilized egg divides into 2

2 cells divide into 4

Morula

Blastocyst

We have learned a lot about how the study of genetics can be useful to mankind. For years, scientists have tried to develop methods to create plants and animals with the exact same genomes so that we can better study and understand genetics. This process is called *cloning*. The concepts of how it works are simple, but it took researchers years to perfect these methods.

How can we clone an animal? An embryo is removed from a female's womb at a very early stage and is broken up into individual cells. The nucleus from each cell is removed and then re-inserted into an unfertilized egg that has already had its nucleus removed. The eggs are bathed in chemical signaling molecules that convince the eggs that they have been fertilized. These eggs are then transplanted into the wombs of several other females. As a result, many individual organisms with the exact same genes (*genomes*) are born. Today, the agriculture industry uses this method to breed a large number of the same animal, or *clones*, with desirable traits. With just one fertilized egg, many cows (and a lot of beef, leather, and milk) can be produced.

This process may sound a little familiar. It is like identical twins, in that two individual organisms are born from a single fertilized egg. Animals cloned in this way have male and female parents, just like those produced from ordinary breeding. They are not the clones of their parents, but of each other.

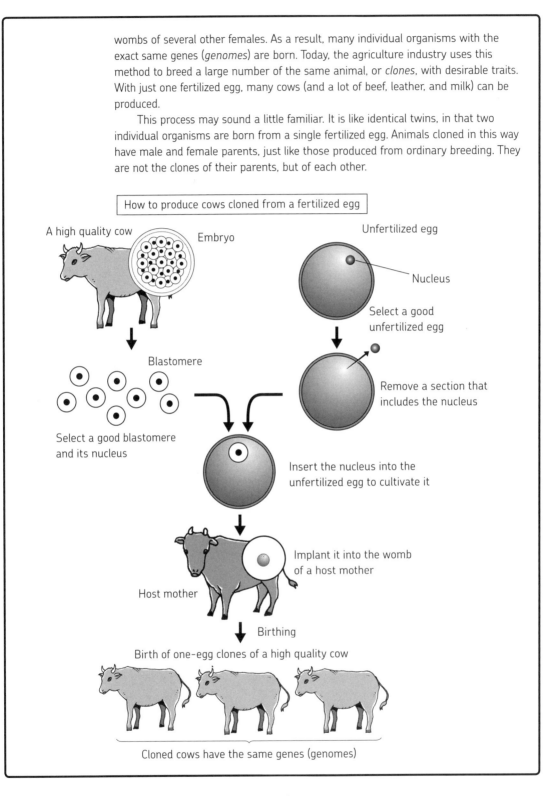

How to produce cows cloned from a fertilized egg

A high quality cow

Embryo

Unfertilized egg

Nucleus

Select a good unfertilized egg

Blastomere

Remove a section that includes the nucleus

Select a good blastomere and its nucleus

Insert the nucleus into the unfertilized egg to cultivate it

Implant it into the womb of a host mother

Host mother

Birthing

Birth of one-egg clones of a high quality cow

Cloned cows have the same genes (genomes)

Scientists have developed more than one way to clone an animal. You may have heard of the sheep, Dolly, who was cloned in a British laboratory in 1996. Dolly was the first cloned mammal to have only one parent. Instead of being cloned from a sex cell (like the eggs mentioned above), Dolly was cloned from a *somatic cell*. Let's take a look at how this kind of cloning works. The process begins by removing the entire nucleus from a breast cell from the parent sheep (the cell was from the *lacteal gland*, an organ that excretes milk). An unfertilized egg is harvested, and the nucleus of the unfertilized egg is replaced with the full nucleus from the breast cell. The egg is then implanted into a host-mother sheep.

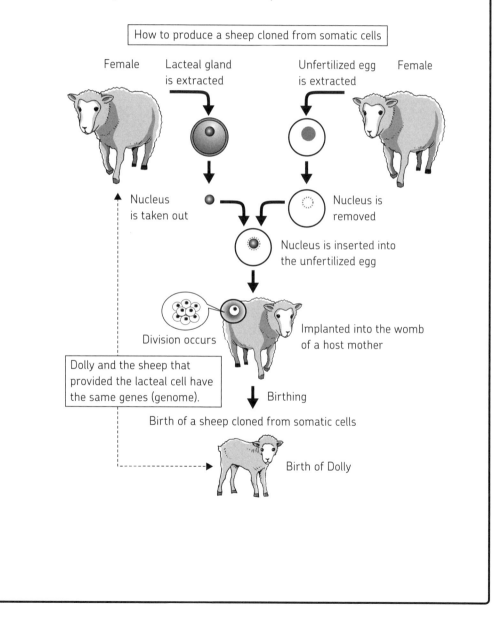

How to produce a sheep cloned from somatic cells

Female

Lacteal gland is extracted

Unfertilized egg is extracted

Female

Nucleus is taken out

Nucleus is removed

Nucleus is inserted into the unfertilized egg

Division occurs

Implanted into the womb of a host mother

Dolly and the sheep that provided the lacteal cell have the same genes (genome).

Birthing

Birth of a sheep cloned from somatic cells

Birth of Dolly

The clone that is born has the exact same genome as its mother, as well as any siblings made in the same way. In this case, Dolly the sheep had no father, just a mother. After the success of Dolly, mice and cows have also been cloned with somatic cells.

Further advances in cloning technology may end up looking a lot like a science fiction novel. It may be possible to produce *cloned humans* using *nuclear transfer*, like Dolly. Theoretically, a person produced using this process will have the same genome as the person (parent) who provided the nucleus. But as we know from twins, people raised in different environments will turn out differently from each other. The idea of human cloning is fraught with legal and moral complications and, while many nations permit studying human stem cells, most nations ban research that would lead to cloning a human.

MOLECULAR EVOLUTION: HOW GENES CAN TELL A STORY

We normally think of evolution as a process that happens in nature. The environment around an organism, from predators to a changing climate, can shape how it evolves by selecting for those who reproduce the most. But what is the exact mechanism of this evolution? Evolution occurs when small (or big) changes in genes occur by chance. If one of these changes helps the organism reproduce (by helping it live longer, catch more food, or tolerate a broader climate, for example) that change can be passed on to the organism's offspring.

Any change to a DNA sequence is called a *mutation*. Mutations occur naturally in DNA due to damage that is repaired improperly or errors in the copying of DNA. Mutations can be small, like switching one base for another, and create little or no change in the protein that the gene codes for. Or mutations can be large; whole sections of the genome can jump from one chromosome to another. In extreme cases, (which can result from a large or small change) the mutation creates a tangible change in the structure, function, or expression of a protein. These changes have more of a chance of altering behavior, survival, and reproduction of the whole organism, and thus can alter the evolution of that organism. The way in which changes in genes correlate to changes in the evolution of an organism is called *molecular evolution*.

Through an in depth analysis of the sequence and structure of genes, it is possible to tell how closely two organisms are related in evolutionary terms. By comparing the sequences of a given set of genes for two different organisms, we can determine how similar the genes are. Those species with very similar sequences are evolutionarily closer than those with very different gene sequences. Because small mutations tend to accumulate at a constant rate over time, these changes can also give us an idea of *when* new species evolved from common ancestors.

Thanks to molecular biology, the study of molecular evolution has progressed to the extent that it is possible to study the history of life on our planet, and evolution itself, at a molecular level.

THE FUTURE OF MOLECULAR BIOLOGY

No one can predict what major advances in molecular biology will bring to mankind. But we can look at the field right now and guess where it is headed. Both genomics and proteomics are huge fields in molecular biology that are just beginning to be researched. *Genomics* is the field of science that aims to explain the function of every gene in our genome, as well as figure out how all of the DNA that doesn't code for genes works within the cell. *Proteomics* is a related field that hopes to describe how every protein functions, as well as how proteins interact with one another.

Stem cells are another tool that molecular biologists use, whose potential is nearly limitless. We spoke early of the ES cell, a totipotent stem cell derived from embryos. Scientists at Kyoto University have discovered how to make a powerful (pluripotent) stem cell from an adult skin cell (called an *induced pluripotent stem cell* or *iPS cell*). While this cell is a little more mature than an ES cell, it can still transform into any other cell or tissue!

Stem cell science is a very exciting field in molecular biology. Using these new technologies, scientists may be able to repair the damage in people with strokes, or those with kidney or heart failure. Induced pluripotent stem cells have the added advantage of being less morally questionable than embryonic stem cells, since they are made from adult tissue instead of embryos.

While molecular biology has the potential to drastically alter the way we treat diseases and live our lives, we cannot count on miraculous breakthroughs. Like any research science, progress in molecular biology is slow. However, all of the processes, techniques, and advances we have learned about come from the combination of creative thought with rigorous scientific methodologies.

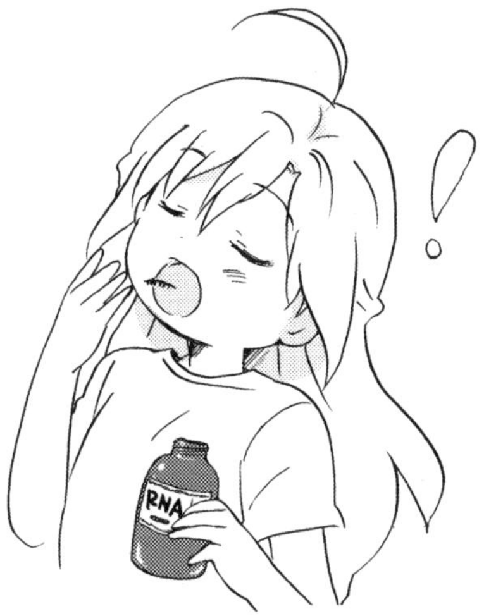
RNA

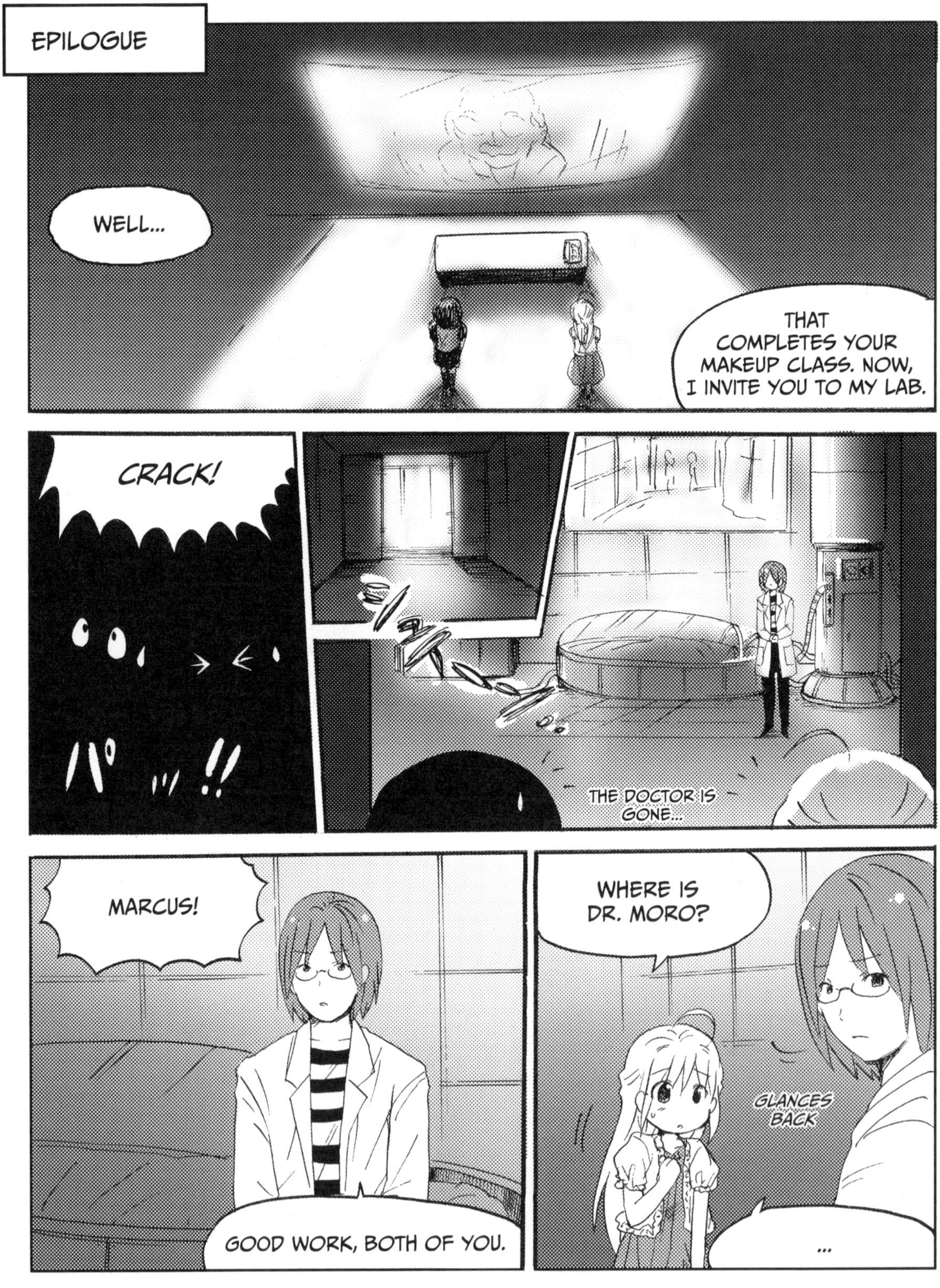
EPILOGUE
WELL...
THAT COMPLETES YOUR MAKEUP CLASS. NOW, I INVITE YOU TO MY LAB.
CRACK!
パッ!!
THE DOCTOR IS GONE...
MARCUS!
GOOD WORK, BOTH OF YOU.
WHERE IS DR. MORO?
GLANCES BACK
...

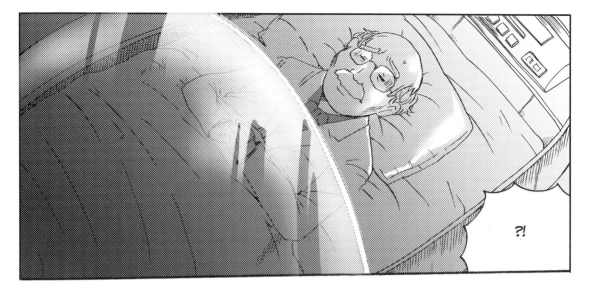

?!

ARE YOU TAKING A NAP?

HE WOKE UP.

BANG!

COUGH

NO, NO!

DON'T GET EXCITED.

COUGH
COUGH
COUGH

DR. MORO LOOKS PALE.

YES, YOU SEE...

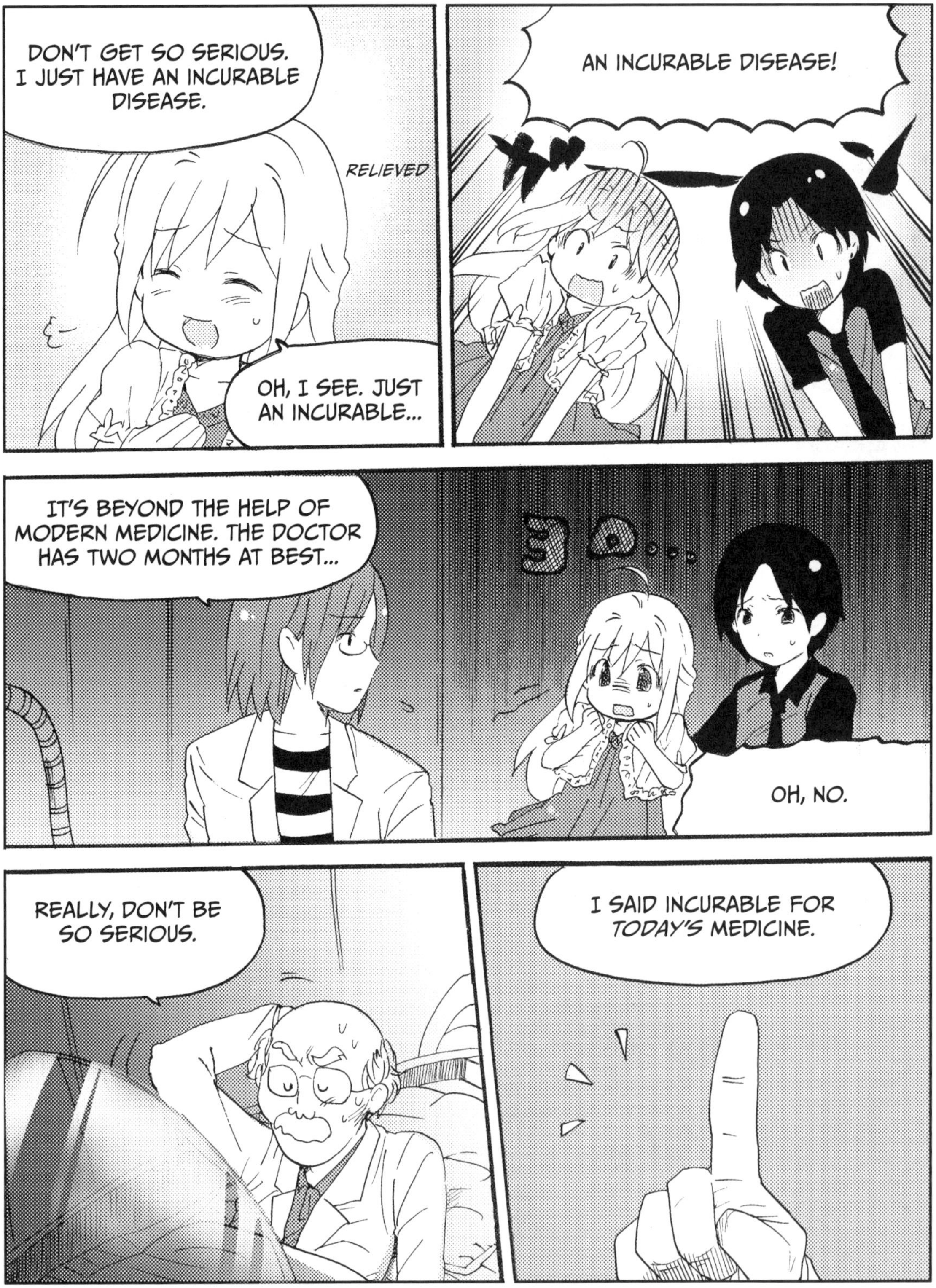
DON'T GET SO SERIOUS. I JUST HAVE AN INCURABLE DISEASE.
RELIEVED
OH, I SEE. JUST AN INCURABLE...
AN INCURABLE DISEASE!
IT'S BEYOND THE HELP OF MODERN MEDICINE. THE DOCTOR HAS TWO MONTHS AT BEST...
OH, NO.
REALLY, DON'T BE SO SERIOUS.
I SAID INCURABLE FOR *TODAY'S* MEDICINE.

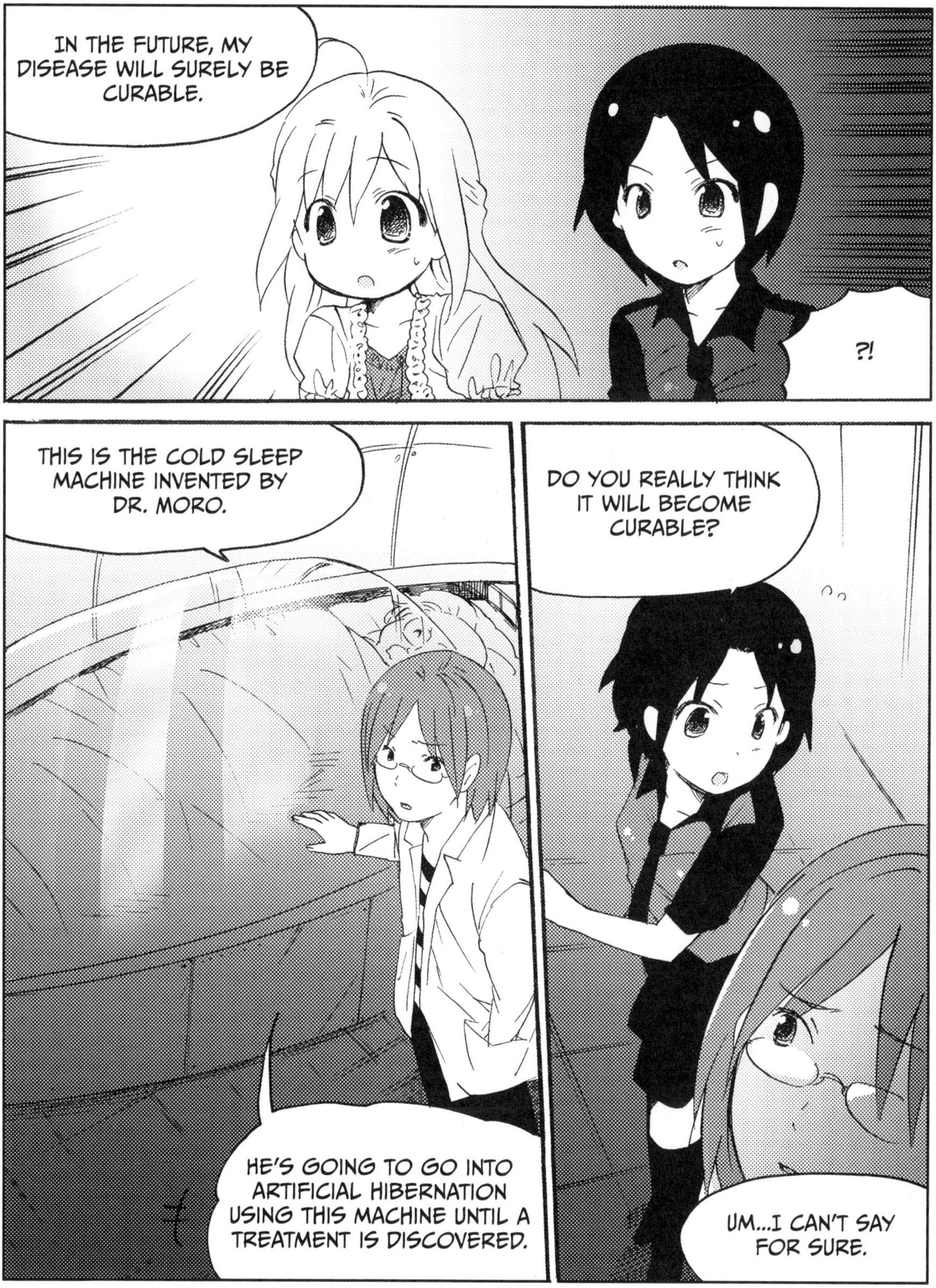
IN THE FUTURE, MY DISEASE WILL SURELY BE CURABLE.
?!
THIS IS THE COLD SLEEP MACHINE INVENTED BY DR. MORO.
HE'S GOING TO GO INTO ARTIFICIAL HIBERNATION USING THIS MACHINE UNTIL A TREATMENT IS DISCOVERED.
DO YOU REALLY THINK IT WILL BECOME CURABLE?
UM...I CAN'T SAY FOR SURE.

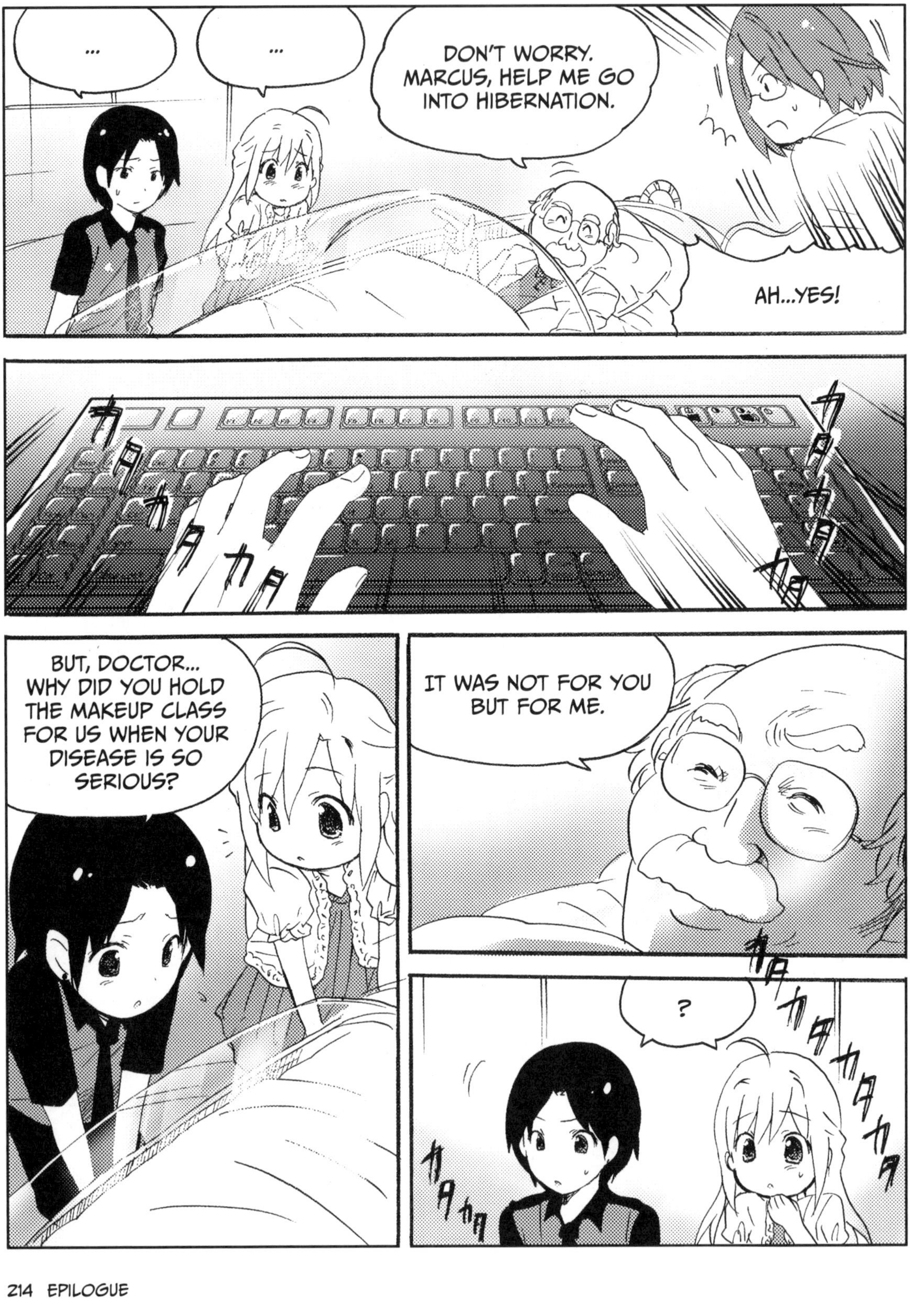
...
...
DON'T WORRY. MARCUS, HELP ME GO INTO HIBERNATION.
AH...YES!
カタカタカタカタ
カタカタカタ
BUT, DOCTOR... WHY DID YOU HOLD THE MAKEUP CLASS FOR US WHEN YOUR DISEASE IS SO SERIOUS?
IT WAS NOT FOR YOU BUT FOR ME.
?
カタカタカタカタ
カタカタ

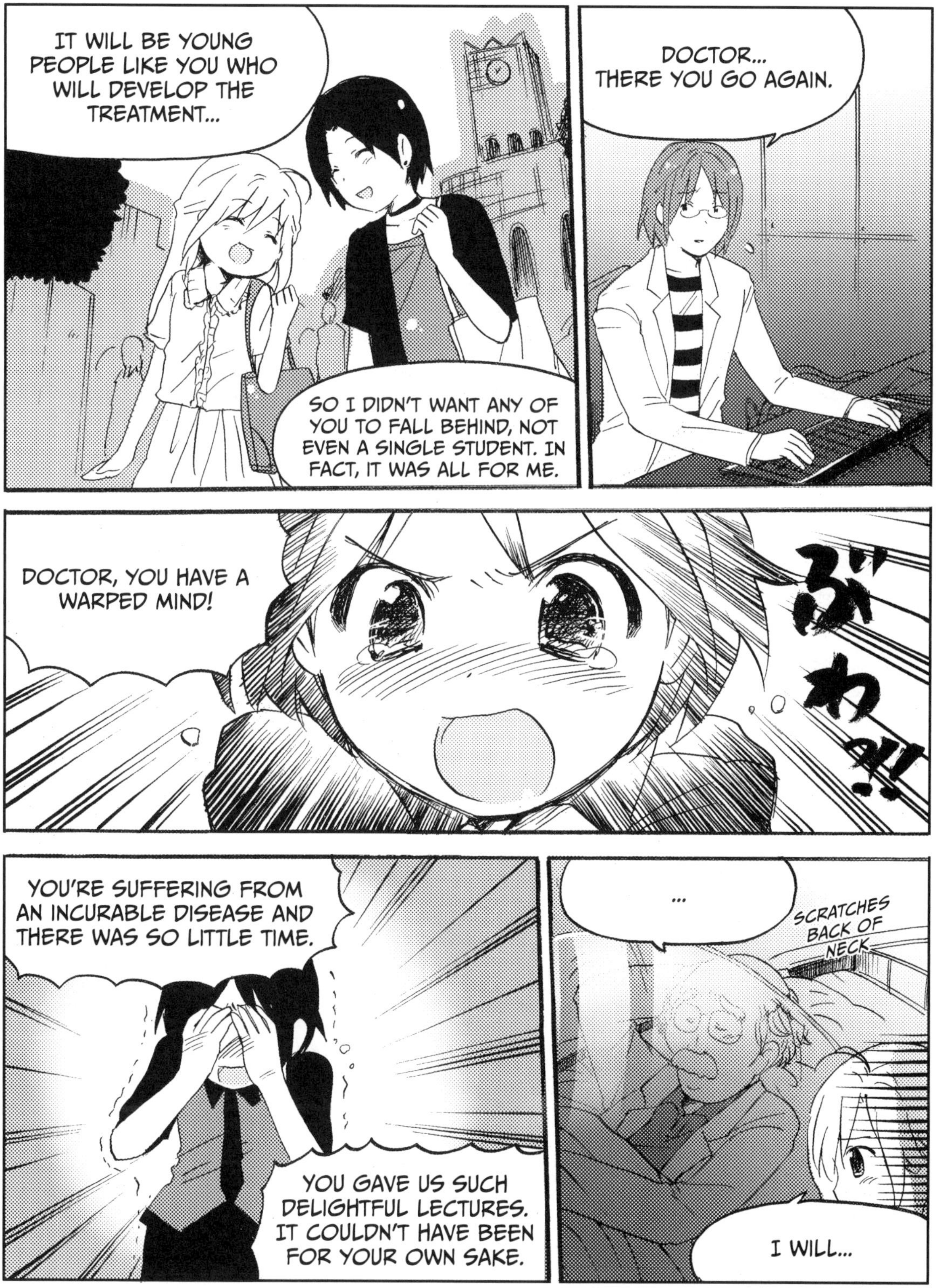
IT WILL BE YOUNG PEOPLE LIKE YOU WHO WILL DEVELOP THE TREATMENT...
SO I DIDN'T WANT ANY OF YOU TO FALL BEHIND, NOT EVEN A SINGLE STUDENT. IN FACT, IT WAS ALL FOR ME.
DOCTOR... THERE YOU GO AGAIN.
DOCTOR, YOU HAVE A WARPED MIND!
ぶわっ!!
YOU'RE SUFFERING FROM AN INCURABLE DISEASE AND THERE WAS SO LITTLE TIME.
YOU GAVE US SUCH DELIGHTFUL LECTURES. IT COULDN'T HAVE BEEN FOR YOUR OWN SAKE.
...
SCRATCHES BACK OF NECK
I WILL...

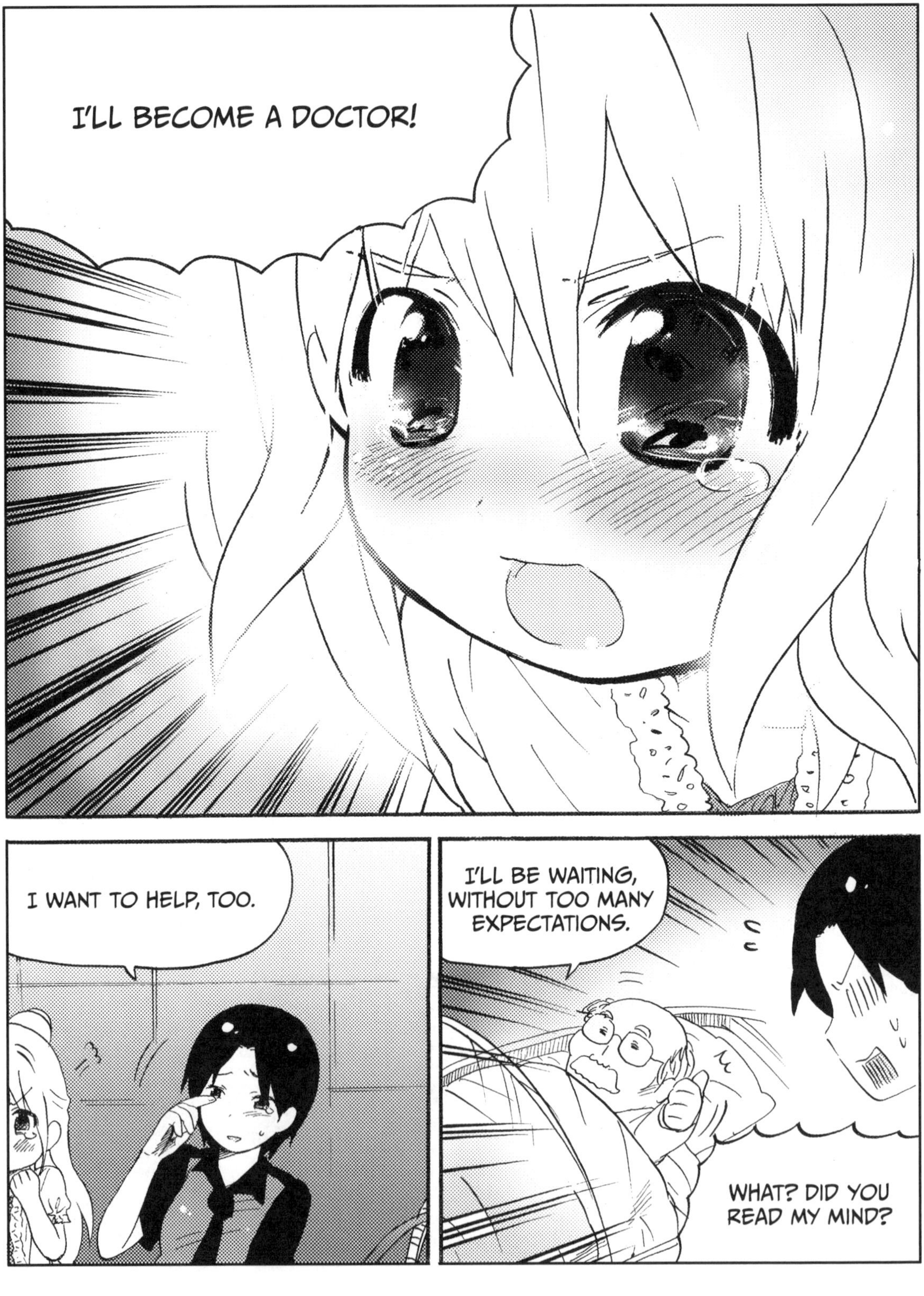
I'LL BECOME A DOCTOR!
I WANT TO HELP, TOO.
I'LL BE WAITING, WITHOUT TOO MANY EXPECTATIONS.
WHAT? DID YOU READ MY MIND?

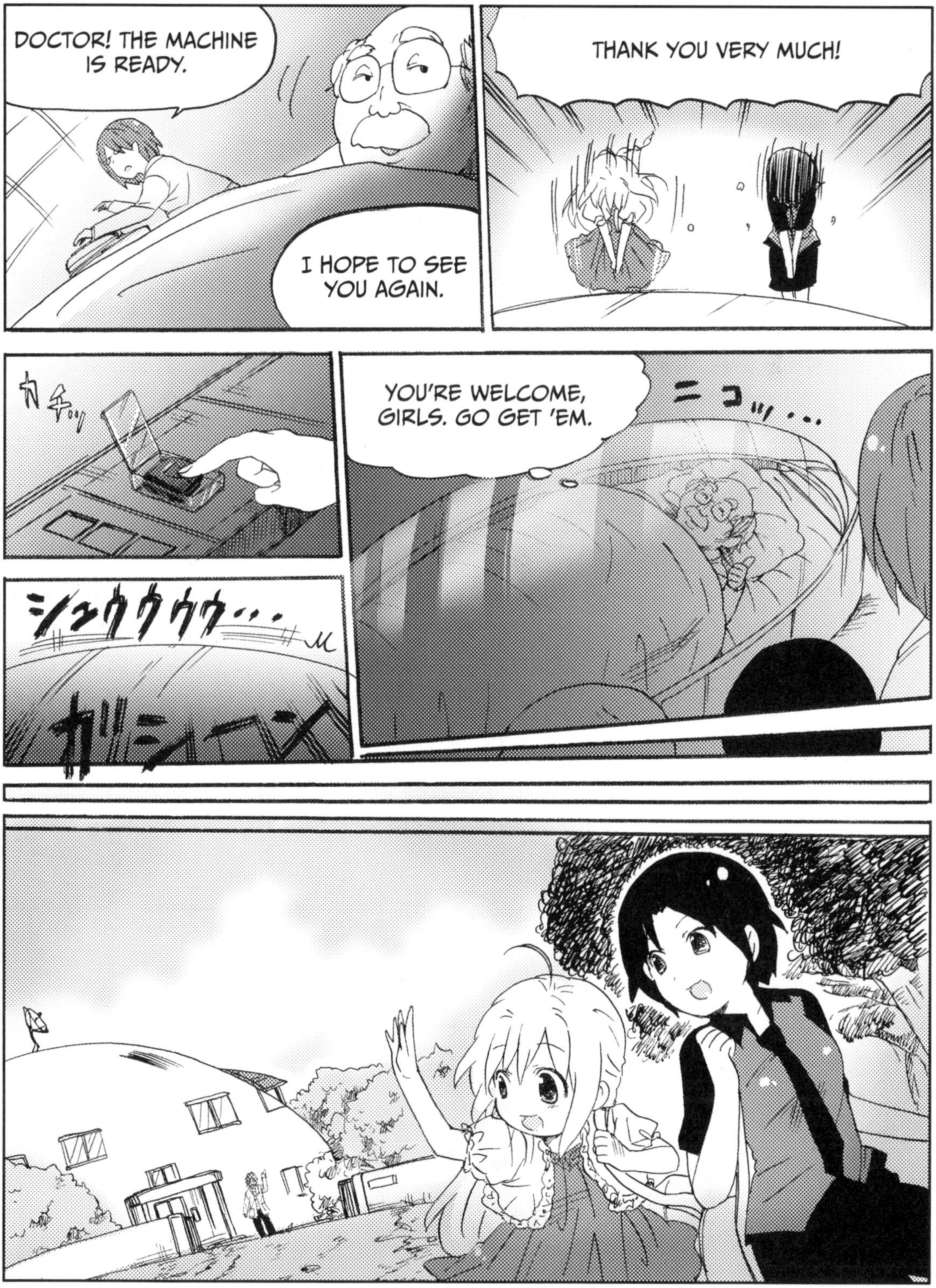
DOCTOR! THE MACHINE IS READY.
I HOPE TO SEE YOU AGAIN.
THANK YOU VERY MUCH!
カチッ
シュウウウウ…
ガシーン
YOU'RE WELCOME, GIRLS. GO GET 'EM.
ニコッ…

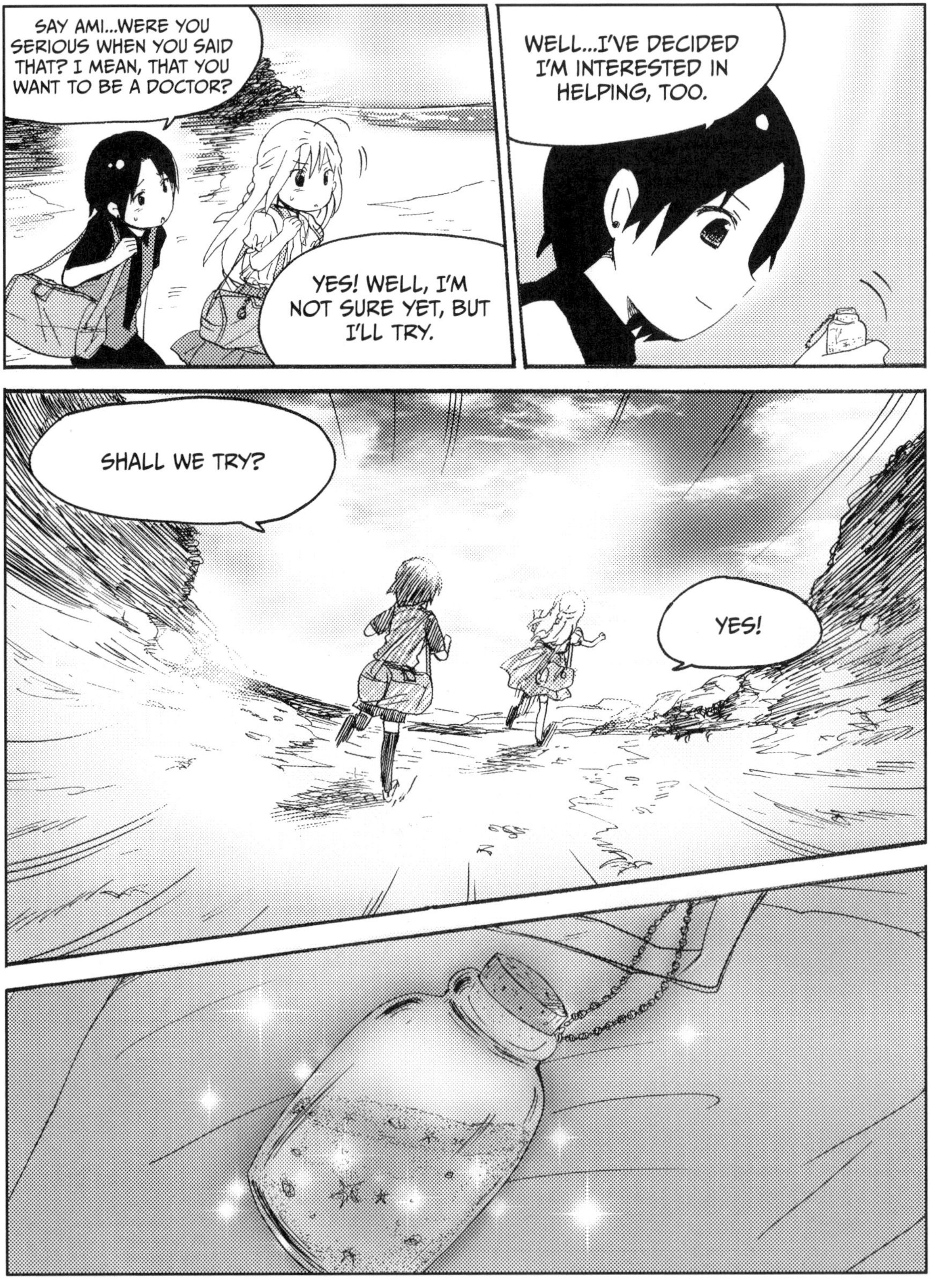
SAY AMI...WERE YOU SERIOUS WHEN YOU SAID THAT? I MEAN, THAT YOU WANT TO BE A DOCTOR?
YES! WELL, I'M NOT SURE YET, BUT I'LL TRY.
WELL...I'VE DECIDED I'M INTERESTED IN HELPING, TOO.
SHALL WE TRY?
YES!

INDEX

D

E

F

G

proteins, *continued*
    DNA and, 42–45, 56–62, 70,
        72, 114–115
    DNA replication and, 114–115
    enzymes and, 61–70
    folded forms, 174, 174n
    functions of, 56–60, 72
    genes and building proteins,
        77–80
    hemoglobin and, 72,
        75–76, 138
    muscular contraction and, 57,
        60, 71–72
    in nucleus, 37–39
    properties of, 77
    role in cell division, 70
    starch breakdown and, 61, 62
    structure and shape of, 73–76,
        86, 161, 174, 174n
    synthesis, 46–47,
        165–167, 202
proteins, types of, 57, 69, 72
    actin, 71, 72
    ADA, 199–200
    globins, 76
    histones, 42, 43, 122, 123,
        144
    insulin, 72
    myosin, 57, 71, 72
    subunits, 76, 165–166
proteomics, 209
protozoa, 48, 52, 100, 178
pseudogenes, 142–143
pure DNA, 191–192
pyruvic acid, 33

Q

quantitative PCR, 204. *See also*
    PCR (polymerase chain
    reaction)

R

replication. *See* DNA replication
reproduction, cell. *See* cell division
reproduction, human and animal,
    92–97, 101
respiratory system, 51

restriction enzymes, 188,
    204–205
ribonucleotides, 159
ribose, 159
ribosomal RNA (rRNA), 161, 165
ribosomes, 31, 32, 34, 38, 45,
    80, 114, 152
    in translation, 165–166,
        170–173
ribozymes, 202
RISC (RNA-Induced Silencing
    Complex), 202
RNA. *See also* mRNA; tRNA
    about, 11–12, 37, 43, 44, 45,
        56, 57, 156–162
    bases/characters of, 156–158
    base sequences of, 141,
        146–150, 153, 160–161,
        170–171, 203
    characters of/bases, 156–158
    flexibility of, 160–161, 164
    messenger (mRNA), 152,
        153, 161
    miRNA (microRNA), 202
    molecules, 202
    ribonucleotides, 159
    siRNA (short interfering RNA),
        201–203
    structure of, 160–161, 164
    sugar-phosphate backbone of,
        158–159
    types of, 161–162
RNA-based medical treatments,
    203
RNA-Induced Silencing Complex
    (RISC), 202
RNA interference, 201–203
RNA polymerase, 148–154
RNA primers, 115–116, 188, 204
RNA renaissance, 201
RNA sequences, 58
rough endoplasmic reticulum, 25
rRNA (ribosomal RNA), 161, 165
RT-PCR, 204. *See also* PCR
    (polymerase chain reaction)

S

SELEX method (Systematic
    Evaluation of Ligands by
    Exponential Enrichment),
    203
sensory epithelium, 49
serine, 74, 75, 154, 167–169
sex chromosomes, 123
short interfering RNA (siRNA),
    201–203
sickle-cell disease, 76
single-celled organisms, 19, 22,
    24, 48–52, 97–100
siRNA (short interfering RNA),
    201–203
skeletal muscle, 50
skin, 49, 57, 102–103, 128, 209
skin cancer, 194
small intestine, 48, 51, 59, 61
smooth endoplasmic
    reticulum, 25
somatic cells, 207–208
S phase, 128–129
spindle apparatus, 126
spindle fibers, 125–126
spliceosomes, 154
splicing, 154, 155
starch, 23, 61
stem cells
    about, 193, 208, 209
    embryonic stem (ES) cells,
        193, 208, 209
    induced pluripotent stem cells
        (iPS cells), 209
    totipotent cells, 193, 209
steric structure, 76
sticky ends, 188, 205
stomach, 48–51, 59
stop codons, 169, 170, 172–173
stratum basale, 103
substrates, 64
subunits, 76, 165–166
sugar-phosphate backbones,
    107, 158–159, 191
sugars
    blood, 61–62, 185, 196
    DNA vs. RNA, 158–159
    glucose, 23, 33, 61–62, 69

ABOUT THE AUTHOR

Masaharu Takemura, PhD, has worked as a research associate at the Nagoya University School of Medicine (Japan), a visiting fellow at the University of Oxford (England), and a research associate at Mie University (Japan). He is currently an associate professor in the Faculty of Science at the Tokyo University of Science. He has published a number of research papers and books on biology-related subjects, including molecular biology, life science, and biology education.

PRODUCTION TEAM FOR THE JAPANESE EDITION

Production: Becom Co., Ltd.

Since its foundation in 1998 as an editorial and design production studio, Becom has produced many books and magazines in the fields of medicine, education, and communication. In 2001, Becom established a team of comic designers, and the company has been actively involved in many projects using manga, such as corporate guides and product manuals, as well as manga books. More information about the company is available at its website, *http://www.becom.jp/*.

Yurin Bldg 5F, 2-40-7 Kanda-Jinbocho, Chiyoda-ku, Tokyo, Japan 101-0051

Tel: 03-3262-1161; Fax: 03-3262-1162

Scenario writer: Masayoshi Maeda

Illustrator: Sakura

Figure Creation: Smith Suzuki, Tsutada Design Planning

Cover Artist: Ogii Hagiwara